Handbook of Automotive Human Factors

Edited by:

Motoyuki Akamatsu

Automotive Human Factors Research Center, AIST, Tsukuba, Japan

For:

Society of Automotive Engineers of Japan, Inc.

CRC Press
Taylor & Francis Group
Boca Raton London New York

CRC Press is an imprint of the
Taylor & Francis Group, an **informa** business

A SCIENCE PUBLISHERS BOOK

CRC Press
Taylor & Francis Group
6000 Broken Sound Parkway NW, Suite 300
Boca Raton, FL 33487-2742

First issued in paperback 2020

© 2019 by Taylor & Francis Group, LLC
CRC Press is an imprint of Taylor & Francis Group, an Informa business

No claim to original U.S. Government works

ISBN-13: 978-0-367-20357-3 (hbk)
ISBN-13: 978-0-367-77962-7 (pbk)

Library of Congress Cataloging-in-Publication Data

Names: Akamatsu, Motoyuki, 1955- editor.
Title: Handbook of automotive human factors / editor Motoyuki Akamatsu
    (Automotive Human Factors Research Center, AIST, Tsukuba, Japan, for
    Society of Automotive Engineers of Japan, Inc.).
Description: Boca Raton, FL : CRC Press, Taylor & Francis Group, 2019. | "A
    science publishers book." | Includes bibliographical references and index.
Identifiers: LCCN 2019010287 | ISBN 9780367203573 (hardback)
Subjects: LCSH: Automobile driving--Human factors. | Automobile
    driving--Physiological aspects. | Automobiles--Design and construction. |
    Human engineering.
Classification: LCC TL250 .H285 2019 | DDC 629.2/31--dc23
LC record available at https://lccn.loc.gov/2019010287

Visit the Taylor & Francis Web site at
http://www.taylorandfrancis.com

and the CRC Press Web site at
http://www.crcpress.com

Foreword

Motoyuki Akamatsu goes by *Moto*, an apt name given his interest in motor vehicles. Over the decades I have known Moto, I have continued to be impressed with his vast knowledge of the history of motor vehicle development, his grasp of fundamental human factors issues, his insights into how Japanese culture influences research and implementation, and with the rigor of his research. About 10 years ago Moto and I collaborated (with others) on a strategic review of human factors activities at Nissan and Renault. After one of those meetings, our hosts took us to a car museum near Paris, and the background knowledge Moto provided was impressive. Several years ago, we collaborated on a paper on the history of automotive factors (Akamatsu et al., 2013). Collectively, these interactions set up high expectations, and of course, Moto's book delivers as expected.

Moto's book contains 6 chapters that complement the existing literature. The first chapter is an overview of automotive human factors, emphasizing contemporary issues. The second chapter concerns the automotive development process, primarily from the view of a researcher. It emphasizes commonly used tools and methods (surveys, simulators, on-road testing). This material is extremely valuable, particularly to researchers beginning in this field. Those interested in the development process should also look at the literature on user experience methods (journey maps, personas, use cases, etc.) given the widespread use in the auto industry. The third chapter concerns occupant comfort and interior quality, a topic that receives scant attention in automotive human factors conferences or journals, but is of great internal interest to automotive manufacturers and suppliers, as comfort and interior quality often determine what customers buy. Those interested in this topic should look at the literature on Quality Function Deployment (QFD)/House of Quality, literature so substantial that it would have been difficult to cram into this chapter. Chapter 4 concerns driver state, which now is a critical topic given increasing interest in the automation of driving. A well-studied issue is transfer of control, a topic for which this chapter provides valuable background. Chapter 5 concerns 4 topics: driver workload, driver interfaces, distraction measurement, and driver assistance systems, and is the longest chapter in the book. Chapter 6, concerning driver models, includes some important models that are not commonly covered in automotive human factors texts.

In a foreword, one is encouraged to proclaim that a tome is the greatest volume since the Bible/Quran/etc. Moto's book is quite useful, but not at that biblical level. Probably the best overall reference on automotive human factors is Bhise's (2016) book (Ergonomics in the Automotive Design Process), which incorporates much of what Bhise learned during his long career at Ford. Moto's book was written to complement Bhise's book, which it does quite well. Moto's book also complements Peacock and Karwowski (1993), a now dated and less comprehensive alternative to Bhise's book.

So, what are the alternatives to Moto's book? Quite frankly, there is nothing that blends applications and research as this book does except for Gkikas (2013), which is a useful edited collection of chapters, but not as well integrated as Moto's book, as it is not of a single mind.

There are also several books that focus on contemporary issues. They include Cacciabue's (2007) edited book on modeling driver performance and 3 books concerned with driver distraction—Regan et al. (2009)—the widely cited classic, Rupp's 2010 book on distraction measurement, and Regan et al. (2013), one of the more recent collections. Somewhat related is the Walker et al. (2015) edited volume that also deals with advanced vehicle technology. Finally, 2 other references worthy of note are Fisher et al. (2011), an encyclopedic book on driving simulation and Smiley (2015), one of many books on the forensic aspects of traffic safety from Lawyers and Judges Publishing.

So, what should one have on their bookshelf and read? I would start with Bhise and then read Moto Akamatsu's book. After that, it would depend on what one is interested in—new technology in general, distraction, driving simulation, or some other topic. In addition, I would also strongly encourage everyone to read SAE Recommended Practice J2944 (Operational Definitions of Driving Performance Measures and Statistics).

References

Akamatsu, M., P. Green and K. Bengler. (2013). Automotive technology and human factors research: Past, present, and future. International Journal of Vehicular Technology.

Bhise, V.D. (2016). Ergonomics in the Automotive Design Process. Boca Raton, FL: CRC Press.

Cacciabue, P.C. (2007). Modelling Driver Behaviour in Automotive Environments: Critical Issues in Driver Interactions with Intelligent Transport Systems. New York, NY: Springer-Verlag New York.

Gkikas, N. (2016). Automotive Ergonomics: Driver-Vehicle Interaction. Boca Raton, FL: CRC Press.

Peacock, B. and W. Karwowski (eds.). (1993). Automotive Ergonomics. London, England: Taylor & Francis.

Regan, M.A., J.D. Lee and K. Young. (2008). Driver Distraction: Theory, Effects, and Mitigation. Boca Raton, FL: CRC Press.

Regan, M.A., J.D. Lee and T.W. Victor. (2013). Driver Distraction and Inattention. Ashgate.

Rupp, G. (2010). Performance Metrics for Assessing Driver Distraction. Warrendale, PA: Society of Automotive Engineers.

Smiley, A. (2015). Human Factors in Traffic Safety (3rd ed.). Tucson, AZ: Lawyers and Judges.

Society of Automotive Engineers. (2015). Operational Definitions of Driving Performance Measures and Statistics (SAE Recommended Practice J2944), Warrendale, PA: Society of Automotive Engineers.

Walker, G.H. and N.A. Stanton. (2017). Human Factors in Automotive Engineering and Technology. Boca Raton, FL: CRC Press.

Walker, G.H., N.A. Stanton and P.M. Salmon. (2015). Human Factors in Automotive Engineering and Technology. Ashgate.

Paul Green
University of Michigan

Preface

This book comes about as a result of the translation of a part of the Automotive Human Factors of Automotive Engineering Handbook published by the Society of Automotive Engineers of Japan (JSAE) in March 2016 in Japanese. It is financially supported by the Automotive Human Factors Research Center (AHFRC) of the National Institute of Advanced Science and Technology (AIST) of Japan. The selecting, compiling and editing works have been accomplished by the editor who is a prime senior researcher of AHFRC. The original articles were written by 37 Japanese authors who are experts on automotive human factors in research institutes, universities and industries.

Human-centered design is a key issue in automotive technology nowadays. Improvement in technology enables us to design driver assistance systems and in-vehicle infotainment systems, cabin interiors and drive ability with greater flexibility to meet human needs, functionalities and activities (human-centered design). In order to achieve it, a wide range of knowledge about humans is required in the design and development process. Knowledge about humans includes human sensory and motor functions, cognitive functions, physiology and homeostatic functions, mental state, behavioral characteristics and so on. However, it is not easy for automotive engineers to learn and master all related disciplines and understand how to apply them by only reading relevant textbooks. Therefore, there is a need to have a complete book that compiles appropriate topics.

The Handbook of Automotive Human Factors aims to provide basic knowledge about measurement and modeling of human functions and research and practices related to automotive human factors. This book offers valuable lessons for researchers, designers and developers for further research and development of human-centered design of future automotive technologies. Some of the contents of the book are well established and some are state-of-the-art. This book covers the following human factors issues: driver state, sensory comfort (Kansei), interactions with systems and driver behavior. It does not cover seating/packaging, displays/controls, visibility and cabin climate because there have already been several books regarding these topics.

Chapter 1 reviews the history of automotive ergonomics and human factors and indicates future issues. Chapter 2 describes the role of ergonomics and human factors in the automobile design process and gives methodologies to understand the driver using questionnaires, measurements using driving simulators and equipped vehicles and sociological approaches. Chapter 3 presents experiences of sound design for vehicles, ride quality and illumination for cabin comfort. Chapter 4 discusses driver state such as fatigue, workload, arousal level and enjoyment of driving, and their measurement methods. Chapter 5 describes HMI design and evaluation of in-vehicle systems and ADAS/automated driving systems. Chapter 6 explains characteristics and modeling of drivers' behavior that will be key issues in the future for designing human-centered ADAS and other systems to support/assist drivers and extend drivers' ability.

Motoyuki Akamatsu, Editor

January 2019

Contents

Foreword   **iii**

Preface   **v**

1. Overview of Automotive Ergonomics and Human Factors   **1**

   1.1   Ergonomics and Human Factors for Making Products and Systems   1
         Compatible with Humans
   1.2   Beginning of Human-compatible Automobile Design   1
   1.3   Vehicle Cabin Design   3
   1.4   Instruments and Displays   4
        1.4.1   Instrument Arrangement   4
        1.4.2   Meters and Displays   5
        1.4.3   Controls   6
   1.5   Riding Comfort and Fatigue   7
        1.5.1   Fatigue   7
        1.5.2   Vibration   8
        1.5.3   Arousal Level   9
   1.6   Vehicle Interior Environment   9
   1.7   Driving Tasks and Non-driving Tasks   10
        1.7.1   In-vehicle Systems   10
        1.7.2   Non-driving Activities like Using Mobile Phones   11
        1.7.3   Visual Distraction   11
        1.7.4   Mental Workload and Cognitive Distraction   12
   1.8   Driver Model and Driving Behavior Measurement   13
        1.8.1   Driver Model   13
        1.8.2   Driving Behavior Measurement   13
             1.8.2.1   Site-based Measurement   13
             1.8.2.2   Driving Simulators   14
             1.8.2.3   Equipped Vehicles and Naturalistic Driving Study   15
   1.9   Driving-assistance Systems/Automated Driving Systems   15
        1.9.1   ACC/Lane-keeping Systems   15
        1.9.2   Automated Driving Systems   16
   1.10 Elderly Drivers   17
   1.11 Positive Aspects of Automobile Driving   18
        1.11.1 Enjoyment and Growth   18
        1.11.2 Stress Relief   20

    1.12 Future of Automobile Ergonomics: Viewpoint of Service   21
          Engineering for Providing Value to Users
References   22

2. **Ergonomic and Human Factors in Automobile Design and**   **32**
    Development Process
    2.1 Ergonomists' Roles and Responsibilities in Automobile Design   32
         and Development
         2.1.1 Ergonomics for Automobiles   32
         2.1.2 Development Process   33
         2.1.3 Identifying Out User Requirements   33
         2.1.4 Ergonomics in Design Stage   34
         2.1.5 Ergonomics in Assessment Stage   35
         2.1.6 Feedback from Users   35
         2.1.7 Designing User's Manual   36
    2.2 Surveys for Understanding Users in Design Stage   37
         2.2.1 Viewpoints for Considering Target Users   37
         2.2.2 Observation-based Approach   38
                2.2.2.1 Knowing User Requirements   38
                2.2.2.2 Behavior Observation   38
                2.2.2.3 Ethnographical Methods   38
                2.2.2.4 Task Analysis   40
         2.2.3 Questionnaire and Interview Approach   41
                2.2.3.1 Objectives of Questionnaires and Interviews   41
                2.2.3.2 Selecting Survey Methods   41
                2.2.3.3 Designing Paper Questionnaires and Interviews   42
                2.2.3.4 Depth Interview Method   43
                2.2.3.5 Group Interview   44
    2.3 Driving Behavior Measurement   44
         2.3.1 Driving Behavior Measurement Using Driving Simulators   44
                2.3.1.1 Objectives of Using Driving Simulators   44
                2.3.1.2 Basic Configuration of Driving Simulators   45
                2.3.1.3 Classification of Driving Simulators   46
                2.3.1.4 Driving Simulator Sickness   48
                2.3.1.5 Other Tips for Use in Driving Simulators   49
         2.3.2 Driving Behavior Measurement Using Instrumented Vehicles 50
                2.3.2.1 Instrumented Vehicle   50
                2.3.2.2 Measurement Environment   52
                         2.3.2.2.1 Measurement on a Test Track   52
                         2.3.2.2.2 Measurement on Real Roads   52
                2.3.2.3 FOT and NDS   53
         2.3.3 Driving Behavior Analysis Using Drive Recorders   53
                2.3.3.1 Drive Recorder Specifications   53
                2.3.3.2 Recording Driving Behavior   54
                         2.3.3.2.1 Face Direction   54
                         2.3.3.2.2 Recording Traffic Conditions   56

2.3.3.3   Data Recording Methods   56
   2.3.3.3.1 Event Trigger Methods   56
   2.3.3.3.2 Continuous Recording Methods   56
2.3.3.4   Examples of Drive Recorder Data Analysis   56
   2.3.3.4.1 Time Series Analysis Using Variation   56
       Tree Analysis
   2.3.3.4.2 Analyzing a Series of Background Factors   58
References   59

3. Comfort and Quality   **62**

3.1   Occupant Comfort During Vehicle Run   62
 3.1.1   Vibration and Comfort   62
  3.1.1.1   Basic Vibration Measurement and Evaluation Methods   63
  3.1.1.2   Riding Comfort Evaluation by Phenomenon   64
  3.1.1.3   Method for Estimating the Vibration of the Seat   64
      when an Occupant is Sitting
 3.1.2   Comfort of the Seat   65
  3.1.2.1   Seat Structure and Vibration Absorption Properties   66
   3.1.2.1.1 Transmission of Vibration through the Seat   66
   3.1.2.1.2 Issues on the Measurement of the   66
       Vibration of the Seat
   3.1.2.1.3 Seat Structure and Specific Characteristics   66
       of Vibration
   3.1.2.1.4 Vibration Characteristics of the Parts of Seat   67
   3.1.2.1.5 Changes in the Characteristics of Vibrations   68
       on People
  3.1.2.2   Body Movements Caused by Acceleration   68
  3.1.2.3   Support Performance of the Seat   69
   3.1.2.3.1 Lateral Movements   69
   3.1.2.3.2 Movements of the Head   70
   3.1.2.3.3 Support by the Seat during Driving   70
 3.1.3   Vibration and Driving Performance   70
3.2   Acoustic Comfort   75
 3.2.1   Design of the Engine Sound   75
  3.2.1.1   Acoustic Characteristics that Influence Sound Design   75
  3.2.1.2   Order Composition of Sounds   75
   3.2.1.2.1 Orders and Generation Mechanism   75
      (1) Engine sound   75
      (2) Suction sound   76
      (3) Exhaust sound   77
   3.2.1.2.2 Relationship of the Order Composition   77
       and the Impression of the Sound
  3.2.1.3   Control of the Sound   77
   3.2.1.3.1 Method that Uses Components   78
       of the Vehicle
   3.2.1.3.2 Method that Uses Devices for Creating   78
       Sounds

	3.2.1.4	Sound Evaluation Methods	79
3.2.2	Sound of the Door Closing		79
	3.2.2.1	Need for Research on Door Sounds	79
	3.2.2.2	Mechanism of Door Closing Sounds	80
	3.2.2.3	Conditions for Good Door Closing Sound	81
		3.2.2.3.1 Arranging the Distribution of Frequency	81
		3.2.2.3.2 Adding Reverberation Effects: It is Effective to give Two Sounds with the Same Frequency Components	81
	3.2.2.4	How to Realize It	82
		3.2.2.4.1 Method of Producing Sounds of Low Frequency	82
		3.2.2.4.2 How to Produce the Two Successive Sounds	82
	3.2.2.5	Other Considerations	83
3.3	Cabin Air Quality		83
3.3.1	Smells in the Interior of the Vehicle		83
	3.3.1.1	Sensory Evaluation	83
	3.3.1.2	Instrumental Analysis	84
	3.3.1.3	Odor Sensors	85
	3.3.1.4	Odor Control	85
3.3.2	Effects of Fragrance		86
	3.3.2.1	Perception Mechanism of Smells	86
	3.3.2.2	Emotional and Physiological Effects of Fragrances	87
	3.3.2.3	Future of Vehicles and Smells	88
3.4	Visual Environment of Vehicle Interior		88
3.4.1	Function and Design of Vehicle Interior Lighting		88
	3.4.1.1	Types of Lighting	88
	3.4.1.2	Requirements for Functional Lighting Design and a Study Example	89
	3.4.1.3	Map and Reading Lamps	90
	3.4.1.4	Vanity Lamps	90
3.4.2	Comfort Provided by Vehicle Interior Lighting		91
	3.4.2.1	Effect of Shape and Brightness of Light Source on People's Impression of Vehicle Comfort and Spaciousness	91
	3.4.2.2	Poor Visibility of Vehicle Interior from Outside	93
3.5	Interior Materials		94
3.5.1	Evaluation Criteria for Interior Material		94
3.5.2	Gripping Functions		94
	3.5.2.1	Functions of Vehicle Operation System	94
	3.5.2.2	Grips that Support Drivers/Passengers with Physical Stability	95
	3.5.2.3	Gripping Functions of Non-grip Parts	95
3.5.3	Effect of Sweat		95
3.5.4	Difference in Skin Structure Among Body Parts		96
3.5.5	Stickiness		96

3.5.6	Thermal Sensation	97
3.5.7	Breathable Seat Materials and Structures	98
3.5.8	Texture and Durability	98
References		98

4. Driver State — **102**

4.1	Driving Fatigue, Workload, and Stress			102
	4.1.1	Stress and Strain		102
	4.1.2	Driver Fatigue		103
	4.1.3	Mental Workload and Tasks		104
	4.1.4	Mental Workload Described in ISO 10075		105
	4.1.5	Task Demand, Mental Resource and Fatigue		107
	4.1.6	Difference Between the Concept of Mental Workload and the Concept of Stress/Strain		107
	4.1.7	Driver's Stress		108
4.2	Enjoyment Generated by Automobiles			109
	4.2.1	Utility of Automobile Use		109
	4.2.2	Automobiles as a Tool for Stimulating Emotions		110
	4.2.3	Flow Theory of Csikszentmihalyi		110
	4.2.4	Flow and Increase of Skills		112
	4.2.5	Flow and the Zone		113
	4.2.6	Effects of Feelings of Enjoyment		114
	4.2.7	Subjective Well-being and Automobiles		114
4.3	Arousal Level			115
	4.3.1	Arousal Level and Sleepiness		115
	4.3.2	Sleepiness Measurement Methods		117
		4.3.2.1	Sleep Propensity	117
		4.3.2.2	Vigilance	118
		4.3.2.3	Subjective Sleepiness	120
	4.3.3	Arousal Level Measurement		121
		4.3.3.1	Driving Behavior	121
		4.3.3.2	EEG	121
		4.3.3.3	Rating Based on Facial Expressions	121
		4.3.3.4	Pupil Diameter	122
		4.3.3.5	Eye Movement	122
			4.3.3.5.1 Saccade	122
			4.3.3.5.2 Slow Eye Movement	123
			4.3.3.5.3 Vestibulo-ocular reflex (VOR)	123
		4.3.3.6	Eyelid Activity	123
			4.3.3.6.1 PERCLOS	123
			4.3.3.6.2 Integrated Indices of Eye-related Measures	124
		4.3.3.7	Heart Rate	126
		4.3.3.8	Summary	126
	4.3.4	Arousal-enhancing Technology		126
		4.3.4.1	Sleepiness and Arousal Level	126
		4.3.4.2	Counter Measures against Sleepiness, Napping	128

		4.3.4.3	Counter Measure against Sleepiness, Other than Napping	129
		4.3.4.4	Summary	130
	4.4	Techniques for Measuring/Analyzing Physical Conditions		130
		4.4.1	Significance of Introducing Biosignal Measurement	130
			4.4.1.1 Purpose of Biosignal Measurement	130
			4.4.1.2 Activities of an Organism and Biological Systems	131
			4.4.1.3 Advantages and Disadvantages of Biological Measurement	131
			4.4.1.4 Potential of Biosignal Measurement	132
		4.4.2	Indices of Central Nervous System Activity	132
			4.4.2.1 Electroencephalogram (EEG)	133
			4.4.2.2 Functional Magnetic Resonance Imaging (fMRI)	135
			4.4.2.3 Functional Near Infrared Spectroscopy (fNIRS)	135
			4.4.2.4 Critical Flicker Fusion Frequency (CFF)	139
		4.4.3	Indices Relating to the Visual System	139
			4.4.3.1 Eye Movement	140
			4.4.3.2 Visual Field	142
			4.4.3.3 Eye Blink	143
			4.4.3.4 Pupil	143
		4.4.4	Indices of Autonomic Nervous System Activity	144
			4.4.4.1 Heart Rate	144
			4.4.4.2 Heart Rate Variability (HRV) Indices	145
			4.4.4.3 Blood Pressure and Pulse Waves	146
			4.4.4.4 Respiration	146
			4.4.4.5 Electrodermal Activity	147
			4.4.4.6 Skin Temperature	148
		4.4.5	Facial Expression	148
			4.4.5.1 Anatomy of Mimetic Muscles	148
			4.4.5.2 Relationship Between Facial Expression and Emotion	149
			4.4.5.3 Techniques for Estimating Emotions Based on Facial Images	151
			4.4.5.4 Relationship Between Facial Expression and Driver States	152
			4.4.5.5 Application of Facial Expressions to Automobile and Future Challenges	152
		4.4.6	Biochemical Reactions	152
	References			155
5.	**Driver and System Interaction**			**162**
	5.1	Mental Workload while Using In-vehicle System		162
		5.1.1	Workload Measurement Using Questionnaires	162
			5.1.1.1 Cooper-Harper Rating Scale	162
			5.1.1.2 NASA-TLX	162
			5.1.1.3 SWAT	164
			5.1.1.4 Workload Profile Method (WP)	166
			5.1.1.5 Rating Scale Mental Effort (RSME)	167

| | 5.1.2 | Mental Workload Assessment Using the Subsidiary Task Method | 168 |

5.1.2   Mental Workload Assessment Using the Subsidiary   168
       Task Method
       5.1.2.1   Two Types of Subsidiary Tasks   169
       5.1.2.2   Psychological Concepts Related to the Subsidiary   169
               Task Method
       5.1.2.3   Example of Application of Subsidiary Task Method   171
5.1.3   Workload Measurement Based on Driving Performance   172
       5.1.3.1   Overview   172
       5.1.3.2   Steering Entropy (SE) Method   173
       5.1.3.3   Real-time Steering Entropy (RSE) Method   175
       5.1.3.4   Summary   178
5.2   HMI of In-car Information Systems   178
     5.2.1   Interaction with a System   178
         5.2.1.1   Design of Interaction   179
         5.2.1.2   Tactile Feedback   179
         5.2.1.3   Audio Interface   179
         5.2.1.4   Integrated Controller   180
         5.2.1.5   Internet Connection of In-car Devices   180
     5.2.2   Route Navigation and Map Display   181
         5.2.2.1   Volume of Graphic Information   181
         5.2.2.2   Mental Map   181
         5.2.2.3   Expression of Maps   183
         5.2.2.4   Displaying Roads   184
         5.2.2.5   Displaying Background   184
         5.2.2.6   Presenting Text   185
         5.2.2.7   Presenting Landmarks   186
         5.2.2.8   Displaying Remaining Distance/Estimated   186
               Required Time
         5.2.2.9   Displaying Routes   187
               5.2.2.9.1   Turn by Turn Display   187
               5.2.2.9.2   Route Display   187
               5.2.2.9.3   Traffic Lane Display   188
               5.2.2.9.4   Crossing Macrograph   188
               5.2.2.9.5   Highway Map   188
               5.2.2.9.6   Manoeuver List   188
               5.2.2.9.7   Guide Information to Support Safe Driving   190
         5.2.2.10   Display of Traffic Information   190
     5.2.3   Design of Menus   191
         5.2.3.1   Menu-based Interaction   192
               5.2.3.1.1   Fundamental Principles   192
               5.2.3.1.2   Presentation and Selection of Menu Items   193
               5.2.3.1.3   Strengths and Weaknesses of Menu-based   193
                       Interaction
         5.2.3.2   Design Guidelines   193
         5.2.3.3   Evaluation Methods for Menu Designs   194
5.3   Assessment of Driver Distraction   194
     5.3.1   Definition of Distraction   194
         5.3.1.1   Characteristics of Attention and Related Definitions   195

5.3.1.2	Distraction	196
5.3.1.2.1	Suggested Definitions	196
5.3.1.2.2	Relation to Inattention	196
5.3.1.2.3	Relation to Arousal Level and Workload	196
5.3.1.3	Conclusion	197
5.3.2	Assumptions for Distraction Assessment	197
5.3.2.1	Information Processing and Distraction	197
5.3.2.2	Ideas and Types of Assessment Methods	198
5.3.2.2.1	Requirements for Assessment Methods	198
5.3.2.2.2	Types of Assessment Methods	199
(1)	Primary task measurement and secondary (subsidiary) task measurement	199
(2)	Assumptions and notes for the secondary task measurement	199
(3)	Secondary task measurement and dual task measurement	199
(4)	Primary task and subsidiary/additional task	200
5.3.2.2.3	Conclusion	200
5.3.3	Visual-Manual Distraction Assessment	201
5.3.3.1	Direct Assessment	201
5.3.3.1.1	Visual Behavior	201
5.3.3.1.2	Driving Performance	203
5.3.3.2	Occlusion Method	205
5.3.4	Cognitive Distraction Assessment	206
5.3.4.1	Lane Change Test (LCT Method)	206
5.3.4.2	Detection Response Task (DRT Method)	208
5.3.4.3	Physiological Index	209
5.3.5	Reference Tasks in Distraction Assessment	210
5.3.5.1	Item Recognition Task	210
5.3.5.2	N-back Task	211
5.3.5.3	Calibration Task	212
5.3.5.4	Conclusion	213
5.3.6	Use of Cellular Phone while Driving	213
5.4	Interaction with Advanced Driver Assistance Systems	215
5.4.1	Presentation and Management of Information	215
5.4.1.1	Design of Warning Signal	215
5.4.1.1.1	Warning	215
5.4.1.1.2	Warning Compliance	215
5.4.1.1.3	Expected Driver's Response	216
5.4.1.1.4	Warning Level and Warning Design	217
(1)	Criticality and urgency	217
(2)	Warning level	217
5.4.1.1.5	Basic Requirements for Warning Designs	218
(1)	Visual presentation of warnings	218
(2)	Impression given by the design of warning signals	219

5.4.1.2   Influence of the Warning Signal on the Driver   221
Behavior
    5.4.1.2.1   Assessment of Effectiveness of the   221
                Warning System on the Avoidance
                of Danger
    5.4.1.2.2   Hazard Avoidance Scenarios of   222
                Experiments
           (1) Effectiveness of warning systems   222
           (2) Assessment of the warning signal   222
    5.4.1.2.3   Assessment of the Compliance with   223
                Warning/alerting Systems
           (1) Compliance   223
           (2) Example of assessment of   223
                 effectiveness of seat belts reminders
5.4.1.3   Priority and Management of In-vehicle Information   225
    5.4.1.3.1   Need for Information Management   225
    5.4.1.3.2   Information Importance   226
    5.4.1.3.3   Message Management   227
           (1) Selection and integration of the   227
                 message to be presented (priority
                 management)
           (2) Design consistency between   229
                 messages from different systems
           (3) Display management   229
           (4) Presentation style management   229
           (5) Time management   229
5.4.1.4   Estimation of the Driving Demand or Workload   229
for Message Management
    5.4.1.4.1   Workload Manager in Information   229
                Management
    5.4.1.4.2   Estimation of the Driving Demand   230
                based on the Road Traffic Environment
    5.4.1.4.3   Estimation based on Automotive Sensor   231
                Signals of Driving Demand in Road
                Traffic Environment
    5.4.1.4.4   Estimation of the Driving Workload in   233
                Real-time based on Sensor Signals
5.4.2   Systems and Drivers   234
    5.4.2.1   Levels of Automation of Systems and Drivers   234
        5.4.2.1.1   Automation of Systems   234
        5.4.2.1.2   Levels of Automation   234
        5.4.2.1.3   Examples of Level 1 to 3   234
        5.4.2.1.4   Examples of Level 4 to 6   235
        5.4.2.1.5   Examples of Level 6.5   235
        5.4.2.1.6   Examples of Level 7   236
    5.4.2.2   Over-trust and Overdependence   236
    5.4.2.3   Monitoring of the System Status by the Driver   239

5.4.2.3.1	Supervisory Control	239
5.4.2.3.2	HMI in Driving Supporting Systems Using V2X communication	239
	(1) Verification of operating status	239
	(2) Easy to understand	241
	(3) Communication certainty	241
	(4) Easy understanding of criticality	241
	(5) Prevention of over-trust and distrust	241
5.4.2.3.3	More General HMI in Driving Support/ automated Driving Systems	242
5.4.2.4	Changes in Driver's Behavior Caused by Introduction of the System	242
5.4.2.4.1	Driving Behavior Induced by the System	242
5.4.2.4.2	Definition of Road/traffic Factors Influencing Driving Behavior	243
5.4.2.4.3	Example of Analysis of Behavioral Changes Caused by the System	243
5.4.2.5	Compatibility of the System with Drivers' Behavior	244
5.4.2.5.1	Distance without the System and Distance with the ACC System	244
5.4.2.5.2	Relation Between Drivers' Characteristics, Driving Behavior and the Distance Selected in the ACC	245
5.4.2.6	Human Factors in Automated Driving Systems	246
5.4.2.6.1	Intersection Between Automated Driving Systems and Humans	246
5.4.2.6.2	Understanding of the System	247
	(1) Understanding of system's functions	247
	(2) Understanding of the system status	247
	(3) Understanding of the system operation	248
	(4) Understanding of the behavior of the system	248
5.4.2.6.3	State of the Driver	248
	(1) State of the driver when using automated driving systems	248
	(2) Gap in the transition to the state where the driver is able to execute driving tasks	248
5.4.2.6.4	Value of Automated Driving Systems for Humans	249
5.4.2.6.5	Interaction Between the Car and other Traffic Participant	249
	(1) Communication between traffic participants	249
	(2) Communication functions that automated vehicles must have	250
References		250

6. Driver Behavior **257**

6.1 Human Characteristics Related to Driver Behavior 257
    6.1.1 Visual Cognitive Functions 257
        6.1.1.1 Visual Attention and Its Psychological Measurements 257
            6.1.1.1.1 Shift of Attention 257
            6.1.1.1.2 Selection of Visual Information at a Fixation Point 258
            6.1.1.1.3 Useful Field of View 259
        6.1.1.2 Physiological Measurement of Attention 262
            6.1.1.2.1 Attentional Resource Allocation and Event-related Potentials 262
            6.1.1.2.2 Evaluation of Visual Attentional Resource Allocation using Eye-fixation-related Potentials 263
            6.1.1.2.3 Evaluation of Attentional Resource Allocation Using Probe Methods 264
        6.1.1.3 Visual Attentional Models 265
            6.1.1.3.1 Saliency Model of Itti and Koch 266
            6.1.1.3.2 Models that take Account of Top-down Factors 267
            6.1.1.3.3 Application of Models to Moving Images 268
    6.1.2 Information Processing and Cognitive Models for Humans 268
        6.1.2.1 Driver Information Processing Models 268
            6.1.2.1.1 Basic Three-stage Information Processing Models for Humans 268
            6.1.2.1.2 Information-processing Model taking Account of Memory and Attention 269
            6.1.2.1.3 Norman's Seven-stage Action Model 270
            6.1.2.1.4 Situation Awareness Model 271
            6.1.2.1.5 Hierarchical Model of Driving Behavior 272
            6.1.2.1.6 Rasmussen's Skills-rules-knowledge (SRK) Model 273
            6.1.2.1.7 Relationship Among Different Human Information-processing Models 274
            6.1.2.1.8 Extended Contextual Control Model (E-COM) 274
        6.1.2.2 Task-capability Interface Model 275
6.2 Driving Performance 277
    6.2.1 Driving Performance Measures 277
        6.2.1.1 Longitudinal Driving Performance 277
            6.2.1.1.1 Velocity, Acceleration, and Jerk 278
            6.2.1.1.2 Response Time 278
            6.2.1.1.3 Headway Distance and Time 279
        6.2.1.2 Lateral Driving Performance 280
            6.2.1.2.1 Steering Operation 281
            6.2.1.2.2 Steering Reversal 281
            6.2.1.2.3 Steering Entropy 281

	6.2.1.2.4 Lane Position of a Vehicle	282
	6.2.1.2.5 Standard Deviation of Lane Position (SDLP)	282
	6.2.1.2.6 Time to Line Crossing (TLC)	282
6.2.1.3	Parking Maneuver	283
	6.2.1.3.1 Cognitive Function Necessary for Parking Maneuver	283
	6.2.1.3.2 Prediction of One's Capability for Park Maneuver based on the Psycho-motor Tests	283
6.2.1.4	Situation Awareness Evaluation Methods	284
	6.2.1.4.1 Situation Awareness Global Assessment Technique (SAGAT)	285
	6.2.1.4.2 Real-time Probe Technique	285
	6.2.1.4.3 Subjective Rating (SART: Situation Awareness Rating Technique)	286
6.2.2	Driving Ability Evaluation for Elderly Drivers	287
6.2.2.1	Ability Evaluation of Driving Behavior	287
6.2.2.2	Evaluation of Perceptual-Motor Coordination	288
6.2.2.3	Evaluation of Cognitive Functions	289
	6.2.2.3.1 Neuro-psychological Tests and Driving Ability	289
	6.2.2.3.2 Screening Test for Elderly Drivers	291
6.2.2.4	Models of Driving Ability for Elderly People	291
	6.2.2.4.1 Multifactorial Model for Enabling Driving Safety	291
	6.2.2.4.2 Adaptive Driving Behavior of Elderly People	292
6.3	Driver's Behavior Models	293
6.3.1	Driving Behavior Models	293
6.3.1.1	Driver Steering Control Models	293
	6.3.1.1.1 Basics of Modeling	294
	6.3.1.1.2 Major Examples of Driver Steering Control Models	295
	(1) Preview-predictive model	295
	(2) Describing function model	296
	(3) Pursuit control model	297
	(4) Other models	297
6.3.1.2	Model of Visual Recognition During Driving	297
	6.3.1.2.1 Perception of Direction of Travel	297
	6.3.1.2.2 Use of Tangent Points	298
	6.3.1.2.3 Use of Information on Near and Far Areas	299
	6.3.1.2.4 Effect of Gaze Direction	301
6.3.2	Information-processing Models Related to Driver's Behavior	301
6.3.2.1	Information-processing Models for Drivers Using Car Navigation System	301

6.3.2.1.1 Information-processing Models for Drivers 302
using a Digital Map System with Self-
localization Function
6.3.2.1.2 Information-processing Models for Drivers 304
using a Turn-by-turn Navigation System
6.3.2.1.3 Information-processing Models for 305
Drivers using a Navigation System
Capable of Displaying an Enlarged
View of Intersection
6.3.2.2 ACT-R (Adaptive Control of Thought-Rational) 307
Model of Driving Behavior
6.3.2.2.1 Driving Behavior and Integrated 307
Driver Models with an ETA Framework
Viewpoint
6.3.2.2.2 Integrated Driver Model using the 308
ACT-R Cognitive Architecture
(1) Control 308
(2) Monitoring 310
(3) Decision-making 310
(4) Component integration and 311
multitasking
(5) Parameter values 311
6.3.2.2.4 Validation and Application ACT-R Model 312
of Driving Behavior
6.3.3 Statistical Behavior Models 312
6.3.3.1 Structural Equation Models for Driving Behavior 312
6.3.3.1.1 Structural Equation Models (SEM) 313
6.3.3.1.2 Structural Equation Model of Driving 315
Behavior for Making a Turn
6.3.3.1.3 Application of Structural Equation Model 316
to Theory of Planed Behavior
6.3.3.2 Bayesian Network Models for Driving Behavior 316
6.3.3.2.1 Bayesian Network Model 316
6.3.3.2.2 Dynamic Bayesian Network Model 317
6.3.3.3 Modeling Driving Behavior Using Hidden 320
Markov Models
6.3.3.3.1 Theoretical Background of Modeling 320
Driving Behavior Using HMM
6.3.3.3.2 Example of Constructing a Driving 321
Behavior Model Using Discrete HMM
6.3.3.3.3 Estimation of Road Shape and Driving 322
Behavior Using Continuous HMM
(1) Collection of driving signals and 322
creation of corpus
(2) Estimation of driving behavior in 323
relation to specific road shape

6.3.3.3.4 Estimating Driving Behavior Using   323
                  HMM and other Applications
              (1) Prediction of driving behavior   323
              (2) Estimating characteristics of   324
                  individuals
              (3) Future direction and issues   324
References   325

Index   **333**

1

Overview of Automotive Ergonomics and Human Factors

1.1 Ergonomics and Human Factors for Making Products and Systems Compatible with Humans

Ergonomics and human factors is a study for designing products, systems and environment that are compatible with human characteristics. It is an interdisciplinary field integrating psychology, physiology, engineering and design. Ergonomics originated with concern for occupational health in the latter part of the 19th century (Jastrzebowski, 1857). 'Human factors' entailed focus on human abilities while using a certain system, mainly in the United States and the United Kingdom around the time of World War II. It was recognized that in order to improve the performance of an overall system, it was necessary to take into consideration the human characteristics which are part of the factors constituting the system. It is generally considered that ergonomics/human factors developed in mid 20th century, though the ergonomic design of automobiles was born at the end of the 19th century (Akamatsu et al., 2013).

As automobiles came into wide use after World War I, the issue of automobile accidents drew widespread attention. At the end of the 1920s, a German psychologist, Narziss Ach, proposed an ergonomic concept called *psychotechnik* (i.e., psychological engineering), emphasizing the need for technology based on psychological studies (Ach, 1929). Meanwhile, at the end of the 1930s, T.W. Forbes in the United States pointed out that accidents are caused not by specific people, but by normal people, and asserted that safety is related to the limits of human abilities, such as visual ability and response time, and therefore human factors, such as psychological and physiological characteristics, should be taken into consideration when designing automobiles (Forbes, 1939). Around this time, it was decided that human characteristics should be understood scientifically so that automobiles are designed according to them.

1.2 Beginning of Human-compatible Automobile Design

At the end of the 19th century, the gasoline-fuelled internal combustion engine was invented, leading to the manufacture of automobiles as machines that move the vehicle body by using power. They were machines fully controlled by humans, whether moving forward, stopping, or turning right or left. Improvements were made to enable humans to operate them; gradually they developed into the current form of automobiles.

The signature improvements were the circular steering wheel and foot pedals. In the early days, a bar handle called a 'tiller' was used as the steering mechanism in order to make the structure simple (Fig. 1.1, left), but it was found that the handle would shake wildly when driving fast on an unpaved bumpy road and the driver would face trouble in controlling it. Therefore, a bar handle with grips on both sides was developed to allow the driver to hold the grips with both hands before the round steering wheel using a gear was introduced.

In a horse-drawn carriage, the speed was decreased by making the horse to slow down its walk. So, it was sufficient to have a parking brake by which the driver pressed the shoe to a carriage wheel or axle with the help of a hand lever. However, since the force applied by a hand lever was insufficient to stop a fast automobile (Fig. 1.1, right), it was replaced with a foot pedal that could apply a greater force. The foot pedal was initially positioned on the vehicle floor close to the seat, but it gradually came to be positioned towards the front. In an ergonomics textbook, you will notice that the maximum force is applied by a foot when it is applied towards the front direction. Pioneering automobile developers empirically discovered the ergonomic pedal position for hard braking.

The 1930s saw the emergence of many designs and functions aimed at safety or driving ease. For visibility, originally hand-operated wipers were replaced by air-driven or electric wipers, and such new equipment as a defroster (Fig. 1.2) and a mirror that could avoid glare at night time driving were developed. Also, in the 1930s, the direction-indicator switch was positioned at an easily reachable part of the instrument panel or on a lever protruding from the steering column to make it easier for the driver to operate the switch (Fig. 1.3).

The steering with the help of a round handle and the acceleration/deceleration control by a foot pedal did not change for 100 years. Of course, keeping the manner of operation consistent was a basic ergonomic design principle, so no need was felt for making unnecessary changes. Furthermore, no tool better than a round handle that can be operated with both hands was discovered for making both slight correctional steering operation for maintaining straight running and large turns at intersections, and for doing these smoothly. The manner of operation that has changed is the introduction of the shift lever. The drive-by-wire technology increased the degree of freedom of design. Whereas the gearshift on the centre control still remains due to maintenance of consistency in the manner of operation, the paddle shift was positioned close to the steering for the same reason as the direction indicator.

Although the introduction of drive-by-wire technology has increased the degree of freedom, no dramatic change is expected in the future unless there is a notable change in the concept of operation, such as changing the tyre direction or changing the pressing force of the brake pad. If the concept of operation were to change, it would be by progress in automated driving technology (see 5.4.2.6). If automobiles gain intelligence, they may become machines that would be controlled by commands or communications, rather than manipulations. Then there would be a need for conducting additional research and development.

1.3 Vehicle Cabin Design

Until around 1920, the positional relationship between the steering wheel/pedal and the seat had been fixed after being designed. Automobiles were not only driven by men, but were also driven by active women. Such women must have had trouble with the layout designed for men. Around the 1930s, the seat-sliding mechanism was introduced (Fig. 1.4).

A group led by R.A. McFarland in the United States collected data on body sizes and published the data in a series as documents of the Society of Automotive Engineers (SAE) from around 1950 (McFarland et al., 1955). He presented the percentile of measurement, which is an important concept in ergonomic design. Understanding that vehicle interior design based on average body measurement values would only satisfy a limited number of people whose body measurements are close to the average values, he indicated the 5th percentile as the minimum value and the 95th percentile as the maximum value (Fig. 1.5).

After around 1930s, when rounded aerodynamic body shape was introduced, it was considered that automobile collision accidents resulted from oversight and that it was important to secure visibility in order to prevent accidents. Therefore, researchers worked on improving both direct and indirect visibility through mirrors. A method to evaluate the range of the driver's visibility was developed based on the range of areas illuminated by lamps positioned at locations that correspond to the driver's eyes (Hunt, 1937). The problem of visibility does not only involve the problem of the size and position of the windows and pillars, but is strongly related to the size of the human body. Since visibility depends on the positions of the eyes, the seat position that decides the eye position and the driver's body size become an issue.

A concept that was introduced for considering the eye position is eyellipse. When considering the variations in human body size, variations in the seating location on the seat, and differences in the seat position, the distribution of eye locations can be approximated to a three-dimensional ellipse. Therefore, an ellipse based on the percentile values of the three-dimensional distribution of eye locations was named 'eyellipse'. These were published by the SAE as SAE standards by the end of the 1950s and served as the basis for the cabin design (SAE Recommended Practice J941, 1965).

In the design process, such human body size is difficult to deal with if there are only numerical data. So, in the 1960s, 2-D manikins and 3-D manikins were developed and standardized as SAE J826 (SAE Recommended practice J826, 1962). In the 1970s, computer technology made progress and the development of CAD manikins, which are manikins on computer, started. Chrysler developed CYBERMAN followed by SAMMIE which was developed in the United Kingdom, RAMSIS in Germany, and Jack in the United States. CAD manikins came to be used in the actual design process by being incorporated in CAD applications for designing automobiles, such as CATIA.

While 2-D and 3-D manikins are measurement models, CAD manikins can reproduce a body of three-dimensional shape, which require not only data of the human body size, but also data of the three-dimensional shape. Therefore, the manikins need technology for three-dimensional shape measurement and for modeling based on the obtained data. Because it is possible to move the manikin on the computer screen, dynamic human movements, such as getting in or out of the vehicle, can be evaluated. However, with regard to reproduction of human body movement, the main approach currently used is to reproduce the data measured through motion capture, and no modeling technology has been established yet for reproducing any intended motion. When evaluating the ease of getting in or out of the vehicle, not only the interference (contact) between the human body and the vehicle, but also the motion stress must be evaluated, and studies are under way to evaluate these by using CAD manikins.

Due to an increase in the number of elderly drivers, consideration should also be given, not only to the differences in the body sizes of the driver and passengers, but also to the influence of their age. Research and development are conducted for CAD manikins that take into account the change in posture, loss of muscle strength, and decrease in the range of joint motion in line with aging. A person's motion is not decided by physical and spatial restrictions alone; it also depends on how the person perceives that space. Therefore, there is a need for CAD manikins in vehicle interior design, taking into account elderly people's perceptual/cognitive characteristics.

1.4 Instruments and Displays

1.4.1 Instrument Arrangement

With regard to the visibility of instruments, there were few instruments until around the 1900s and up to around early 1910 when they were attached to the bulkhead separating the engine and the cabin, providing poor visibility (Fig. 1.6, left). In the late 1910s, an instrument panel began to be installed and an array of meters was arranged on a luxury car (Fig. 1.6, right). In the 1930s, a meter cluster was installed near the steering wheel, in order to improve the visibility for the driver. Incidentally, in terms of visibility, the first instrument that was positioned at an easily visible location was the thermometer attached to the radiator head and which emerged in the 1910s. This was later replaced by the radiator mascot.

Instruments came to be positioned in front of the steering wheel, where it is easy for the driver to see, but sufficient consideration had not been made with regard to their interference with the steering wheel. The use of eyellipse made it possible to design instruments at positions where their visibility was not obstructed by the steering wheel. From the 1950s, the meter cluster was moved up to the position on the upper part of the instrument panel. While the fascia was lowered to secure the view, the visibility of the meters was ensured. In the 1970s, non-reflecting glass came to be used for the surface of meters, thus reducing the glare and improving their visibility. In the latter 1990s, the center meter, which is positioned in the middle of the dashboard, appeared (Fig. 1.7). It was a position where the amount of the gaze shifted from the road scene ahead to the meters is small. The visual recognition time was experimentally measured, and the position was found to be notably effective, particularly for elderly people (Atsumi et al., 1999).

As new instrument technology, the head-up display (HUD) was introduced. HUD was originally developed for aircraft. Because it had a great advantage in reducing the amount of gaze shift, studies began in the 1970s to apply HUD to automobiles (Rutley, 1975). HUD was first mounted on mass-produced cars in 1988 on a GM car and a Nissan car (Okabayashi et al., 1989). Although HUD was proved to be ergonomically effective (Ward and Parkes, 1994), it did not come into wide use because the benefit was not large when only the speedometer was displayed.

With the introduction of the driver assistance system, the information to be displayed during driving increased. Accordingly, HUD became more useful and came to be used more widely. Moreover, when automated driving technology becomes established, further use of HUD should also be considered. When the driver does not have to control the traveling course or distance between cars in real time due to the use of automated driving technology, it will be possible to display on a large HUD the road scenery with information overlapped on it by augmented reality (AR) technology. It may also be possible to use an AR display that would call the driver's attention or support and enhance the driver's situation awareness during automobile driving.

1.4.2 *Meters and Displays*

With regard to the meter design, bobbin-type speedometers using the centrifugal force of a rotating cable were used until around 1920, but later the clock-type rotary meters came to be widely used. From the 1930s, lettering in meters was artistically designed but some font designs were difficult to read and in low contrast. This tendency continued until around 1960, but easy-to-read fonts started to be used in the 1970s.

The digital number display that started with neon tubes developed into seven-segment liquid crystal displays came to be extensively used in calculators. Digital speedometers were introduced in automobiles in the 1980s (Ishii, 1980). The advantages of digital speedometers are that because only numerical values are displayed, they can be visually recognized quickly and do not interfere much with the steering wheel due to their small surface area. On the other hand, they have challenges including flickering and glare on account of being self-luminous. Therefore, development of technology for preventing flickering and studies on easy-to-read brightness and contrast were conducted.

Reading analog meters can be quick using pattern recognition, while reading digital meters requires symbol processing. Analog meters are convenient for grasping the approximate values and their changes. Digital displays indicate accurate values, but the changes in values are difficult to grasp. Digital displays are suitable for cases where accurate values are required, such as in the case of cruise control or speed limiter. An appropriate display method must be considered, depending on its functions (Merker, 1966).

With the progress of automated driving technology, the information to be displayed becomes different from the past. It is necessary to display the distance from the preceding vehicle, position within the lane, positions of the surrounding vehicles, and relative speed. Furthermore, the operation status of the system must be displayed, such as the ON/OFF/READY status of the system, system error, and system mode (see 5.4.2.6). These are currently designed through trial and error. Since many elements would need to be displayed, the challenge is how to display them in an integrated manner. Although they are presented, sometimes it is difficult to detect the changes in the elements from slight differences (Fig. 1.8). The design needs to be based on the user's mental model of the system and its operation.

1.4.3 Controls

Many early automobiles used lever-turning switches, which were for activating the linkage mechanism. Later, when wire began to be used, pull-type switches were introduced. Then, when electricity started to be used, see-saw-type switches came into use (Fig. 1.9). The advantage of see-saw-type switches is that they are directly visible whether the switch is turned on or not. The driver can confirm that the operation has been securely conducted. In the case of switches for which the operation status is not directly visible, an ON/OFF indicator is needed (Nevett, 1972).

Initially, there was no indication as to which switch was to be used for which function. The driver read the instruction manual and memorized the functions of the switches. Then, the functions were indicated using a printed word on the switches, and later initial letters of the functions were also used. From around 1960, symbols came to be indicated in automobiles, mainly in Europe, where multiple languages are used (Fig. 1.9). Since symbols cannot be understood unless they are standardized, standardization efforts were initiated with publication of *SAE J1048* in 1974 (Heard, 1974). The symbols are specified internationally in *ISO 2575*. P. Green pointed out that many of them were created from an engineer's viewpoint (for example, the brake symbol expresses the structure of a drum brake) and are not necessarily understandable for the general public (Green and Pew, 1978).

At times when switches were independent and had the same shape, it was important to distinguish them by indicating their functions. However, when switches were incorporated into a touch panel, they could be designed with more freedom. Their display size was made variable and their layout could be changed. They could be expressed in various colors and gradations, but care had to be taken to avoid a design that would distract the driver while driving. We must not forget that it is different from PCs, smartphones, or home electric appliances, which are used on their own.

Touch panels offer freedom of design and because the display and operation coincide, their operability is somewhat like the conventional switches. A visual change or a sound is used as feedback in response to an operation. Furthermore, technology to give tactile and kinesthetic feedback, resembling a mechanical switch, has made progress. However, a display device that is looked at while driving, such as a car navigation system, should desirably be installed at a location close to the front-road view. Also, since elderly people have a poor near-distance vision, it is advisable to place the display device at a distant location from such a viewpoint. If the display device is placed at a highly visible location for such reason, the input device cannot be reached by hand, and a touch panel cannot be used.

In that case, the display and the controls will have to be positioned at distant locations, and here the type of input device to be used becomes an issue. Such devices as joysticks, touch-pads, and rotary knobs are used. While joysticks and touch-pads allow two-dimensional operation, which adopts a human machine interface (HMI) similar to the concept of the graphical user interface (GUI), rotary knobs allow only one-dimensional operation. Therefore, some indication or metaphor is required to make clear that the menu is to be selected by one-dimensional rotating operation (see 5.2.1).

There are also expectations for controlling the system using gestures, due to the advantage of not having to reach toward switches, etc. Gestures that can be used for operation must be those that can be used intuitively. A system that requires the driver to learn new gestures for operation will hardly be acceptable. The feature of not having to reach toward switches may seem to be an advantage, but moving a hand by holding it in midair causes muscle load. Taking these into account, the number of potentially usable gestures would be limited, being unable to cover the variation of operation instructions needed for an in-vehicle system. Further, even if intuitive, people's gestures are often not the same though they may be common to a certain extent. This point also presents a problem.

1.5 Riding Comfort and Fatigue

Just after the birth of the automobile, speed was slow, and suspension technology that had been used in horse-drawn carriages was employed. Soon after, vibration became a big problem as speed increased. At the beginning of the 1900s, Frederic Lanchester in the UK, who attacked the issue of uncomfortable vibration, postulated that vibration during driving should be the same as that during walking, which people were accustomed to. He adopted a design policy to stabilize the visual field based on the oculo-motor control for compensating the head movement caused by vehicle vibration (i.e., VOR or vestibulo-ocular reflex). This was pioneering work in automobile design considering the compatibility with human functions. A man's walking pitch is about 2 Hz irrespective of the walking speed. In order to achieve this, a linkage mechanism was devised whereby the leaf spring was attached upside down and the axle was attached at the rear end to make it move in the vertical direction with a link (Clark, 1995). Still today, the frequency of vibration of upspring mass is about 2–3 Hz, and this originates from Lanchester's study.

Although vehicles were designed so as to prevent vehicle vibration, driving actually required drivers/passengers to have physical strength and patience due to the underdeveloped road conditions. In the 1930s, the term 'ride comfort' was used to cover both fatigue and ride quality. Figure 1.10 shows the factors related to ride comfort which R.W. Brown illustrated in 1938. It indicates that Brown had considered vehicle vibration due to uneven road surfaces to be a major factor (Brown, 1938).

1.5.1 Fatigue

Fatigue was regarded as a natural phenomenon accompanying daily activities in the 19th century (Dhers, 1924). As working conditions changed, for instance, introduction of work hours, in European countries at the end of 19th century, fatigue came to be regarded as an unpleasant human state that was to be removed. A simple definition of fatigue is deterioration in work performance resulting from continuous work for long hours. However, apparent changes in driving performance often cannot be observed even after continuous driving. A pioneering study conducted by F.A. Moss in 1925 failed to show any change in driving performances after 60 hours of sleeplessness (http://siarchives). From this experience, Moss postulated that driver fatigue needs to be objectively identified and that physiological indices, such as heart rate, gravity center fluctuation, metabolism, response time, and blood components should be measured (Moss, 1930). Since then, there have been many studies (see 4.1).

Crawford stated difficulties in studying driving fatigue in 1961 (Crawford, 1961). One of the difficulties is to define the driving performance, and thus to measure it. Contrary to our general impression, it was difficult to show a clear relationship between hours of driving and the number of accidents. In an experiment of long-haul truck drivers by McFarland and Mosely in 1954, about a half of near-accidents occurred in the first two hours and far fewer in the latter stage of trip (McFarland and Moseley, 1954). Another thing that made it difficult is that driving and fatigue are interrelated. Whereas fatigue is caused by driving, the influence of fatigue shows up in driving, so the cause and the result cannot be separated. After 20 years, in 1979, Broadbent stated in his paper, 'Is a Fatigue Test Now Possible?' but still we cannot solve these difficulties (Broadbent, 1979). While changes in physiological indices were observed in many studies, they were not seen in some other studies (see 4.4). Psychological fatigue and physiological fatigue should be considered separately. Furthermore, I.D. Brown pointed out the need to consider, not only the duration of the driving task, but also circadian rhythms and sleep records, especially in fatigue of truck drivers (Brown, 1994).

If the driving performance deteriorates, the risk of accidents may increase. In order to avoid the increase of risk, the driving performance should be maintained to meet the driving demand (Fuller, 2005). This can be a reason that the driving performance maintains a certain level even in fatigued conditions.

1.5.2 Vibration

The issue of vibration, drew attention since the 1930s regarding riding comfort and was studied as an issue of biomechanical load that triggers occupational drivers' back pain and a decline in performance caused by vibration. Vibration of the body includes whole-body vibration if the frequency is low, but if the frequency becomes higher, vibration has a different influence on various parts of the body, not only as physiological influence, but also as a health disorder. However, vibration dampens on the respective organs of the body when the frequency is 300 Hz or more. The resonance frequencies of the respective body organs were obtained, such as about 20 Hz for the head. Regarding whole-body vibration, human sensitivity to vertical vibration frequencies was measured. This was published in 1974 as *ISO 2631*, and came to be widely cited (*ISO 2631*).

ISO 2631 covers various kinds of whole-body vibration, but only such vibrations of a specific frequency range are applicable to automobiles. An issue specific to automobiles is ride comfort. With regard to this issue, the relationship between a linguistically expressed impression of ride quality and frequency was obtained (see 3.1.1).

Vibration was regarded as a negative factor that had to be eliminated, but if vibration was reduced, there were less somato-sensory cues about vehicle behavior. The perception of vehicle behavior from vibration of 1 Hz or less and the gripping feeling of the tyres from high frequency vibration was sensory feedback that was necessary for identifying vehicle behavior (see 3.1.3), so optimum vibration needed to be considered.

1.5.3 Arousal Level

The issue of drowsy driving was addressed as part of the fatigue problem relating to occupational drivers who drove long distances over long hours (see 4.3.1). Due to working long hours, the resultant deprivation of sleeping hours, and the night work, occupational drivers tended to become sleepy and submit to drowsy driving. The fluctuation of the arousal level caused by circadian rhythms was robust, and the arousal level reduced at around 4:00 a.m. in early morning and during 2:00 p.m. to 3:00 p.m. in the afternoon. Accident statistics showed that many accidents occurred during these hours. The influence of sleeping hours was also strong. In the case of working at night due to shift work, sleeping hours tended to become short as a result of not being able to sleep enough during daytime. Moreover, there was a growing tendency to stay up late, with so many people chronically lacking sleep. In such a case, the arousal level easily reduced while driving.

Attempts were made to develop devices for preventing drowsy driving since the 1980s (Fig. 1.11). To that end, there was a need to detect reduction in the arousal level. Changes in physiological indices, including the electroencephalogram (EEG), heart rate, respiration rate, and galvanic skin response (GSR), were observed at the low arousal level (Erwin et al., 1976). Since sensing devices that required contact with the skin surface to measure the physiological indices were difficult to install in vehicles, there were studies on non-contact measurement methods (Dingus et al., 1985). Among these indices, facial expressions, particularly the state of eyelids, was found to be highly sensitive in detecting sleepiness. Thus, an index called PERCLOS (PERcentage of eyelid CLOSure time) that can be detected by a camera was developed and has become a standard index (Wierwille, 1994) (see 4.3.3).

Even if sleepiness can be detected, drowsy driving cannot be prevented unless the arousal level is raised. Various stimuli, such as light, sound, smell, and vibration have been studied, but a decisively effective method has yet to be found (see 4.3.4). A definitely effective factor is a nap. However, a driver cannot always take a nap.

1.6 Vehicle Interior Environment

In the early days, automobiles had no roof. The driver and passengers protected themselves against the cold by wearing warm clothing. Vehicle's cabin heating was started in the 1920s. Interior air was heated by drawing the engine heat into the cabin (Fig. 1.12). Studies on temperature and humidity have a long history. In 1904, Willis Carrier created a psychometric chart that illustrates the relationship between wet-bulb temperature, dry-bulb temperature, humidity, dew point, etc. People's thermal comfort was decided by air temperature, radiation temperature, wind velocity, humidity, metabolism rate, and articles of clothing and their heat insulation property. It was difficult to measure these factors within an automobile cabin, but it became possible after NASA developed a thermal manikin to be used in a spacesuit in 1966. The thermal manikin was introduced to assess the thermal comfort in the vehicle cabin (Wyon et al., 1989).

Thermal comfort is regarded to be better at a thermal equilibrium state, but in summer, it feels more comfortable when the temperature is lower than the equilibrium state, and in winter, it is more comfortable when the temperature is higher than the equilibrium state. Since the state of sitting in an automobile seat is more complicated than the state of merely wearing clothes, the conditions of the contact with the seat surface and the seat material also affect the level of comfort. Also, the conditions that are found to be comfortable differ between the time immediately after getting into cabin and after driving for long hours. Efforts were made to create thermal comfort by controlling the interior air temperature. However, considering energy efficiency, there was a need to create a comfortable state efficiently by using local heating or local cooling.

1.7 Driving Tasks and Non-driving Tasks

1.7.1 In-vehicle Systems

When people drove automobiles to reach a destination, they used a map and a magnetic compass. When automobiles were first invented, road maps had already been published. They were made for traveling on bicycles (Fig. 1.13). People first kept a map in their pocket and later in the automobiles' glove box or door pocket (Fig. 1.14).

There were automobiles equipped with a magnetic compass, but it was not until electronic technology made progress in the 1980s that the car navigation system came to be used. Honda's Gyrocator appeared in 1981, and by around 1985, in-vehicle devices that displayed the distance to and the direction of the destination were released under the names of 'drive computer', invoking the image of mounting a computer to an automobile (Fig. 1.15). The current car navigation style using a digital map was introduced by Toyota's Electro Multi-Vision in 1987 (Fig. 1.16). Car navigation systems began to be popularized in the 1990s with the use of the global positioning system (GPS) and development of digital maps (Ikeda et al., 2010).

With the introduction of in-vehicle systems, a new human factor issue was raised, which was the dual task condition where a driver performed the car driving task and the task of using the in-vehicle device at the same time. While using a car navigation system, the driver would often check the present location or which intersection to turn at, resulting in frequent interactions with the system. When using a paper map, the driver would stop the automobile and check the map, but with the car navigation system, the driver was able to obtain information for the route even while driving the automobile.

The route guide was useful for driving and it also contributed to safety since the driver did not have to decide which route to choose at an intersection and thus abrupt lane changes were avoided. At the same time, however, there were concerns that the continuous use of an in-vehicle information device during driving would increase the burden on the driver. This concern had been recognized since the development phase, so ergonomic studies were carried out in parallel to the release of the system. In Europe, studies on HMI and information management were conducted in the PROMETHEUS project, which started at the end of the 1980s, and the successive projects of DRIVE (see Fig. 5.58), GIDS, HASTE, and AIDE (Michon and Smiley, 1993; Parkes and Franzen, 1993). In response, guidelines for achieving a balance between safety and usability (JAMA guidelines [Japan], UMTRI guidelines [US], HARDIE guidelines [Europe], etc.) and relevant international standards were actively developed (Heinrich and Conti, 2015).

1.7.2 Non-driving Activities like Using Mobile Phones

Concurrently with the spread of car navigation systems, mobile phones became common around the mid-1990s. Soon drivers started to bring mobile phones into automobiles. The issue of using a phone while driving became a big social problem and the handheld mobile-phone use while driving was prohibited in many countries. Because hands-free phone calls were allowed, a hands-free kit or a system became available. However, with the spread of smartphones in recent years, the desire to operate a smartphone other than for talking, such as checking e-mails and browsing the internet, grew among the users. How this problem should be treated was also an issue.

Concerns developed at a driver's engagement in an activity other than driving, not only using a mobile phone but also eating, drinking, or the like (non-driving-related activity) would disturb driving and impair safety. In the US, the National Highway Traffic Safety Administration (NHTSA) conducted the 100-Car Naturalistic Driving Study to investigate the actual situation. Around this time, the term 'distraction' came to be used widely. Distraction is defined as a diversion driver's attention away from activities for safe driving toward competing activities (Regan et al., 2008). It is distinguished from a disturbance of driving caused by low arousal level.

When a driver conducts a non-driving task during a driving task, there will be two distraction aspects—one is the issue of looking aside while driving, and the other is the issue of limits in the capacity of mental resources to perform two tasks simultaneously. A distraction caused by a non-driving-related visual task or visual-manual task is called 'visual distraction', and a distraction caused by mental activity which does not involve a visual task is called 'cognitive distraction'. These are collectively referred to as 'driver distraction'.

1.7.3 Visual Distraction

A similar issue of looking aside while driving already existed in relation to car audio. Rockwell measured the driver's visual behavior toward an audio device when operating it while driving. Rockwell revealed that the average time per glance for selecting a radio station or controlling the sound was almost the same at 1–1.5 seconds. While the driver glanced only once to adjust the sound volume, the driver needed to glance four to five times to select a radio station (Rockwell, 1988). In other words, the time of looking away from the road while driving was limited to within 1.5 seconds, and this was repeated in order to complete the task while driving. Based on this, Zwahlen proposed a guideline that a task which can be completed by repeating a glance of about one second less than three times is acceptable while driving, but a task that requires visual recognition of two seconds or more each time and a task that needs to repeat the glance four times or more is not acceptable while driving (Zwahlen et al., 1988) (Fig. 1.17). Studies began on how to determine the allowable extent of visual-manual tasks of in-vehicle devices while driving.

In the 1960s, J.W. Senders developed the occlusion method to measure how much visual information was required (visual demand) to perform a driving task. Senders used a helmet with a smoked visor which opened when pressing a switch to secure a view, and closed when releasing the switch to shut off the view (Fig. 1.18). He conducted an experiment, in which participants were asked operate the visor to the extent necessary, to see how much visual demand a driver needed to drive on an actual road (Senders et al., 1967). While this was a study for estimating driving demand, it turned out to be applied for estimating the visual demand of in-vehicle tasks while driving.

By applying this occlusion method to in-vehicle tasks, a method using liquid crystal shutter goggles was developed. In Senders' study, the time during which the view was shut off corresponded to the time of looking aside. In the case of an occlusion method for an in-vehicle task, the time when the view was obtained corresponded to the time for operating an in-vehicle device, and the time when the view was shut off corresponded to the time of looking at the road head. The occlusion method for estimating the influence of an in-vehicle task on driving in advance was developed around 2000 and was applied to the JAMA guidelines and the AAM guidelines. The method was also internationally standardized (*ISO 16673*).

1.7.4 Mental Workload and Cognitive Distraction

Automobile driving, where by the driver controls the automobile by obtaining mainly visual information from the road traffic conditions, is an information-processing task (mental work) (see 6.1.2.1), and the use of an in-vehicle device, such as a car navigation system or a communication device, is also an information-processing task. Performing these tasks simultaneously means having to process a large amount of information while driving.

Study of a human's information-processing ability or mental capacity (or attention resource capacity) was started in applied psychological research around 1950. Since there are limits to conducting many tasks at the same time, a person's information processing capacity was considered to be limited. Such limit came to be called 'mental capacity'. The capacity remaining after consuming the part used for a certain mental work is called the 'spare capacity'.

The load of mental work, in other words the mental workload, is considered to be the amount of resources used for performing a task relative to the person's total mental capacity. In the case of muscle work, the exerted force can be physically defined, and the load can be measured based on the task itself. In the case of mental work, however, the load size cannot be measured directly. Studies to measure mental workload were conducted using physiological measures, subjective measures, and performance measures. The subsidiary method, which is one of methods using the performance measures, was established in the process of research on mental workload (see 5.1).

Assuming that the total mental capacity is constant, the person uses a part of the resource for performing a task. The amount of use will be large in the case of a high-demand task, and small in the case of a low-demand task. Therefore, the reserved capacity (amount of remaining resource) during the task will be small in the former case and large in the latter case. The subsidiary task technique is a method that asks the subject to perform a subsidiary task by using the entirety of the reserved capacity, and estimates the amount of resources required for the main task (primary task) from the decline in performance of the subsidiary task (Fig. 1.19). This experiment technique was developed by German psychologist Bornemann in the beginning of 1942 (Bornemann and Teil, 1942). It was a technique to experimentally show that, while an action is conducted almost automatically when it becomes a skill, a subject will devote mental resource to conduct another task.

I.D. Brown et al. conducted an experiment to present eight-digit numerical sequences to a driver at a constant interval during driving, and have the driver say which digit was a different value from the previous numerical sequence, and demonstrated that a driver with higher skill exhibited a higher subsidiary task score (Brown and Poulton, 1961). In research on pilots conducted at NASA, a simple visual task whereby the subject pressed a button when a small lamp lighted up was used as a subsidiary task. T. Miura et al. applied this method for subjects while driving an automobile or a motorcycle, and measured the workload during the driving (Miura, 1986). The subsidiary task method was widely used for tasks of operators, pilots, and drivers (see 5.1.2).

While the subsidiary task method was introduced to measure the workload of a driving task itself, given the concern regarding use of in-vehicle devices while driving, the technique came to be used for evaluating the influence of such devices (cognitive distraction) (see 5.3).

1.8 Driver Model and Driving Behavior Measurement

1.8.1 Driver Model

Norbert Wiener's cybernetics is an idea inspired by the feedback mechanism in human motion control. It is an interdisciplinary research field of control, communication and living organisms (animal and human). This was also applied to automobile driving, and studies on driving control models that regard steering as a feedback system started in the 1960s (McRuer and Weir, 1969) (see 6.3.1.1). In order to investigate man's control characteristics, tracking tasks were used as abstracted steering tasks. At the simplest level, the steering task was approximated by a system with a time-delay plus first order lag, and driving was regarded as preview control. It was also demonstrated that steering in curve driving can be modeled by a combination of feed-forward control, pursuit control, and compensatory feedback control (McRuer et al., 1977).

These were theoretical studies, but they are applied to automated driving technology, such as adaptive cruise control (ACC) and lane-keeping system. In order to establish an assistance system that will be accepted by drivers, the system should provide control similar to control by humans, and studies on driver models will be essential for achieving such control. Meanwhile, the ideas of feedback and feed-forward were also handled in studies on driving skills. Since racing drivers gaze far ahead, measures for acquiring visual information as input information were studied (Land and Tatler, 2001) (see 6.3.1.2).

1.8.2 Driving Behavior Measurement

1.8.2.1 Site-based Measurement

Many ergonomics studies were conducted in laboratories. However, in studies on automobile driving, driving behavior on real roads needed to be investigated. Such studies serve as basic data for designing laboratory experiments that appropriately reproduce actual driving behavior.

Before the development of an equipped vehicle with various sensors, site-based measurements on roads were conducted. Driving behavior was recorded by filming a vehicle traveling on a road. However, due to restrictions on the installation of a camera on a real road and lack of image-processing technology, it was not easy to obtain quantitative indices. Therefore, various types of on-road sensors were investigated. For example, in the 1960s, the traveling speeds of vehicles and headways were obtained by using a device that detected a vehicle by changes in the pressure inside the rubber tube laid across the road (Sonntag, 1963). In Japan, traffic congestion in the Tokyo metropolitan area became a social issue in the high-growth period because of which Japan was quick in installing traffic counters using ultrasonic sensors.

1.8.2.2 Driving Simulators

On real roads, various uncontrollable factors influence driving behavior. Thus, driving simulators came to be used as devices for measuring driving behavior under controlled conditions. A kind of driving simulator was introduced in driving schools in Japan since the 1950s. Student drivers took driving lessons by sitting in a mock driver's seat and looking at the road view displayed on a screen. In the 1960s, General Motors developed a simulator that has a gimbal structure for reproducing the acceleration and deceleration feelings accompanying driving operations (Beinke and Williams, 1968).

In the 1970s, when computer graphics (CG) were not well developed, simulators using diorama were developed in various countries, like MIT in the US, TNO in the Netherlands, the Mechanical Engineering Laboratory in Japan, etc. (Fig. 1.20) (Kikuchi et al., 1976). In order to reproduce the changes in road views according to driving operations, these simulators displayed images sent from a camera shooting the road scenery fixed above a moving belt at the driver's eye position. The computational power of computers gradually increased from the 1970s, and studies on simulators using CG were carried out, but it took time to achieve the level needed for practically applying them as experimental devices in automotive human factors research.

Simulators started to draw attention from around 1985, when Mercedes-Benz's driving simulator was developed (Drosdol and Panik, 1985). This simulator was developed for studies on vehicles' dynamic characteristics rather than for measuring driving behavior. Since the 1990s, research institutions and automobile manufacturers have been developing various types of simulators. From 2000 onward, the use of simulators spread wide as experimental devices for measuring driving behavior since they allowed experiments of driver assistance systems, such as collision warning systems, without exposing the participant to danger.

Since the tracking tasks used in the studies of man-machine systems simulated driving, a tracking task device was sometimes called a 'simulator', whereas there were also large-scale simulators, such as those of the National Advanced Driving Simulator (NADS) and Toyota. Driving behavior is considered to be influenced not only by hardware conditions, such as motion-base, size of the screen, and force feedback from the steering, but also by software conditions, such as the vehicle model, road structure, and generation of other traffic (see 2.3.1).

Results can be obtained after conducting an experiment with a simulator, but whether or not the results can be applied to real vehicles or systems depends on what the simulator can simulate and what it cannot simulate. A difficult point about a simulator is that the driving behavior of the participant is incorporated into the overall scenario. Although a simulator can control experimental conditions, such as timing of warning signal, there are still uncertain factors compared to control in psychological experiments. Therefore, experiments should be carefully designed, such as the traffic scenario, instructions to the experiment participant, and vehicle speed control.

1.8.2.3 Equipped Vehicles and Naturalistic Driving Study

Studies using experimental vehicles with measuring instruments had been conducted since the 1930s, but these were no more than experimental studies. Due to the development of sensor technology, reduction of camera prices, and introduction of higher-speed and larger-capacity PCs for recording data from around the end of the 1990s, it became possible to identify the driving behavior in the real road/traffic environment. A famous study is the 100-Car Naturalistic Driving Study conducted by NHTSA (Klauer et al., 2006). This project was originally conducted to address driver distraction, but the study was continued in the successive project, SHARP2, and a large amount of crash data and near-crash data was accumulated, which contributed to clarifying how people behave in the real world.

Naturalistic driving studies have been promoted not only in the US, but also in Canada, Australia, and Europe. In Japan, 'driver recorders' are in widespread use, and efforts are made to identify the cause of crashes and near-crashes through these drive recorders (see 2.3.3). Also, in order to see the effects of ADASs, field operational tests (FOT) using vehicles with systems installed have been conducted (see 2.3.2.3).

While accident data is needed for developing a safety-related system, it is also important to identify normal driving behavior as background data for the development of various driver-assistance systems. It is not easy to handle large amounts of naturalistic driving data, which is recorded under extremely diverse conditions. Driving behavior depends on the road/traffic environment; therefore, it needs to be annotated based on the road/traffic condition. Standardization of annotation of safety-related naturalistic driving data has been discussed (ISO TR 21974, 2018).

1.9 Driving-assistance Systems/Automated Driving Systems

1.9.1 ACC/Lane-keeping Systems

There has been progress in automation system technology that makes machines perform tasks which the driver performs. With automatic transmission using the torque converter, the driver no longer needs to conduct clutch operation and the gear-change operation. There are also automatic wipers and automatic lighting. The first system that automated automobile travelling itself was cruise control. It began to be used in the 1950s in the US, where drivers often traveled long distances at a constant speed.

Cruise control is a device for maintaining speed. A device called a 'governor' for keeping the engine rotation constant was used for it (Fig. 1.21). The governor itself was used as technology for maintaining the engine rotation during the warm-up immediately after starting the engine. When cruise control is activated while driving at a certain speed, that speed can be maintained and the driver can take his/her foot off from the accelerator pedal and be released from the work of continuing to press the accelerator pedal with a constant pressure.

Later, the engine became electronically controlled, allowing precise speed setting. Cruise control with a mechanism to keep a certain distance to the lead vehicle is adaptive cruise control (ACC), which appeared around 2000. It measures the distance to the lead vehicle by using a laser or radio waves, and controls the speed to keep the distance constant, allowing an automobile to follow the lead vehicle. Even if the automobile catches up with the lead vehicle, the driver no longer needs to conduct the pedal operation. At around the same time, the lane-keeping system was developed. The system detects the markers on the left and right sides of the lane by using an image sensor and controls the steering so that the automobile does not run over the markers.

The ACC raised a new human-factor issue. It was an issue of whether the user could correctly understand the functions, conditions, and limits of the system and appropriately use it. For example, when traveling on a sharp curve, the sensor may fail to detect the lead vehicle and the system, which understands that there is no lead vehicle, may start to accelerate up to the set speed. The driver may be surprised by this system action (i.e., automation surprise) if he/she has believed that the automobile will follow the lead vehicle as long as there is one. Meanwhile, if the driver thinks that the system has a functional role in automatic deceleration, rather than in just following the preceding vehicle, he/she may not realize that he/she must decelerate by him/herself when approaching a toll gate. In a lane-keeping system, when there are several system states (modes), such as a state where the system can detect the white line on the right but cannot detect the line on the left, there is an issue of whether the driver can correctly judge which mode the system is in. On the assumption that users do not read instruction manuals, it is necessary to design HMI for making it possible for users, who do not know the mechanism of the system, to correctly understand the system (see 5.4.2).

In order for ACC to be accepted by users, drivers need to be able to correctly understand the system, and the control conducted by the system must not cause anxiety or discomfort to the driver. The driver may feel anxiety if the automobile does not decelerate at the timing when the driver would have decelerated, or feel discomfort if the automobile decelerates when the driver does not see the need to decelerate. Since drivers have different ideas about the comfortable distance between vehicles and approach timing, the anxiety and discomfort can be eliminated by providing control compatible with each driver's characteristics (Sato et al., 2006) (see 5.4.2.5).

This is an issue not only for the driver driving the automobile, but also for the drivers of surrounding automobiles. If the brake lamps frequently turn on and off when accelerating and decelerating to keep a constant distance between vehicles, the driver of the following vehicle may feel discomfort and may misunderstand that such an operation is being conducted intentionally with ill intent. Also, a normal driver would shorten the distance from the preceding vehicle if the traffic flow speeds down, but since the ACC tries to keep the distance constant, stronger deceleration is conducted at an earlier timing than a normal driver. The driver of the following vehicle may not have expected such behavior and may cause a rear-end collision. An understanding of humans' driving behavior and studies on driver models are important for the system behavior to be acceptable to both the driver and other surrounding road users.

1.9.2 Automated Driving Systems

Automated driving systems can be realized as the sensing accuracy and reliability of the ACC and lane-keeping systems improve. Human factor challenges of automated driving systems include not only the above-mentioned understanding of the system and acceptability of the systems' behavior for the drivers and other road users, but also the issue of the drivers' states.

The systems have limited sensing abilities and cannot provide stable control for some conditions. Accordingly, the systems often do not work appropriately, and in such cases, the drivers override the driving task (SAE level-2 system). When using this level of systems, drivers are constantly monitoring the system conditions and the surrounding traffic conditions so as to see the need for overriding (see 5.4.2.6). When the systems improve and overriding is rarely needed but the operation still cannot be totally left to the systems, the sole role of the drivers is monitoring. Monitoring is a task where one needs to continue paying attention although nothing happens. Humans are not good at a task requiring such vigilance. If a state absent of stimuli continues, the arousal level will be lowered (see 4.3). However, if the driver is not allowed to sleep, the decreased arousal needs to be prevented. As discussed earlier, technology for raising the arousal level has not been established; so this is a big challenge.

The human factors, challenge to driver-assistance systems, such as in-vehicle information systems, was the issue of mental workload whereby the driver must conduct multiple tasks in parallel, but such a challenge for automated driving systems would be the issue of switching tasks. If the use of an automated driving system is limited to expressways, it is necessary to switch from the automated driving system to the driver's own operation when exiting an expressway or coming to the end of an expressway section. It is the switching from the state of conducting a monitoring task or non-driving-related task to the state of performing a driving task. Normally, when we drive an automobile, we start the engine, do some preparations, and begin to travel. As we gradually increase the speed, we gradually adapt a driving behavior that suits the vehicle characteristics. In contrast, when we use an automated driving system, we face a situation where we suddenly have to perform the driving operation of the vehicle which is traveling at a certain speed from the state of not being involved in the driving operation. The issue of such switching presents a new human factor issue to be resolved (see 5.4.2.6).

1.10 Elderly Drivers

In many countries, the rate of the aging population is increasing. Elderly people experience a decline in various functions, including the sensory function, motor function, sensori-motor association such as response time and cognitive function. Meanwhile, analyses have been made of the characteristics of the accidents of elderly drivers. It has been found that crossing collisions, driver-side turn accidents, and red-light violation accidents are frequent, whereas rear-end accidents are infrequent in Japan. It has also been revealed that elderly people find it difficult to predict the time headway or to predict the arrival time of an approaching vehicle, and often make predictions mainly based on the distance without taking into consideration the vehicle speed.

However, which functional impairment accounts for the accidents caused by elderly drivers has not yet been clarified. Driving behavior is generated as a result of integrating various functions, and the relationship between the measured functions and the behavior cannot be simply identified (see 6.2.2). Also, we should keep in mind that, even in cases where driving characteristics of elderly driver were to be identified in an experimental environment, elderly people cannot immediately adapt to the experimental environment, and often show unnatural behavior. Furthermore, a functional impairment may simply seem to cause an impairment in performance, but humans behave adaptively according to functional impairments; in other words, they perform compensation behavior or coping behavior and act in a manner that does not cause a problem as a result (Akamatsu et al., 2006). The characteristics of elderly drivers need to be understood by taking such factors into consideration.

Elderly people do not exhibit the same characteristics, even if they are of the same age. Individual differences among elderly people are larger as compared to those among non-elderly people. This is considered to be because the combination of functions that have changed with age differs individually. Accordingly, studies that categorize elderly drivers merely by age will only reveal average tendencies and are unlikely to lead to assistance technology that will be useful for individual elderly drivers. It is necessary to focus on individuals and evaluate their functions as well as identify their driving behavior, including compensation behavior, on a real road (see 6.2.2.4.2). In spite of the individual differences, categorization would be possible to some extent; so an appropriate categorization technique and assistance technologies that correspond to the categorized characteristics should be developed. Moreover, it is necessary to develop not only technologies for ensuring the safety of elderly drivers, but also technologies for increasing their motivation, such as encouraging them to go out to engage in social activities. The technology can work as a safeguard for social activities.

With regard to elderly drivers, not only their driving behavior, but also their health status is an issue. Although there are no accurate statistics, a fairly large portion of accidents caused by elderly drivers are considered to have been caused due to health problems. Accidents have occurred as a result of loss of consciousness due to cerebral infarction, cardiac failure, or other attacks during driving. Since other people are involved in such accidents, this is not a problem for elderly people alone. It is hoped that an early warning will be given or automatic stop will be activated by monitoring the health status of the driver through driver-monitoring technology. Such an incident cannot be sufficiently prevented through monitoring using an in-vehicle system alone. The accuracy of prediction is expected to be more precise by integrating information from the monitoring physiological conditions and behavior at home before driving.

1.11 Positive Aspects of Automobile Driving

1.11.1 Enjoyment and Growth

With ergonomics and human factors researches, products and systems are improved by clarifying the problems occurring where the product or system is found to be incompatible with human characteristics. It is an approach to eliminate negative aspects of technology, but it does not necessarily increase the value of the technology. It is desirable to develop positive ergonomic technology that does not only eliminate negative aspects, but can help people enjoy or develop their skills.

Some people call their automobiles their beloved cars, and consider automobiles to be an object of passion; automobiles are regarded as technology that has a positive aspect for humans. In a questionnaire survey targeting elderly drivers, nearly 70% of both males and females in their 60s and 70s answered that they liked driving very much or somewhat liked driving. 'Like driving' may give an impression of thrilling driving, but these respondents were unlikely to drive in such a way. It should be considered that one can enjoy even ordinary automobile driving itself.

One of the theories of happiness is the flow theory of Mihaly Csikszentmihalyi (Csikszentmihalyi, 1990). People with high skills, such as automobile racers, professional athletes, chess players, and brain surgeons, sometimes feel a sensation of deep joy, a kind of ecstasy, when concentrating and being devoted to their task. In such a state, their consciousness becomes clear, their body automatically moves to perform the required actions, they receive clear sensory feedback from the object, and feel that everything is under their control. Csikszentmihalyi studied how such a state occurs, and called such a state 'flow', which refers to the spontaneous feeling of enjoyment (see 4.2.3).

The requirements for obtaining this feeling are that the subject feels he/she is tackling a challenge by exerting his/her own skills, and that the behavior is not carried out for someone else but is the purpose in itself, that is, the behavior itself is the purpose (autotelic). In such a state, the subject receives clear feedback from the action, and as a result, experiences an immense amount of satisfaction at performing an appropriate action. Moreover, as the action is integrated with the object, the subject feels in control over his/her situation, including the behavior and environment.

The flow state appears when the subject tackles a challenge by exerting his/her skills (see 4.2.4). Here, it indicates that the subject's skills and the challenge need to be balanced in order to feel a flow. Also, this coincides with what is indicated by R. Fuller's task-capability interface model, which is one of the driving behavior models (see 6.1.2.2). Fuller's model suggests that, in automobile driving, the task demand from the driving conditions and the driving performance are balanced. If we regard that driving performance is exertion of one's driving skills and demand is a task to be performed, which is a tackling of the given conditions, it would be the same view as the balance between skill and challenge of the flow theory. In the flow state, the subject will be able to acquire higher abilities and tackle more difficult challenges while acquiring a higher flow as a result and improving his/her skills. In automobile driving as well, if the driving skill improves through driving experience, the driver will be able to manage more difficult conditions.

The subject feels he/she has control over the conditions in a flow state, whereas the level of demand in automobile driving can be either raised or lowered by speed or the headway distance selected by the driver. In other words, the driver is able to control his/her driving difficulties by him/herself. Also, in automobile driving, the driver receives somato-sensory/visual feedback from the reaction force or vehicle behavior that arises in response to the steering wheel or pedal operation, which corresponds to the feedback in the flow theory. Furthermore, people have opportunities of 'going out for driving' where the act of driving becomes the purpose in itself. In this manner, the flow theory and automobile driving coincide in many respects, and it is possible to use the flow theory to explain the fact that people enjoy automobile driving.

People feel enjoyment in tackling challenges and improving their skills probably because organisms have a mechanism to actively adapt to a new environment. An organism that stays in the same environment is in a stable state, adapted to the environment. However, the environment eventually changes. The change may result from an external cause or from the organism itself eating up all the food. Then, the organism needs to jump into a different environment. By being able to feel enjoyment in tackling challenges in a new environment, the organism can jump into the new environment without escaping from the challenge. Vehicle driving is a dynamic environment that does not exist in daily living at home, but helps in acquiring driving skills while adapting to that environment. It can be said that people tackle challenges in order to adapt to a new environment, which in this case is the road traffic environment, which serves as an opportunity to experience growth themselves. There are expectations for research and development in automobile technologies that allow drivers to feel more flow experience, such as the design of reaction force and vehicle response, which are feedback actions, and technologies that support demand control.

1.11.2 Stress Relief

People often say that driving automobiles relieves one of stress. Does this mean that people feel refreshed when they drive at a high speed? Stress as a scientific term is strongly related to the above-mentioned demand control. The term 'stress', spread by Hans Selye, refers to the adaptation syndrome (Selye, 1937). If a person cannot adapt to an environment and is in an escapist condition, it has an adverse effect on the body and increases the risk of contracting a disease. Meanwhile, Lazarus focused on the balance between load and task performance, and explained that psychological stress arises when the person feels that his/her performance is insufficient for meeting the task demand (Lazarus, 1966). Furthermore, Karasek showed that a person with high demand and little decision latitude faces a high risk of cardiovascular disease, but if the person has high decision latitude, there is no such risk even with a high demand (Karasek, 1979) (see 4.1.7). In other words, a person does not feel stress if he/she can control the task demand.

Lazarus's viewpoint of evaluating the balance between demand and task performance is same as the balance between driving demand and performance in the task-capability interface model (see 6.1.2.2). The driver feels stress if his/her performance is insufficient for the driving demand, but not if these are balanced. The driver's ability to control the demand by changing the headway distance to the lead vehicles or the vehicle speed corresponds to the decision latitude for a job in Karasek's stress theory. Thus, automobile driving can be regarded as a task that can be performed with decision latitude. It is considered that people can relieve stress through automobile driving because they can gain decision latitude, which is difficult to gain in a job.

Stress is not a mere load, but an issue of whether the subject is able to appropriately cope with the situation. People who feel stress in automobile driving are not able to demonstrate the performance corresponding to the situation or the situation by themselves. It is hoped that, by identifying the characteristics of such people's driving behavior and providing them with information support or driving hints accordingly, they can come to drive more actively, and that automobile driving becomes a positive act for humans.

1.12   Future of Automobile Ergonomics: Viewpoint of Service Engineering for Providing Value to Users

Automobile ergonomics originated as a study for making automobiles compatible with humans' muscle strength and body dimensions; this in turn developed into a study for ensuring safety and comfort. To date, ergonomics and human factors have mainly contributed to resolving issues that are hard, uncomfortable, or difficult for humans; that is, for eliminating aspects that are negative for humans. However, value cannot be provided to users by merely eliminating negative aspects. In order to raise the value of products, such as automobiles, there are expectations for ergonomic technologies to create aspects that are positive for humans. While the previous section focused on the positive aspects of automobile driving, it is also necessary to consider providing value in users' daily lives.

The concept of service engineering emerged around 2005 as a method to provide value to a person who has different needs, depending on the situation. Every person is different, but the idea of regarding all people as same underlies the practice of providing products through mass production, as in the case of automobiles. However, with the progress of ICT, it has become possible to create diversity in the same hardware by using software and also to measure or detect users, due to progress in various sensor technologies. Therefore, the idea of individual compatibility is spreading also in automobile technology. However, according to the idea of service engineering, it is not enough to regard each person as different, but take into account the fact that each person also needs to be regarded as having diverse behaviors. A person may find value in different things, depending on the situation. The same person may eat at a fast-food stand on one occasion, but go to a star-rated restaurant another time. The choice is made as a result of the value determination at the time. Even in the case of using the same restaurant, the user expects different services when using it for a job meeting and when using it for a family dinner. What is necessary for providing value according to human conditions, that are so widely varied and that change according to situations, is to constantly observe and analyze the user and design and apply what is suitable for the user (Fig. 1.22).

It is not easy to know what humans find value in which type of a situation. Manufacturers often develop a product on the assumption of what the users are like. This is especially the case where the developers and users rarely come in direct contact with each other, such as in the manufacturing industry. Developers must not forget open-mindedness in order to learn about users. To this end, behavior observation and ethnographic methods can be applied without prejudgment (see 2.2.2). We can construct a hypothesis by such a method, act on users according to it, and observe once again whether it could actually make the users happy. If not, we should not only doubt the hypothesis itself, but review the requirements for the hypothesis, which may not have been understood in detail, and refine the hypothesis. While data mining can be used for constructing a hypothesis, the data available is often not sufficient for understanding the conditions of each individual user. It is important to directly see the actual situation through behavior observation, etc. If we forget that people and society change constantly, we will not be able to provide value to people. The essential step toward providing value to ever-changing people is to continue the loop of creating a hypothesis based on observation or measurement and applying it to the actual settings, especially in the era of MaaS (Mobility as a Service) and CASE (Connected, Autonomous, Shared, Electric).

References

Ach, N. (1929). Psychologie und technik bei bekampfung von autounallen. Industrielle Psychotechnik, 6: 87–105.

Akamatsu, M., K. Hayama, A. Iwasaki, J. Takahashi and H. Daigo. (2006). Cognitive and physical factors in changes to the automobile driving ability of elderly people and their mobile life questionnaire survey in various regions of Japan. IATSS Research, 30(1): 38–51.

Akamatsu, M., P. Green and K. Benglar. (2013). Automotive technology and human factors research— Past present and future. International Journal of Vehicular Technology, Article ID526180, 27 pages.

Atsumi, B., K. Kimura and Y. Ohsumi. (1999). Study of meter cluster location for visibility performance. JSAE Review, 20(3): 369–374.

Beinke, R.E. and J.K. Williams. (1968). Driving simulator. In: Proceedings of Automotive Safety Seminar, 24, General Motors, Warren, Michigan, USA.

Bornemann, E. and I. Teil. (1942). Untersuchungen uber den Grad der geistigen Beanspruchung, I. Teil. Ausarbeitung der Methode, Arbeitsphysiologie, 12(2): 142–172.

Broadbent, D.E. (1979). Is a fatigue test new possible? Ergonomics, 22(12): 1277–1290.

Brown, I.D. and E.C. Poulton. (1961). Measuring the spare "mental capacity"of car drivers by a subsidiary task. Ergonomics, 4(1): 35–40.

Brown, I.D. (1994). Driving fatigue. Human Factors, 36(2): 298–314.

Brown, R.W. (1938). Riding comfort requirements. SAE Technical Paper 390036.

Clark, C.S. (1995). The Lanchester Legacy: A Trilogy of Lanchester Works, 1: 1895–1931, Coventry University Enterprise, UK.

Crawford, A. (1961). Fatigue and driving. Ergonomics, 4(2): 143–154.

Csikszentmihalyi, M. (1990). Flow: The Psychology of Optimal Experience, Harper and Row, New York, USA.

Dhers, V. (1924). Les Tests de Fatigue. Essai de Critique Theorique. J.-B. Baillie're et fils, Paris, France.

Dingus, T.A., L.H. Hardee and W.W. Wierwille. (1985). Detection of drowsy and intoxicated drivers based on highway driving performance measures. IEOR Department Report #8402, Virginia Tech. Department of Industrial Engineering and Operations Research Blacksburg, VA USA.

Drosdol, J. and F. Panik. (1985). The Daimler-Benz driving simulator: a tool for vehicle development. SAE Technical Paper 850334.

Erwin, C.W., J.W. Hartwell, M.R. Volow and G.S. Alberti. (1976). Electrodermal change as a predictor of sleep. In: C.W. Erwin (ed.). (1976, February), Studies of Drowsiness (Final Report); The National Driving Center, Durham, North Carolina.

Forbes, T.W. (1939). The normal automobile driver as a traffic problem. Journal of General Psychology, 20: 471–474.

Fuller, R. (2005). Towards a general theory of driving behavior. Accident Analysis and Prevention, 37(3): 461–472.

Green, P. and R.W. Pew. (1978). Evaluating pictographic symbols—An automotive application. Human Factors, 20(1): 103–114.

Heard, E.A. (1974). Symbol study—1972. SAE Technical Paper 740304.

Heinrich, C. and A.S. Conti. (2015). Fighting driver distraction International Standards. Zeitschrift fur Arbeitswissenschaft, 69(2): 98–103.

http://siarchives.si.edu/blog/science-service-close-sleeplessness-study-part-1-insomniacs.

Hunt, J.H. (1937). Automobile design and safety. SAE Journal, 41(2): 349–369.

Ikeda, H., Y. Kobayashi and K. Hirano. (2010). How car navigation systems have been put into practical use—Development management and commercialization processs. Synthesiplogy, English Edition, 3(4): 280–289.

Ishii, I. (1980). Comparison of visual recognition time of analogue and digital displays in automobiles. SAE Technical Paper 800354.

ISO 2631 (1974). Mechanical vibration and shock. Guide for the Evaluation of Human Exposure to Whole Body Vibration.

ISO TR 21974 (2018). Naturalistic Driving Studies—Defining and Annotating Safety Critical Events.

Jastrzebowski, W. (1857). An outline of ergonomics, on the science of work based upon the truths drawn from the science of nature. Ochorny Pracy, Central Institute for Labour Protection, Warsaw, Poland.

Karasek Jr, R.A. (1979). Job demands, job decision latitude and mental strain: Implications for job redesign. Administrative Science Quarterly, 24(2): 285–308.

Kikuchi, E., T. Matsumoto, S. Inomata, M. Masaki, T. Yatabe and T. Hirose. (1976). Development and application of high-speed automobile driving simulator. Technical Report of Mechanical Engineering Laboratory 89 (in Japanese).

Klauer, S.G., T.D. Dingus, V.L. Neale, J.D. Sudweeks and D.J. Ramsey. (2006). The impact of driver inattention on near-crash/crash risk: An analysis using the 100-car naturalistic driving study data. Technical Report DOT HS 810 594, U.S. Department of Transportation. National Highway Traffic Safety Administration. Washington. DC. USA.

Land, N.F. and B.W. Tatler. (2001). Steering with the head: The visual strategy of a racing driver. Current Biology, 11(15): 1215–1220.

Lazarus, R.S. (1966). Psychological Stress and the Coping Process, McGraw-Hill, New York, USA.

McFarland, R.A. and A.L. Moseley. (1954). Human Factors in Highway Transport Safety, Harvard School of Public Health, Boston, USA.

McFarland, R.A., A. Damon and H.W. Stoudt. (1955). Human body size and passenger vehicle design. SAE SP-142.

McRuer, D.T. and D.H. Weir. (1969). Theory of manual vehicular control. Ergonomics, 12(4): 599–633.

McRuer, D.T., R.W. Allen, D.H. Weir and R.H. Klein. (1977). New results in driver steering control models. Human Factors, 19(4): 381–397.

Merker, H.M. (1966). Imperial sentry signal warning system. SAE Technical Paper 660045.

Michon, J.A. and A. Smiley. (1993). Generic Intelligent Driver Support, Taylor and Francis, London, UK.

Miura, T. (1986). Coping with situational demands: A study of eye movements and peripheral vision performance. pp. 205–216. *In*: A.G. Gale, M.H. Freeman, C.M. Haslegrave, P. Smith and S.P. Taylor (eds.). Vision in Vehicles I, North-Holland, Amsterdam, The Netherlands.

Moss, F.A. (1930). Measurement of comfort in automobile riding. SAE Technical Paper 300002.

Nevett, L.J. (1972). Human engineering applied to the design and grouping of electrical controls in the motor vehicle. SAE Technical Paper 720233.

Okabayashi, S., M. Sakata, M. Furukawa and T. Hatada. (1989). How head-up display affects recognition of objects in foreground in automobile use. pp. 283–296. *In*: R.E. Fischer, H.M. Pollicove, W.J. Smith (eds.). Current Developments in Optical Engineering and Commercial Optics, Vol. 1168, International Society for Optical Engineering,San Diego, CA: SPIE.

Parkes, A.M. and S. Franzen (eds.) (1993). Driving Future Vehicles, Taylor and Francis, London, UK.

Regan, M., J. Lee and K. Young (eds.) (2008). Driver Distraction Theory: Effects and Mitigation, CRC Press, Boca Raton, FL., USA.

Rockwell, T.H. (1988). Spare visual capacity in driving. pp. 317–324. *In*: M.H. Freeman, P. Smith, A.G. Gale, S.P. Taylor and C.M. Haslegrave (eds.). Vision in Vehicle II, Elsevier, North Holland, Amsterdam, The Netherlands.

Rutley, K.S. (1975). Control of driver's speed by means other than enforcement. Ergonomics, 18(1): 89–100.

SAE Recommended Practice J826 (1962). Manikins for use in defining vehicle seating accommodation.

SAE Recommended Practice J941 (1965). Motor vehicle driver's eye range.

Sato, T., M. Akamatsu, A. Takahashi, Y. Takae, N. Kuge and T. Yamamura. (2006). Influence of ACC on frequency of situations experienced while driving on highway, 6 pages. In: Proceedings of 16th World Congress on Ergonomics (IEA 2006).

Selye, H. (1937). The significance of the adrenals for adaptation. Science, 85(2201): 247–8.

Senders, J.W., A.B. Kristofferson, W.H. Levison, C.W. Dietrich and J.L. Ward. (1967). The attentional demand of automobile driving. Transportation Research Board, 15–33.

Sonntag, R.C. (1963). Some traffic flow relationships on two-lane urban streets. Joint Highway Research Record, (6).

Ward, N.J. and A. Parkes. (1994). Head-up displays and their automotive application: An overview of human factors issues affective safety. Accident Analysis and Prevention, 26(6): 703–717.

Wierwille, W.W. (1994). Research on vehicle-based driver status/performance monitoring; Development, validation and refinement of algorithms for detection of driver drowsiness. Technical Report DOT HS 808 247, U.S. Department of Transportation, Washington DC, USA.

Wyon, D.P., S. Larsson, B. Forsgren and I. Lundgren. (1989). Standard procedures for assessing vehicle climate with a thermal manikin. SAE Technical Paper Series, 890049: 1–11.

Zwahlen, H.T., C.C. Adams and D.P. DeBals. (1988). Safety aspects of CRT touch panel controls in automobiles. pp. 335–344. In: M.H. Freeman, P. Smith, A.G. Gale, S.P. Taylor and C.M. Haslegrave (eds.). Vision in Vehicle II, Elsevier, North Holland, Amsterdam, The Netherlands.

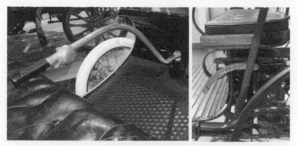

Fig. 1.1: Oldsmobile 1902 (left) and Benz Patent Motor Vehicle (replica) (right) (Toyota Automobile Museum).

Fig. 1.2: Cord 810 (1936) (Toyota Automobile Museum).

Fig. 1.3: Morris Eight Series (1937) (Toyota Automobile Museum).

Fig. 1.4: A seat adjuster level in the seat base as seen in the Mercedes Benz 500k (1935) (Toyota Automobile Museum).

Body Measurement	5th Percentile	50th Percentile	95th Percentile	Range 5th-95th Percentile
1. Height	64.6	68.4	72.5	7.9
2. Weight (lb)	129.0	163.7	212.8	83.8
3. Abdominal Depth	7.9	9.5	12.1	4.3
4. Arm Reach, Anterior	32.9	35.7	38.4	5.5
5. Arm Span, Total	66.5	70.9	75.5	9.0
6. Buttock-Knee Length	22.1	23.8	25.8	3.7
7. Calf Circumference (average both)	12.6	14.1	16.1	3.5
8. Chest Breadth	10.2	11.8	13.5	3.3
9. Chest Circumference	34.1	38.3	44.2	10.1
10. Chest Depth				
11. Elbow Breadth				
12. Elbow Span				
13. Elbow-Middle Fir				
14. Foot Breadth				
15. Foot Length				
16. Hand Breadth				
17. Hand Circumferer				
18. Hand Length				
19. Head Circumferer				
20. Head Height				
21. Hip Breadth				

Fig. 1.5: Human body size and passenger vehicle design (Reprinted with permission of SAE International from McFarland, R.A., A. Damon and H.W. Stoudt. (1955). The application of human body size data to vehicular design. SAE Technical Paper 550320; Permission Conveyed through Copyright Clearance Center, Inc.).

Fig. 1.6: Hispano-Suiza Alfonso XIII (1912) (left), Duesenberg Model J. (1929) (right) (Toyota Automobile Museum).

Fig. 1.7: Center meter from brochure of Toyota Prius (TH00017-0005, 2000) with permission.

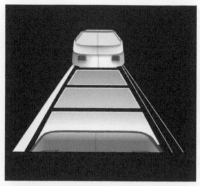

Fig. 1.8: Example of HMI of ACC and lane keeping system.

Fig. 1.9: Alpha Romeo Tipo 103 (prototype) (1955) (Museo Storico Alfa Romeo [Alfa Romeo Museum]).

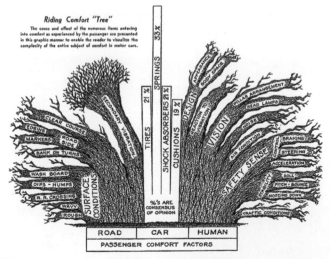

Fig. 1.10: Riding comfort tree (Reprinted with permission of SAE International from Brown, R.W. (1938). Riding comfort requirements. SAE Technical Paper 390036, Permission Conveyed through Copyright Clearance Center, Inc.).

Fig. 1.11: Device that encourages the driver to a rest (from brochure of Nissan Bluebird (3101AMM, 1983) with permission).

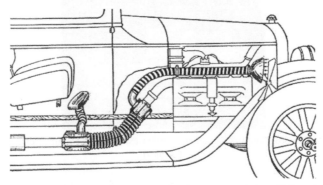

Fig. 1.12: Heater in the 1920s (from Automotive Equipment Association, Universal Catalogue D, 1927).

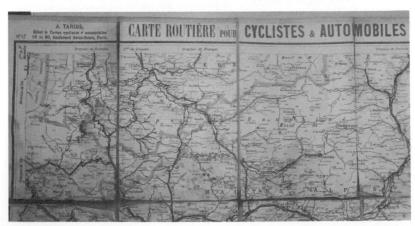

Fig. 1.13: Road map published by Taride in France (1910). The words 'for cyclists and automobiles' are seen at the top.

Fig. 1.14: Small size guide and altas (left: first Michin Guide published 1900 (reproduction), right: Sur Route Altas-Guide de Poche ("Pocket guide" in French) published in 1905).

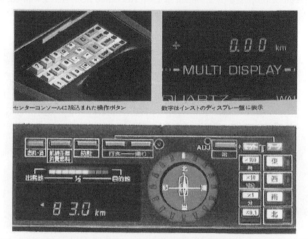

Fig. 1.15: 'Drive Computer' in 1980s (from brochure of Nissan Silvia (0084K908, 1980) (above) and Toyota Crown (101010-5906, 1984) (below) with permission).

Fig. 1.16: Toyota Electro Multi-vision from brochure of Toyota Crown (101024-6209, 1987) with permission.

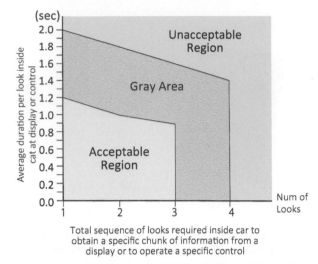

Fig. 1.17: Design guide proposed by Zwahlen in 1988 (modified from H.T. Zwahlen, Safety Aspects of CRT Touch Panel Controls In Automobiles, Vision In Vehicle II, 1988).

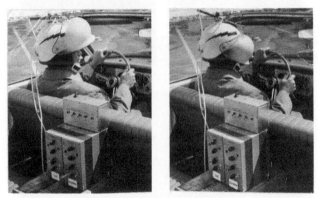

Fig. 1.18: Measuring visual demand by the occlusion method (1967) (photos: courtesy of J.W. Senders).

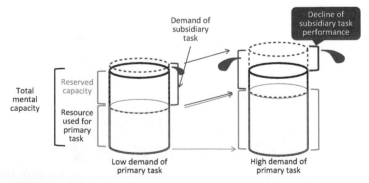

Fig. 1.19: Model of mental capacity (resource) and a subsidiary task method.

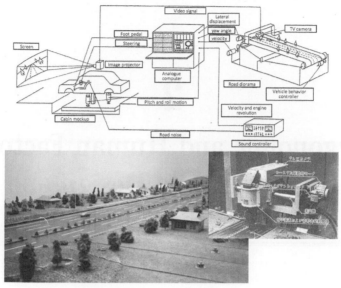

Fig. 1.20: Simulator using diorama of Mechanical Engineering Laboratory of AIST first developed around 1970 (from Technical Report of MEL, No.89, 1976).

Fig. 1.21: Cruise control of Cadillac Eldorado (1959) Left: The cruise control switch on the left side of the meter; Right: The governor in the engine room (Toyota Automobile Museum).

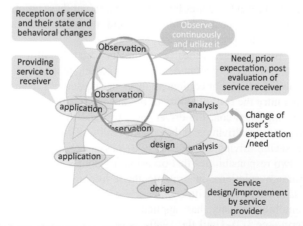

Fig. 1.22: Loop of observation, analysis, design, and application in service engineering.

2

Ergonomic and Human Factors in Automobile Design and Development Process

2.1 Ergonomists' Roles and Responsibilities in Automobile Design and Development

2.1.1 Ergonomics for Automobiles

The International Ergonomics Society defines ergonomics as 'the scientific discipline concerned with the understanding of interactions among humans and other elements of a system, and the profession that applies theory, principles, data and methods to design in order to optimize human well-being and overall system performance.' Automobiles felicitate range of movement for individuals and have great influences on the safety, security, comfort and health of drivers and passengers, and on the society surrounding the automobiles. Automobiles' function and their influences are far greater than those of any other durable consumer good. Automakers are strongly and socially required to design and manufacture automobiles that take into account these points. Ergonomists' role is to realize safe, secure, comfortable and health-friendly automobiles in cooperation with engineers and designers.

There are two responsibilities ergonomists should fulfill for automobile development—the first is to guarantee interior comfort, the ease of getting in and out, and the usability of interior equipment to secure passengers' comfort and health; second is to guarantee the controllability of the driving positions and control systems for accurate driving operation, the driver visibility and the instruments for getting information required for driving and the controllability of equipment for safe driving to ensure safe, secure driving and the safety of passengers and the society. In order to fulfill these two responsibilities that occasionally conflict each other, ergonomists must find optimum solutions, incorporate them into the design and verify their appropriateness in cooperation with engineers and designers.

The automobile industry has applied ergonomics to vehicle development over a long term (see 1.2). Until the 1980s, automobile ergonomics had developed for human sizes, posture, muscle load, and had mainly related to performance concerning interior packaging, such as interior comfort, driving positions, layout of controls, and driver visibility. Since the 1990s, however, cognitive usability has become a major challenge for ergonomics as navigation and other information systems are commonly used today. In recent years, ergonomics has been applied to the development of human machine interfaces (HMI) for advanced driving assistance systems (ADAS).

2.1.2 Development Process

* *Setting concept and target*: Product planners determine the concept of a new automobile for development by setting target users and deciding values to be provided to them. Ergonomists analyze user needs/requirements and clarify targets for values to be provided.

* *Designing to realize concept*: To decide ways and means for achieving concept, ergonomists work for ergonomic considerations in cooperation with engineers and designers. Ergonomists exploit the past research results and design tools. As necessary, ergonomists conduct ergonomic experiments and analyze relations between test conditions and data regarding human behaviors and actions as well as physical and mental workload, senses and sensibility to clarify the required conditions for achieving the goals. They then provide these conditions to engineers and designers.

* *Verification*: After the design is completed, a prototype of the new automobile is built before mass production. The prototype is used for ergonomic tests to confirm 'whether the preset goals are achieved' and 'whether there are any ergonomic problems'. If the goals fail to be achieved or some problems are extracted, ergonomists may conduct ergonomic tests to identify the causes of failure or problems and consider potential solutions to clarify design conditions for necessary modifications.

To play their role in the automobile development process, ergonomists must always study and increase their knowledge about basic matters, including what comfort means for humans and how humans can drive automobiles safely. They also must make tools for various considerations to be prepared to conduct efficient considerations in the development process. To prevent any test conditions or assessment viewpoints from being missed, ergonomists must always check and figure out how automobiles are actually used and what actual problems emerge. This is very important for ergonomics as a practical science and indispensable for ergonomists' activities.

2.1.3 Identifying Out User Requirements

When new automobiles are developed, it is important to accurately figure out the requirements of target users of the vehicles. User requirements differ, depending on physical size, gender, age and other human characteristics. Particularly, user requirements for automobiles vary, depending on the purposes, modes, frequency and environments for use of automobiles that are personal products made for global sales. Therefore, it is particularly important to clarify the requirements of target users of new automobiles to be developed.

User requirement survey methods include the internet and mail-in, behavior observation at clinics and interviews. Behavior observation and interviews are useful for figuring out specific and detailed user requirements, but their coverage is limited. Therefore, it is desirable for behavior observation and interviews to be used along with questionnaires that can collect massive data over a wide region (see 2.2). Developers of automobiles for large-scale global sales must survey the requirements of various users in a wide region and use various survey methods to collect user requirement data.

2.1.4 Ergonomics in Design Stage

Designers are required to design automobiles that meet various legal conditions, critical design requirements, cost goals and achieve ergonomic performance targets. To make designing efficient, ergonomists must prepare tools for ergonomic considerations which predict human perception and behavior/actions in the design drawing and check the achievement of ergonomic conditions.

For example, cabin designing cannot start without human body size-related tools that predict the seating posture, eye and shoulder locations of passengers and their arm lengths. For these tools, the SAE J1100 (motor vehicle dimensions), J941 (motor vehicle drivers' eye locations) and other standards are frequently used. However, there are cases where these standards are not sufficient; here the ergonomists create tools as are necessary. Tools for predicting human senses of space for passengers to feel the space to be roomy and those for estimating physical load while operation controls are used for designing the cabin and cockpit to meet the targets.

Given, however, that it is difficult to comprehensively predict complex human perception, behaviors and senses and that new equipment using new technologies are being introduced one after another, it is impossible to create tools for covering all the assessment items included in the drawings. Therefore, cockpit and cabin mockups are frequently built for ergonomic tests to examine detailed design conditions (Fig. 2.1). To improve the accuracy of considerations, a cockpit mockup is combined with a driving simulator and virtual reality to conduct tests in simulated driving conditions in order to acquire more accurate results (Fig. 2.2). In a CAD (computer-aided design) process, digital mannequins may be used for virtual tests (Fig. 2.3).

Fig. 2.1: Work space evaluation in cockpit mockup.

Fig. 2.2: Driving simulator.

Fig. 2.3: Work space evaluation by digital mannequin.

Although digital mannequins make it easy to conduct quantitative assessment, the latter must be confirmed with ergonomic tests using human participants.

Based on these test results, ergonomists give feedback to engineers and designers to help them determine the final design specifications.

2.1.5 Ergonomics in Assessment Stage

After the design is completed, a prototype is built for ergonomic tests to confirm the achievement of ergonomic targets and extract problems before production vehicles are manufactured. It is important to conduct the tests for a comprehensive assessment. Checklists are used to avoid missing out any assessment points. Ergonomic experts take leadership in assessment. As necessary, a clinic is conducted with potential users. To shorten the development process, digital CAD data may be verified with digital mannequins before a prototype is fabricated.

If the tests find that the targets are not achieved or that unexpected problems have arisen, the test results may have to be fed back to designers for design modifications. To determine details for the feedback, ergonomic tests are conducted to clarify what counter measures can solve problems to achieve the target.

2.1.6 Feedback from Users

After the production vehicles, which have been manufactured based on the final design specifications, are put into the market, information such as actual usages and user satisfaction are always collected for minor changes and for future development of new automobiles. Users' opinions are directly given to customer-service offices and dealers or collected through questionnaire surveys to check any problems and their causes. Users' opinions are often fragmentary, making it difficult to find the cause of the problems. Therefore, more data must be successively collected and/or compared with other vehicles being analyzed to extract problems and identify their causes. Results of these analyses are precious data for clarifying user requirements for new automobiles to be developed.

2.1.7 Designing User's Manual

As for the present automobiles, whose functions are sophisticated and complex, it is important to use instruction manuals to inform users on how to use these functions and precautions. If all matters considered in the design stage were to be included in manuals, they would become too massive for users to read. Therefore, manuals must provide the necessary explanations in an easy-to-understand manner and be so compact in volume that users are tempted to pick up. Particularly for devices with new functions, a few-page quick reference guide covering the minimum necessary information that users need to understand may be attached to detailed manuals. In the recent trend, car manufacturers have provided systems allowing users to read manuals on in-vehicle monitors and digitized manuals to read with mobile terminals, such as smartphones.

When manuals are designed, necessary information for users must be strictly selected and described in a manner that is easy to search and read. Manuals need to explain operation procedures for each function of the device or for achieving the goal of each user task. As for new functions that users may face for the first time, the latter approach may be easy for users to understand (Fig. 2.4). However, this approach may tend to produce a big volume and make the necessary search difficult. An effective

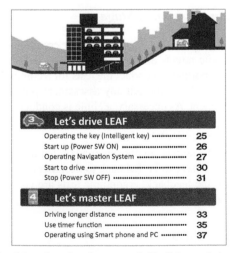

Fig. 2.4: Example of task-based explanation manual (English translation of *Nissan Leaf User's Manual*, 2010 with permission).

way is to separate a small-volume booklet for task-based explanations from the main search-based manual. Various factors may be balanced when designing manuals.

2.2 Surveys for Understanding Users in Design Stage

2.2.1 Viewpoints for Considering Target Users

In the product planning and function/specification design stages, it is important to figure out what the users require from automobiles. User requirements may range from those that can be expressed in a particular language to those at deep psychological levels that users do not clearly understand by themselves. Product planners and engineers must skillfully obtain these requirements from the users. The following viewpoints for considering target users before user requirements are figured out are as follows:

* If target users are drivers, a method to set target users in reference to personal characteristics of the driver may be conceivable. Sex, age, residence, educational background, occupation, income and family makeup are called demographic attributes, which are the basic attributes used in various areas.

* Drivers may be classified by their driving practices or experience as inexperienced, beginner or skilled. Driving practices broadly involve two viewpoints—driving frequency and annual driving mileage. There is not necessarily a positive correlation between driving frequency and mileage, as there are drivers who frequently drive cars to certain shops or stations close to their homes, while limiting their annual driving mileage. As there are drivers who do not drive cars for a long time after getting a driver's license, we must clarify whether driving experience means the number of years after getting a driver's license or the periods of time when drivers actually drive cars.

* Regional characteristic is an important viewpoint because user characteristics differ from region to region. This may depend on cultural or lifestyle differences; for example, the meanings and what is conjured by certain colors reportedly differ from country to country or from region to region. Apple iOS has particularly high smartphone market shares in Japan and the United States, while Android commands high market shares in other regions. This may be a key point when considering what users require to HMI of information devices. The places where the user lives is related to the frequency of driving and driving characteristics. What users require from driving may differ, depending on the road traffic environments, such as traffic-limited suburban roads and urban districts plagued with heavy traffic congestion. Road environment differences between rural, urban and suburban regions influence the driving styles.

Target users are not limited to drivers. Understanding is required about front- and rear-seat passengers, for example, in the case of a family vehicle, such as a minivan, any person could become a rear-seat passenger, ranging from a child to an elderly person. There are studies on car-sickness of rear-seat passengers caused by entertainment displays (e.g., Kato and Kitazaki, 2008). In the case of welfare-purpose vans, target users, as passengers, include disabled people using wheelchairs and caretakers, who put wheelchairs on lifts to move them inside. Given that product plans, functions and specifications may be related to decisions to purchase specific vehicles, vehicle planners should consider and take into account how decision-makers would use vehicles and what they require from vehicles.

2.2.2 Observation-based Approach

2.2.2.1 Knowing User Requirements

User requirements are useful for determining the design policies for products and services. Designing without understanding the user requirements may lead to wrong designs which fail to satisfy user requirements or cause users' discontent. Approaches to objectively grab user requirements include behavior observation ethnographic surveys, task analyses questionnaire, and interviews that are expected to give insights.

2.2.2.2 Behavior Observation

Behavior observation is an approach to give insights through observation of samples' behaviors. This approach looks into subjects' involuntary behaviors and is more objective than the questionnaire surveys or group interviews. It is effective to find the potential needs. Behavior observation can extract needs that cannot be expressed in language and can suppress stereotypes. It is useful when new viewpoints are required for developing new products and services.

Behavior observation is summarized by Matsunami as an approach to produce innovations in the following way (Matsunami et al., 2016): When innovative viewpoints are required, traditional ways of thinking and prejudices must be destroyed. Therefore, observers are required to observe behavior in places where subjects act, grasp facts as they are and understand what is happening at the real sites. Based on factual findings, observers analyze why such events have occurred and gain insights into user needs. Observing involuntary behaviors provides hints for clarifying unsatisfied needs. It is difficult to gain such hints with questionnaire surveys.

Updating viewpoints based on gained insights is called 'reframing'. Reframing means creating a new framework that leads to innovation, for example, the conventional idea that vehicles are a means for moving may be reframed to view vehicles as a means for enhancing enrichment of leisure time. Such reframing may hint at new results. Based on gained hypotheses, improvement proposals may be produced and implemented to confirm their effectiveness.

2.2.2.3 Ethnographical Methods

Ethnography is a research method initiated in the sociology field. It was initially used to study cultures, concept of values, lifestyles and habits in a particular community through close contact with the community. This approach is applied for understanding the subject's activities in user surveys.

CCE (cognitive chrono-ethnography) entails applied ethnography for user surveys (Akamatsu and Kitajima, 2012). The CCE is developed to understand behavioral principles of people on the site and lead to ideas about new services and products that raise the experience value. The CCE, though being close to behavioral observation, emphasizes not only observing current behaviors of subjects but also investigating their past experiences and their unconscious cognitive state. It gives priority to understanding the context of subjects' use of particular goods or services, such as living environments and usage situations, and to extracting conditions for and identifying directions for the optimization of goods or services. The CCE seeks to gain insights into cognitive states that are influenced by subjects' life experiences and events through behavioral observation and retrospective interviews on their experiences and to clarify unaware values and needs or discontents. A similar methodology that applies ethnography to human factors researches was developed by Lahlou et al. This method is called 'subjective evidence-based ethnography' (Lahlou et al., 2015).

Information that should be gained through ethnographical methods is how particular people select particular behaviors in particular cases. Therefore, subjects for ethnographical studies are carefully selected according to behavior-characterizing parameters (called 'critical parameters' in CCE) and are called 'elite monitors'. To cover a wide range of behaviors and responses to services, subjects with different critical parameters are recruited. To maximize the information gained, subjects are selected through an audition. The minimum number of subjects with specific parameters may be around three (Table 2.1).

In the behavior observation session, subjects use portable recording equipment to record their behaviors. Video cameras, digital cameras, notebooks and others that can easily record activities without disturbing one's natural behavior are used. Recorded video images and photos are used in interviews as triggers to recall memories. Multiple behavior observation and interview sessions with each subject are effective. The first session is used to acquire and confirm factual data; the second and later sessions are exploited to acquire and confirm background information and historical data. Repeated sessions allow researchers to widen and deepen the data coverage. Repeated reviews of past activities using video images enable subjects to more deeply understand themselves and provide more honest comments and accurate information.

To gain insights into the cognitive state of subjects, researchers visualize information the subjects attributes (personality, experiences and human relationship, etc.) and find relations between information acquired through behavior observation and interviews and the subjects' attributes. It is useful to describe the subject's profile as basic information, behavioral flows reflecting goods/service-accepting processes and a service history including relations between the subject's past and present life and his or her acceptance of goods/services. It is sometimes effective to describe the sociogram of human relations for the subject. From the analysis of relations between

Table 2.1: Comparison Between CCE and Traditional Research Approach.

	CCE	Traditional Ergonomic Research
Research premise	Assuming biases of samples based on a hypothesis	Extracting population averages
Sampling methods	Purpose-oriented sampling	Random sampling
	Screening samples in line with a hypothesis	Random sampling is important (sampling must be random)
Objective for sampling	Securing reproducibility	Minimizing errors
Views on the number of samples	A large number is not important	A large number is desirable
	The number of samples guarantees the reproducibility of results	The number of samples is directly related to the accuracy of data

behavioral patterns and sample's attributes, principles can be extracted for designing functions/services to improve user experiences.

2.2.2.4 Task Analysis

Task is a series of actions and procedures to achieve a subject's purpose or goal. The task analysis is a method to observe the subject's activities to reach to the goal state by operating an equipment/system and extracts system-design problems. The analysis of the user behavior can find problems that cannot be provided through user interviews, such as unaware problems. In task description, the sequence of actions and information accessed by a subject and their procedures are decomposed in detail and described one-by-one for each scene. The analysis clarifies physical or cognitive user requirements regarding the decomposed actions and procedures and results are used for designing user interfaces, work environments and procedures.

In an approach to extract a scene for use, if the systems already exist, a task analysis may be conducted through observation of subjects using the existing systems and interviews with them. In the development phase of new systems, task scenarios are to be specified and actions and procedures to achieve goal may be generated for the analysis. To extract problems, each subject's information acquisition, understanding and decision, and operation for each task or subtask is examined from the viewpoint of layout, visibility, cues and procedures.

Both ideal and realistic solutions to problems extracted are investigated in the task analysis. To draft solutions, researchers make considerations from the viewpoints of structure, operation, systems and human errors while comparing the products in question with others. Solutions are recorded as ideal and realistic ones. If there is a large gap between the ideal and realistic solutions, describing reasons for the gap may lead to solutions. Table 2.2 shows an example of task analysis.

Table 2.2: An Example of Task Analysis.

Task	Extracting Problems			Draft Solutions	
	Information Acquisition	Understanding and Decision	Operation	Realistic	Near-Future
Connecting a power cord on the main body	• It is difficult to reach the power cord connector because it is located in the back. • It is difficult to insert the cord.			Making it easy to insert the cord sideways.	Enabling the power cord receiver to swive.
Inserting a power plug into a socket	• The 1.5-meter cord is too short.			Extending the cord length to 2 meters to make it available for a large sink or dining table.	Wireless power supply system requiring no power cord.
Pushing a lock button	• It is difficult to understand that the unlock button must be pushed for hot-water supply. • It is difficult to know that the temperature has reached 98 degrees. • The sheet switch is difficult to push.			• Using an arrow to indicate the relationship between unlocking and the hot-water supply button. • Adopting a switch that is easier to push.	

2.2.3 Questionnaire and Interview Approach

2.2.3.1 Objectives of Questionnaires and Interviews

Although a questionnaire is based on respondents' subjective views, it should be designed for a questionnaire survey to eliminate any unnecessary prejudice and to extract essential users' desires. To make a survey effective, we must focus on an objective and targets of survey and explain what we want to know through the survey before designing the questionnaire. Given that the time and the number of questions for the questionnaires or interviews are limited, we must give priority to questions which help in achieving the survey objectives. In many cases, the number of questions may increase, but by focusing on key questions based on objectives we may succeed in conducting a better survey.

2.2.3.2 Selecting Survey Methods

A questionnaire survey is one of the quantitative methods to figure out users' opinions, attitudes and realities by asking them to respond to pre-set questions. It is suitable not for an exploratory survey but for verifying a hypothesis or for grasping facts. One questionnaire survey alone fails to collect all the requisite data. It may be combined with a group interview or any other qualitative survey to gain more accurate results. In traditional questionnaire surveys, paper questionnaires are distributed to respondents through postal mail, hand delivery or meetings of respondents. Over recent years, web questionnaires are frequently conducted. Every questionnaire method has its advantages and disadvantages. It is desirable to select a method suitable for fulfilling the survey objective.

Web questionnaires can be aggregated more easily than paper questionnaires. Conditional branching for web questionnaires is easier than for paper questionnaires. However, there are concerns that respondents do not read the questions carefully and give irresponsible responses. While web questionnaires can be distributed on a large scale, respondents may vary. Respondents' answers must be screened. In addition, we must remember that web questionnaires are answered through personal computers and feature a bias to respondents with a high PC literacy.

Conditional branching and other expressions for paper questionnaires (distributed through mail or hand delivery) are more difficult than for web questionnaires. Space is more limited. Therefore, paper questionnaires are more difficult to develop. Another disadvantage of paper questionnaires is that greater time is consumed for aggregation after the collection of questionnaires. Mail questionnaires force respondents to resort to mailing procedures, which may lead to lower collection rates. On the contrary, handwritten-paper questionnaires are effective for collecting data from users with lower PC literacy. Paper questionnaires are distributed to target service subscribers or product users through mail or direct hand delivery at event sites. In facilitating participants to give responses, paper questionnaires excel over web questionnaires.

Another questionnaire method is to gather samples at a site for blanket responses. At event sites, samples can respond to questions while still excited just after the event. However, it may be burdensome for respondents to express their feelings in handwritten language. Facilitators may be required for such questionnaires to improve the accuracy.

Questionnaire planners consider some specific coverage that are required to gain sufficient responses. Preliminary research may be required for considering such coverage. For web questionnaires, it is effective to calculate sample appearance rates or implement simple questionnaires for prior sampling. To analyze respondents by age group or sex, questionnaire planners must determine the total number of samples to secure a sufficient number of samples for analyses in each category after the categorization of samples. The following equation may help those planners who consider some specific questionnaire coverage: Coverage = Collection rate × Appearance rate × Required number of samples.

Web questionnaires use questionnaire systems that distribute questionnaires and collect responses. Paper questionnaires are distributed through mail or direct hand delivery and collected through mail, questionnaire boxes or hand delivery. As people are required for distribution and collection, paper questionnaires may be assigned to some special professionals. Whether companies have in-house monitors or assign questionnaires to professionals, some counter measures may be required to raise the collection rates.

2.2.3.3 Designing Paper Questionnaires and Interviews

While interviews can be conducted flexibly, paper questionnaires are fixed in advance and cannot be changed to suit the situation. This means that paper questionnaire-planners must clarify questionnaire objectives and consider data and analysis methods.

When the overall structure of the paper questionnaire and question orders are considered, it is desirable to itemize the questions and sort them out. With consideration given to the burden on respondents, the number of questions for a web questionnaire should be limited to say, 30 (to be answered in 10 to 20 minutes). An interview should not exceed 90 minutes. A rise in the number of question items increases the burden on respondents, causing a decline in the accuracy of responses.

The question order should begin with easy-to-answer questions. Generally, general questions should come first to be followed by those on specific issues. A questionnaire should start with general questions on a relevant theme and then focus on the prime theme to prod respondents to provide specific responses after remembering the relevant conditions sufficiently. Such an approach can increase the accuracy of responses. A paper questionnaire is recommended to ask respondents to specify their age and sex in the final part because early questions on such items may affect their responses to later questions. The wording and expressions in questions should be devised to avoid misunderstanding, to avoid leading to a desired answer and to facilitate respondents to give accurate responses right till the end. Free descriptions, though useful provision of details should be minimized to prevent the burden on respondents. Successful answering of questions maintains the motivation of respondents who are keen to provide responses.

In an interview, an interviewer should ask open-ended questions to draw more information from a user to explore user requirements before shifting to closed-ended questions which need thinking and figuring out to confirm facts. Open-ended questions include how an interviewee thinks about something and are not answered with a simple 'yes' or 'no', while closed-ended questions are answered with a brief response such as 'yes' or 'no'. To develop topics and deepen thoughts, an interviewer must adopt techniques that successfully combine open-ended and closed-ended questions.

Subjective scales are used to confirm the level of the workload, satisfaction, comfort and feeling of accomplishment. Subjective assessment is given by a subjective judgment based on their feelings. It should be kept in mind that individual subjective assessment may be influenced by their experiences in the use of goods and systems. For subjective assessment, Likert scales and the semantic differential (SD) method are used frequently, for example, the following seven scales are proposed for assessing usability:

7: Very easy to use
6: Easy to use
5: Slightly easy to use
4: Neither easy nor difficult to use
3: Slightly difficult to use
2: Difficult to use
1: Very difficult to use

After the collection of paper questionnaires, the collected results are aggregated and analyzed. Before analyzing the results, it must be checked if the responses have any defects. To increase the accuracy of analyses, wrong responses, skipped responses and errors must be excluded. Questions that confirm contradictory and wrong responses, like scale questions, can be included.

When aggregating data, simple aggregation and drawing of graphs must be done to make it easier to figure out the results. Then, various viewpoints must be introduced to analyze the data and find similar trends. Cross aggregation based on respondents' attributes, scatter diagrams and frequency distribution tables are used to find some tendency. Multivariate analysis and other statistical analyses are also useful, while findings and hypotheses are derived from analysis to answer the survey objectives. Paper questionnaires are suitable for verifying hypotheses and grasping facts. However, it is difficult to gain further insights. Objective behavior observation and other measures should be taken to confirm details.

2.2.3.4 Depth Interview Method

One-to-one interviews are useful to acquire details on personal backgrounds. An interviewer can spend time with an interviewee in a one-to-one interview and extract more information from an interviewee through direct talks than in a group interview. Not only opinions and attitudes but also their reasons and background factors are explored. An effective interview allows an interviewer to conduct an exploratory survey before any hypothesis is established.

Interviewees must be selected and limited to users of products meeting research objectives or people who have preferences regarding such objectives. The standard number of interviewees for an objective may be around eight. If there are multiple attributes for the candidates, three to five interviewees for each attribute may be appropriate.

Objectives of interview research are divided into two categories—one for figuring out a situation for the use of a product and another for assessing the acceptability of a product. In a situational research interview, an interviewer asks an interviewee when and how a target product or service is used in order to grasp the situation used for a product. In an acceptability research interview, an interviewer leads the interviewee to touch a product and explore how and why the interviewee likes the product, in a bid to find what points of the product are acceptable.

A ladder-up approach is used for interviews. The approach to ask questions about upper concepts regarding why an interviewee took some action or made some choice or assessment is combined with a ladder-down approach to question an interviewee about what particular service or product he/she would accept. Such laddering to explore a value structure is useful for understanding the users more deeply.

Given that a one-to-one interview, where an interviewer asks an interviewee to talk about personal details, may exert a great burden on the interviewee, the interviewer must take care to build confidential relations (rapport) with the interviewee and lead the interviewee to talk freely. Basically, the interviewer should first ask easy-to-answer questions. If the interviewee talks about general topics rather than personal experiences or views, the interviewer may have to bring the interviewee back to personal matters by asking him/her to talk about personal views and experiences.

2.2.3.5 Group Interview

Group interviews, where multiple interviewees give opinions, are useful to obtain information at the same time. A group interview has the synergic effect among participants (or group dynamics) and deepens the discussions. In this way, a group interview is suitable for exploratory research before building hypotheses. It is also useful for grasping behaviors of consumers, who act in groups.

In a group interview, a group of five to six people basically hold some three sessions. Group members must be so arranged as to be of the same age, sex or social background. If interviewees' attributes are different, then disputes or viewpoint differences may arise between interviewees, affecting the discussions and making it difficult to acquire the expected information. Synergy effects cannot be expected in such cases. If research is conducted on multiple user groups, such as young and old generations, then attribute-based groups may have to be formed for separate interview sessions.

2.3 Driving Behavior Measurement

2.3.1 Driving Behavior Measurement Using Driving Simulators

2.3.1.1 Objectives of Using Driving Simulators

Driving simulators are used for various objectives: to evaluate changes in accelerator operation for developing automobiles, to assess the workload of human machine interface (HMI) use while driving for developing in-vehicle information system, to assess the effect of warning signals for developing ADAS, to analyze human errors and assess driver characteristics, and to simulate hazardous situations for driving education. Driving simulators are an effective tool for experimentally measuring driving behavior in complex road-and-traffic conditions As indicated in Table 2.3, driving simulators bridge real driving behavior and computer simulation using mathematical models.

An advantage of using driving simulators for ergonomic experiments is that they can generate experimental conditions that are to be reproduced for each participant and do not harm the participants even in a crash situation. While it is almost impossible to generate various road conditions (road width, slope angle, lighting conditions, etc.) for experiments in a real environment to examine their effects on driving behavior, we can easily do so by using driving simulator (Akamatsu et al., 2004; Sato et al., 2008; Akamatsu and Onuki, 2008; Rudolf et al., 2004). Driving simulators have been also used for assessing interior designs at the vehicle design stage (Kemeny, 2009). Driving simulators are also developed for assessing vehicle performance control designs (Zeep, 2010; Yasuno et al., 2014).

Figure 2.5 shows application of driving simulators in the automobile development process for active safety systems. After conducting traffic-accident analyses, driving behavior in dangerous situations is measured using driving simulators and accident mechanisms are investigated. Based on this, performance targets for developing active safety systems are set. Then the development process and the effects of the systems are verified, using driving simulators. In this way, driving simulators play two important roles—analyze driving behavior and assess the effects of systems under development.

Table 2.3: Driving Behavior Measurement Methods for Automobile Development.

Method Element	Real World	Mixed Reality	Driving Simulator	Computer Simulation
Environment	Real	MR	Virtual	Virtual
Vehicle	Real	Real	Virtual	Virtual
Driver	Real	Real	Real	Virtual

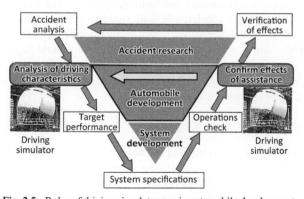

Fig. 2.5: Roles of driving simulator use in automobile development.

2.3.1.2 Basic Configuration of Driving Simulators

Figure 2.6 shows a typical driving simulator configuration. Based on the amount of driver operation measured through the acceleration pedal and steering, the acceleration/speed/displacement are calculated by a vehicle dynamics model. Accordingly the point of view is moved in a simulated three-dimensional space to display a road-environment image. Then, engine sound, wind noise, road surface vibration and other stimuli may be provided to give the feeling of a moving vehicle. Moving bases, such

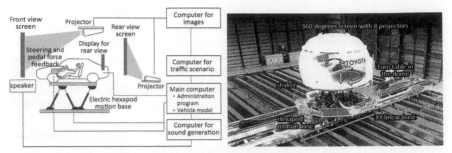

Fig. 2.6: Example of driving simulator configuration.

as a hexapod motion platform may be used to generate the feeling of acceleration based on a calculated vehicle motion. A moving base is effective in improving the sense of reality to make the driving operation seem close to actual driving. It also reduces simulator sickness, a problem with driving simulators.

An image system requires a wide-angle screen for right and left turns at intersections and improves the sense of immersion in stimulated driving. However, it has the problem of increasing the occurrence of simulator sickness. The feeling of acceleration depends heavily on hardware performance of moving bases and is also influenced by software, including motion control algorithms.

An important component of the driving simulator in driving behavior measurement is a driving scenario to generate traffic conditions, such as other vehicles running on a road-environment. The driving scenario is controlled by the location of the subject vehicle for triggering events to secure reproducibility under controlled conditions.

Driving simulators can measure the following: accelerator and brake pedal and steering operation amount; acceleration and running speed; vehicle motions and other vehicle states; vehicle locations, road widths, traffic signal and other traffic environment information; and data on other vehicles' status including inter-vehicular distances and relative positions. In the simulated driving environment, compared to the real vehicle, all sensing data is without error. It can also be used to simulate the use of system with less-accurate sensors or sensor errors.

2.3.1.3 Classification of Driving Simulators

The field of view for image systems and the degree of freedom for motion system are important elements for classifying the driving simulators. The range of applications of driving simulators is influenced by the performance of system components as classified roughly in Fig. 2.7, for example, a simple cockpit with a single-display monitor and a racing wheel for video game is frequently used to demonstrate a dangerous situation and driving assistance system concepts in driving education. A screen that can display only the front view of a monotonous road environment, like a highway, is used for HMI assessments and for distraction research. A wide-angle screen/display, covering a complete periphery of the subject vehicle, is necessary when the driving scenario includes right and left turns at intersections. In order to generate 0.3 G front, back, left and right acceleration covering 80–90 per cent of usual driving, a 35 m × 20 m linear motion system is necessary to simulate the vehicle-generated acceleration. Parks proposed a potential classification for training simulators (Parks, 2013).

Body vibrations, engine sounds and environmental sounds during driving must be reproduced to improve a sense of reality. The force feedback when steering or holding a steering wheel and the feel of brake when operating must be similar to actual vehicles when driving operations are to be measured. There are two types of visual system—fixed and movable screens. Visual systems are further classified by image resolution and by presenting 2D or 3D image. Originally, the display system for three-dimensional stereoscopic vision was used for assessing the vehicle interior and other static settings (Kemeny, 2009). Recent improvement in the visual system performance has allowed their use in driving simulators for assessing dynamic settings. Furthermore, the evolution of virtual reality devices has allowed head mount displays to be used for developing simplified systems (Aykent et al., 2015).

As the ability of indoor vehicle driving simulators is limited, a new approach has been adopted to use a test vehicle that amounts to a driving simulator, which can run in a real environment (Fig. 2.8). The imaging device fixed in front of the driver displays a mixture of real environment images and the virtual ones which are produced with computer graphics (Uchida et al., 2017).

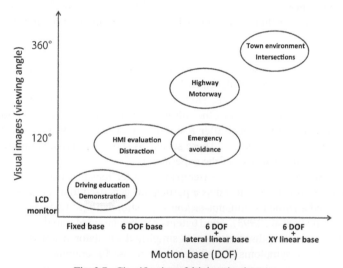

Fig. 2.7: Classification of driving simulators.

Fig. 2.8: An experimental vehicle using augmented reality technology (JARI_ARV) with permission.

2.3.1.4 Driving Simulator Sickness

A driving simulator using a large screen to provide more realistic road scenes can cause simulator sickness, although a simple driving simulator using a personal computer-monitor has no such problem. Simulator sickness occurs mainly by discrepancy between visual information and somato-sensory information obtained from the vestibular system. While visual information changes according to the position of the vehicle in the simulated road space, somato-sensory information reflects the movement of the driver's body, that is, on the seat of the simulator cabin. When a simulator is of fixed-base type, the body does not move even if the driver presses the acceleration pedal to speed the vehicle. Even if simulators are motion based, it is difficult to precisely reproduce the same acceleration as by actual vehicles. There is always a discrepancy between the visual and somato-sensory information in driving simulators. Perception of self-motion caused by visual information is called 'vection' and is induced by optical flow through the movement of visual images. If the visual image is large, then the perceived self-motion becomes large. Therefore, a driving simulator, that has a wide-angle screen and lacks motion system, easily leads to simulator sickness.

Test scenarios, including a right or left turn at an intersection and deceleration to a stop, tend to lead to simulator sickness. As yaw angle speed is high for a right or left turn, image drawing is often delayed and thus the image motion is discontinuous. The absence of a physical sense of yaw rotation is another cause of simulator sickness. Deceleration to a stop is mainly caused by a lack of a somato-sensory feeling of deceleration in fixed-base driving simulators. In moving-base driving simulators, using hexapod motion system or other tilting mechanisms, a motion to return to the initial (horizontal) position just after a stop (a cabin leans forward when a vehicle stops), causes simulator sickness. These scenarios should be avoided unless they are indispensable for the objective of the experiments.

People can adapt to sensory discrepancies after repeated exposure. Therefore, simulator sickness can be reduced as a participant gets used to a simulator. Kennedy et al. developed a 16-item-simulator-sickness questionnaire (SSQ) to assess simulator sickness (Kennedy et al., 1993). The 16 items can be classified into three symptoms—nausea, oculomotor and disorientation. A simplified simulator sickness questionnaire (SSSQ) with these symptoms (Fig. 2.9) is easy to use for ergonomic experiments to assess the changes in sickness symptoms during repeated simulator drives (Akamatsu, 2001). By using the questionnaire, sickness effects and relevant countermeasures have been devised (Table 2.4). The questionnaire can also be used to arrive at a decision

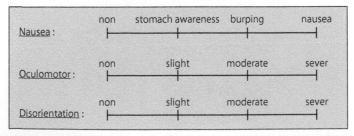

Fig. 2.9: Simplified simulator sickness questionnaire (SSSQ) (Akamatsu 2001).

Table 2.4: Simulator Sickness Effects and Counter Measures.

Sickness Symptoms	Effects on Tests and Counter Measures
Nausea	Repeated driving can monotonously mitigate the symptom. It is effective for participants to get used to the simulator's characteristics by first driving on an expressway or the like, where acceleration and deceleration and steering are less than on other roads.
Oculo-motor	The symptom does not change in one day. However, repeated driving over days can reduce the symptom. Avoid scenarios where steering maneuver, including right or left turns at intersections, is frequent, until they get used to the simulator.
Disorientation	The symptom increases through repeated driving in one day, producing no familiarization effect. As a counter measure, consideration should be given so that the duration of each driving session is not long.

regarding stopping an experiment to avoid further sickness or for screening robust participants.

Given the above, it is recommended that preparatory sessions are held where participants drive the simulator for familiarization purposes before undertaking the main experimental sessions. In such a driving scenario, the participants first drive on an expressway or the like, where acceleration, deceleration and steering are less than on urban and countryside roads, so that sickness is limited. In this way, they may get used to the characteristics of the simulator. It is effective to use the SSSQ to screen out people, who are vulnerable to simulator sickness. Experimenters should gather more participants than required, considering that some of them may be rejected due to simulator sickness. Some people, who get used to the simulator, may feel uncomfortable when actually driving a vehicle. Therefore, participants should refrain from driving a vehicle just after their simulator test. Also, given that people just after eating are susceptible to simulator sickness, participants should not undertake simulator tests just after a meal.

As mentioned above, simulator sickness is a major problem for simulator-using tests. To address the simulator sickness problem, consideration should be given, not only to hardware configuration for the driving simulator, but also to test procedures and participants.

2.3.1.5 Other Tips for Use in Driving Simulators

While projectors are used frequently to project images in the simulator, we must understand that their illumination intensity is far less than actual daylight conditions. Even in a daytime condition on the simulator, it is so dark that participants tend to become sleepy. Drowsy driving is more frequent for the simulator than for a real vehicle, especially in a monotonous driving scenario.

Ensuring safety even in a dangerous driving situation is a great advantage in driving simulator experiments; however, as participants are aware of not meeting with an accident after a crash, there is the possibility of their driving behavior being not the same as in a real road environment. Lack of the sense of risk may also affect the results of the experiment on driver's distraction. In dual task conditions, performing both in-vehicle devise task and driving task, the amount of attention focused on the driving task depends on the perceived demands of the drivers. When participants think there is no risk of accident, they may tend to look more often or longer at the in-vehicle devise. Instructions to participants and driving scenario, including other traffic, should thus be very carefully designed.

2.3.2 Driving Behavior Measurement Using Instrumented Vehicles

To objectively understand the driving behavior, the driver's behavior data must be measured and recorded. An instrumented vehicle is used to measure the driving behavior on real roads or on test tracks. The instrumented vehicle is equipped with sensors to measure the position of the vehicle, the speed, acceleration, angular speed and other vehicle conditions, as well as the driver's operation data which also include the accelerator and brake pedal stepping pressure/stroke and the steering angle. Cameras are also mounted on the instrumented vehicle to record the front or surrounding view, the driver's face and whole body to measure the traffic conditions and the driver's state.

2.3.2.1 Instrumented Vehicle

Measurement of driving behavior data using instrumented vehicles began in the 1960s (Michon and Koutstaal, 1969). At that time, measurement systems were big enough to fill the rear seats and the trunk of a station wagon. Since the 1990s, driving experiments on real roads, using instrumented vehicles, have been conducted to collect driving behavior data (Katz et al., 1995) for statistical modeling of the driving behavior (Pentland and Liu, 1999).

Figure 2.10 shows an example of sensors mounted on an instrumented vehicle and driving behavior data.

Data for vehicle conditions include:

- Speed
- Acceleration (X, Y, and Z directions)
- Angular speed (roll, pitch, and yaw)
- Angle (roll, pitch, and yaw)
- Position (latitude, longitude, direction, and altitude)
- Driving distance

The driver operation data include:

- Accelerator pedal stroke
- Brake pedal stroke
- Steering angle
- Shift position
- Turn signal (left/right) on/off
- Brake lamp
- Head lamp
- Wiper operation

Data for environmental conditions surrounding the vehicle include:

- Distance to a car in front, relative speed
- Distance to a car behind, relative speed

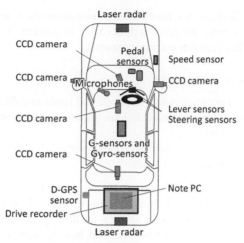

Fig. 2.10: Sensors mounted on an instrumented vehicle to measure driving behavior.

A laser distance sensor may be installed just above the brake and accelerator pedals to measure the presence or absence of the foot on the pedals. The detection of pedal stroke and foot position through the sensor allows measurement of the driver behavior as to when the driver moves a foot on the pedal to prepare for stepping. Instead of distance sensors, cameras that capture pedal operations are also used for this purpose.

Driving behavior depends on the driving conditions; so it is recommended that as many behavior-influencing factors as possible must be measured to understand the driving behavior in a real environment. Surrounding vehicles are one of the factors the influence the driving behavior. In addition to sensors installed on bumpers that are used to detect vehicles in front and behind, some instrumented vehicles have an infrared sensor to detect the surrounding vehicles (e.g., an instrumented vehicle of the University of Southampton (Brackstone et al., 1999)).

In addition to the above-mentioned numerical data, video data are collected with cameras installed in the vehicle; particularly the number of lanes and the vehicle's position in a lane are quantified through camera footage. Generally, front and rear views are recorded with cameras. To detect the driver's states, cameras cover the driver's face, head, feet, and the arms from behind and the sides (steering and in-vehicle system operation). An eye-tracker may be used to measure the driver's visual behavior to understand the perceptual/cognitive conditions (Lappi and Lehtonen, 2013). Physiological sensors may also be used to collect physiological data (Helander and Hagvall, 1976).

The driving behavior data are recorded in a memory device. A sampling frequency range of 30 to 100 Hz is recommended to avoid loss of precise timing of the onset of an event. Sampling frequency of video image is limited, depending on the capacity of memory device and the length of recording duration. The minimum used to be eight frames per second; however, as the memory size increases, higher frequency image is being recorded these days. When data is uploaded using a cloud service through network, instead of recorded it on-board device, the size of data is limited by network conditions. Given that an instrumented vehicle is designed to measure the driver's natural driving behavior, the recording devices should be installed in the trunk and other locations outside the driver's direct view to prevent the driver from being conscious of his or her behavior getting measured.

2.3.2.2 Measurement Environment

Instrumented vehicles are used on test tracks or real roads to collect driving behavior data. This section describes the considerations regarding the assessment tests in each measurement environment.

2.3.2.2.1 Measurement on a Test Track

Driving behavior is influenced by the road environment and traffic conditions. If driving behavior, free from such influence, is to be assessed, measurement on a test track may be effective. For example, studies on driving skills are measured on a test track to investigate the driver's behavior on how to operate the brake pedal to decelerate and stop, on the skill in keeping the vehicle speed constant, and the gaps between skilled and unskilled drivers when controlling the vehicle during circular driving. Eliminating any effect by other vehicles and securing safety when a large space is available are advantages of ergonomic experiments on a test track.

If driving behavior at an intersection is to be measured on a test track, then traffic lights must be installed with pedestrian crosswalks to reproduce an intersection. When a driver's side turn at an intersection is tested, it may be difficult to set the same timing for a vehicle coming from the opposite direction for various conditions. Thus, limits exist on the reproduction of a real-road environment and traffic conditions on a test track.

Furthermore, we must take note of the fact that the same driving behavior as that on a real road may not necessarily be reproduced on a test track as constraints exist on the reproduction of a real-road environment. When the timing of the onset of deceleration is measured at an intersection with a 'Stop' signal reproduced on a test track, for example, the average deceleration start timing on a real road for a driver differs from that on a test track; this is so even for the same driver and the same intersection size. Also, variation of driving behavior, i.e., variance in timings of the application of the brake over repeated driving is smaller on a test track than on a real road.

2.3.2.2.2 Measurement on Real Roads

When an instrumented vehicle is used on real roads to measure driving behavior, the participant may use the instrumented vehicle for daily life or drive the vehicle on a preset route. A participant who frequently uses the vehicle for daily life may be able to provide extensive driving behavior data, which is an advantage. If the participant fails to pass the same locations when driving the vehicle, his driving-behavior data at the same location may be limited. This is a disadvantage if the driving behavior is analyzed statistically. In this kind of measurement, drive recorders are installed on participant-owned vehicles to acquire the relevant data.

In case of a preset route experiment, the driving behavior data at the same location are collected for comparison between the same driver's data on different days and between different drivers' data. Here it is important to give consideration to the driver's familiarization with the instrumented vehicle. As the vehicle used by the driver is not necessarily of the same type as the instrumented vehicle, a period must be set for the driver to become familiar with the instrumented vehicle for which a preset route is about a week that allows the driver's ordinary driving behavior to be observed.

2.3.2.3 FOT and NDS

The field operational test (FOT) and naturalistic driving study (NDS) have been conducted on a large scale in countries around the world under their national projects in recent years. The FOT-Net has been attempted for countries to share FOT information (*FOT-NET Data*). The FOT uses an instrumented vehicle equipped with a driving assistance system to measure the driver's ordinary driving behavior. This system is expected to be commercialized in the future. The FOT aims to assess the acceptability of the system for drivers and the system's influence on the driving behavior (see 5.4.2.4). Methods to plan an FOT, implement it and assess the collected data are compiled in the *FESTA Handbook* (*FESTA* Handbook 2017). The handbook defines the FOT as "a study undertaken to evaluate a function, or functions, under normal operating conditions in road traffic environments typically encountered by the participants using study design so as to identify real-world effects and benefit." Data acquisition guidelines for sensors and data storage are described in this handbook.

The Naturalistic Driving Study (NDS) is a series of studies aimed at collecting driving behavior data with vehicles that people own. Famous NDS programs in the United States are the 100-Car Naturalistic Driving Study (Dingus et al., 2006) and its successor, Strategic Highway Research Program 2 (SHARP2) (Hallmark et al., 2013). 'Naturalistic' means drivers' daily life driving, in a narrow sense. The NDS represents a private car equipped with a drive recorder to collect daily driving data, while in the wider sense, it includes studies using an instrumented vehicle on a preset driving route.

2.3.3 Driving Behavior Analysis Using Drive Recorders

The drive recorder is a device that automatically records video images and vehicle behavior data for dozens of seconds before and after encountering safety critical events (including crashes, near-crashes and crash-relevant event), triggered by strong deceleration. Therefore, the device allows us to objectively analyze factors behind accidents on real roads. In recent years, naturalistic driving studies for collecting crash and near-crash data have been conducted in US, Canada, European countries and Japan to record and analyze traffic conditions, driving operations and vehicle behavior data before and after safety critical events.

This section takes up a naturalistic driving study for accident prevention and safety research, which was conducted under a research contract (FY 2005–2008) awarded by the Japan Automobile Manufacturers Association and provides tips on a driving behavior analysis, using drive recorders (Nagai, 2006).

2.3.3.1 Drive Recorder Specifications

Figure 2.11 shows the system configuration of a drive recorder used in the study. The system consists of CCD cameras, a global positioning system (GPS), vehicle sensors,

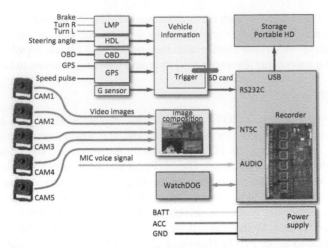

Fig. 2.11: System configuration for a drive recorder for accident prevention and safety study.

a recorder, a power control and other components. The camera image data are put into the system through five channels and synchronized with GPS and vehicle sensor data before being recorded on a portable hard-disk drive. Data recording is triggered by thresholds for acceleration and other vehicle sensors and user-triggered button press data are recorded for 30 seconds, before and after the trigger.

2.3.3.2 Recording Driving Behavior

2.3.3.2.1 Face Direction

The driver's face-image data collected through recording cameras are the most important data for analyzing a driver's visual perception during driving. Although some drive recorders record a wide scope of the vehicle interior, including the driver and his or her surroundings, it is desirable to zoom in on the driver's face as much as possible with head movements within the camera frame to analyze the driver's visual behavior, such as inattentive driving or lack of safety-confirmation behavior. The drive recorder has a CCD camera (Fig. 2.12) installed around the rear-view mirror for capturing the driver's face with a relatively narrow horizontal angle of 53 degrees. With this camera, face images that synthesize five pictures, with the quarter video graphics array (QVGA) standard resolution, allow viewers to find rough eye directions unless there is glass lens reflection (Fig. 2.13). When the face-image data are recorded with the drive recorder, the driver's consent must be obtained in advance with mosaic treatment made for data subject to analysis, to prevent the driver from being identified. Such privacy protection measures are required from the viewpoint of research ethics.

By recording the driver's actions in operating the pedals, including moving the foot from the accelerator to cover the brake pedal, the driver's preparation or readiness for hazardous events around the vehicle can be analyzed. Numerical data about accelerator-pedal operations may be acquired from the OBD 2 on-board diagnostics terminal. Pedal covering must be recorded with various three-dimensional position sensors (including an ultrasonic sensor) or image data acquired through a camera for capturing the feet. Usually the space for a sensor or camera for the driver's feet is limited with little freedom in the installable position. A wide-angle lens (with a horizontal angle of view of 115 degrees) with a CCD camera was fixed at the lower part of the center console (Fig. 2.14) in this study. From the viewpoint of research ethics, the driver's consent for the installment of a camera for recording the movements of the feet should be obtained in advance, as in the case with the driver's face.

Fig. 2.12: Camera for capturing the driver's face.

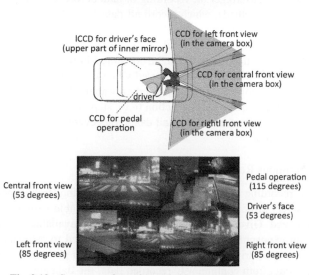

Fig. 2.13: Camera configuration and video-image data acquisition.

Fig. 2.14: Camera for recording the driver's feet.

2.3.3.2.2 Recording Traffic Conditions

To accurately understand driving behavior, we must record driving behavior and traffic conditions with a wide scope around the vehicle simultaneously. This is because outside image is helpful to understand whether the driver focuses on a direction when the oncoming vehicle is approaching or looks at a non-driving related object. When using a wide-angle lens, we must take note of the fact that the recorded subjects are small in the recorded image. In case of a near-crash with pedestrians, pedestrian behavior, such as whether or not he or she is moving and the direction of any movement, is to be identified. Three CCD cameras, with a horizontal scope of more than 190 degrees at a high resolution level for a detailed analysis of traffic conditions on the front, left, and right sides of the vehicle, are used in this study.

2.3.3.3 Data Recording Methods

2.3.3.3.1 Event Trigger Methods

Data recording methods are roughly divided into two types—event trigger methods, which use some event to trigger the recording of data before and after the event, and continuous recording methods, which record all data from the start of the engine to the stop. A general event-trigger method uses acceleration. When braking causes a stronger deceleration than a preset level, for example, data for dozens of seconds before and after the application of the brake are stored in the recording media of the drive recorder. This method is frequently used when the storage capacity is limited. It has an advantage of collecting data efficiently for analyzing critical events. In case the driver does not take quick evasive action (e.g., step on the brake pedal), then the relevant data may not get recorded. Not all dangerous event data during study period are necessarily recorded.

2.3.3.3.2 Continuous Recording Methods

A continuous recording method allows all data, ranging from normal driving conditions to safety critical conditions, are recorded to enable various safety-related data to be analyzed. Given that normal driving data are available as baseline data, for example, an exposure-based analysis can be conducted to estimate the risks of some specific types of driving behavior. Since all recorded images must be observed to extract all events from continuously recorded data, the data analysis may be very heavy for large-scale driving studies. A system to use image-recognition technology to extract relevant events automatically has been under consideration, but it is not yet available for practical use.

2.3.3.4 Examples of Drive Recorder Data Analysis

2.3.3.4.1 Time Series Analysis Using Variation Tree Analysis

A method to describe factors leading to an accident chronologically with a time-series diagram to estimate a process or background factors that cause the accident is called a 'variation tree method' (Ishida and Kanda, 1999). Describing a process leading to a hazardous situation with a variation tree, using near-crash data acquired through the drive recorder, can analyze human factors resulting in an accident.

Figure 2.15 indicates an analysis of data for a near-crash with an oncoming motorcycle during a driver's side turn at an intersection. Analysis results indicate that the driver decided to turn driver-side before the motorcycle came into the driver's view. He or she looked at the direction to turn, took time to detect the motorcycle appearing from a blind area, leading to a near crash with the motorcycle. Figure 2.15 indicates that an analysis can be done on the driver's eye direction and recognition of a hazardous object, on the prediction of danger through pedal operations and on the urgency of the situation as indicated by evasive action.

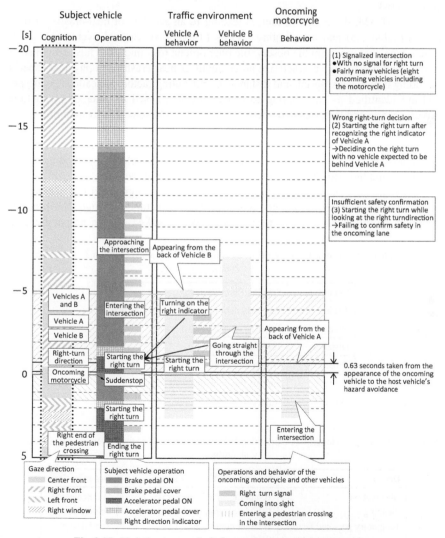

Fig. 2.15: Variation tree analysis for a near-crash with a motorcycle.

2.3.3.4.2 Analyzing a Series of Background Factors

The driver reliability and error analysis method (DREAM) is an analytical method to understand the background factors responsible for accidents (Aust, 2010). As does the variation tree analysis, DREAM tracks the factors involving an accident backward from the accident. Unlike the variation tree analysis that puts events and factors in chronological order, DREAM takes into account the parallelism of human cognitive-information processing and identifies the links between the events and factors. By aggregating analysis results (DREAM chart) for individual near-rash events corresponding to specific event types (e.g., a driver's side turn), DREAM can comprehend an accident scenario and indicate the background factors that increase collision risks.

In DREAM, there are two factors leading to an accident or near-crash—critical events (Table 2.5) and contributing factors (Table 2.6). Critical events are called 'phenotype' that represent the observable phenomena resulting from false response to traffic conditions. They characterize an accident or near-crash from physical aspects including timing gaps, speed, and distance. Contributing factors are called 'genotype' and are identified as contributing to critical events, based on data. They include driver, vehicle and traffic environment factors. DREAM selects corresponding items (subcategories into which categories in Tables 2.5 and 2.6 are broken down) provided in the DREAM manual to analyze a series of accident factors (Warner et al., 2008). A critical event characterizing an accident or near-crash is selected with contributing factors behind the critical event divided into direct and indirect ones.

After the analysis of individual cases, results were aggregated to consider the occurrence rates of critical events and contributing factors. Through this process, common causation patterns for near-crashes were chalked out. In a specific data

Table 2.5: Phenotypes Categories (Critical Events) in DREAM Manual (Version 3.0) (Warner et al., 2008).

Phenotypes	Specific Phenotypes
Timing	Too early action; too late action; no action
Speed	Too high speed; too low speed
Distance	Too short distance
Direction	Wrong direction
Force	Surplus force; insufficient force
Object	Adjacent object

Table 2.6: Genotype Categories (Contribution Factors) in DREAM Manual (version 3.0) (Warner et al., 2008).

Driver	Vehicle and Environment	Organization
Observation	Temporary HMI problems	Maintenance
Planning	Vehicle equipment failure	Vehicle design
Temporary personal factors	Weather conditions	Road design
Permanent personal factors	Obstruction of view due to object	
	State of road	

analysis, we selected the most appropriate one from the preset categories (Tables 2.5 and 2.6) in the manual while viewing video images to describe the links between critical events and contributing factors.

Two or more researchers should analyze one case. If differences arise, they should coordinate opinions through group discussions to produce the final result. As an analysis result, the causation chart shown in Fig. 2.16 (summary of 13 driver-side turn near-crashes at intersections) indicates the following features of an increase in risks of collisions with pedestrians upon a driver's side turn at an intersection:

- In about 70 per cent of near-crash cases, starting a turn too early (A1.1) fell into a critical event attributable to wrong situational assessments (C2) (e.g., taking a turn despite the presence of pedestrians around a pedestrian crossing).

- Two main factors contributing to wrong situational assessments (C2) observed in all cases were failure to find pedestrians (B1) and the driver's attention allocation (E2). The two were identified in more than 90 per cent of the cases.

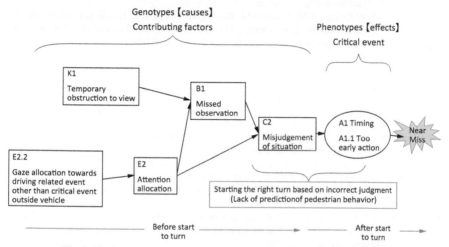

Fig. 2.16: Summarized DREAM analysis result for intersection right turns.

- In most cases, the reason for a driver's allocation of attention (E2) other than pedestrians was the confirmation of other traffic participants (E2.2).

- Major factors contributing to the failure to find pedestrians (B1) were the driver's allocation of attention to the above-mentioned safety confirmation (linking E2.2 to E2) and temporary obstruction of view caused by other vehicles (K1).

References

Akamatsu, M. (2001). Development of hi-fidelity driving simulator for measuring driving behavior. Journal of Robotics and Mechatronics, 13(4): 409–418.

Akamatsu, M., A. Yamaya, N. Imacho, H. Ushiro-oka and T. Hamanaka. (2004). Driving simulator study for designing light emitting delineators for tunnels. Proceedings of the DSC, 2004, Europe, 81–90.

Akamatsu, M. and M. Onuki. (2008). Trends in technologies for representing the real world in driving simulator environments. Review of Automotive Engineering, 29: 611–618.

Akamatsu, M. and M. Kitajima. (2012). Designing products and services based on understanding human cognitive behavior—Development of cognitive ethnography for synthesiological research. Synthesiology, English Edition, 4(3): 144–155.

Aust, M.L. (2010). Generalization of case studies in road traffic when defining pre-crash scenarios for active safety function evaluation. Accident Analysis and Prevention, 42(4): 1172–1183.

Aykent, B., F. Merienne and A. Kemeny. (2015). Effect of VR Device-HMD and screen display on the sickness for driving simulation. Proceedings of the Driving Simulation Conference Europe, 2015: 235–236.

Brackstone, M., M. McDonald and B. Sultan. (1999). Dynamic behavioral data collection using an instrumented vehicle. Transportation Research Record, 1689: 9–17.

Dingus, T.A., S.G. Klauer, V.L. Neale, A. Petersen, S.E. Lee, J. Sudweeks, M.A. Perez, J. Hankey, D. Ramsey, S. Gupta, C. Bucher, Z.R. Doerzaph, J. Jermeland and R.R. Knipling. (2006). The 100-Car Naturalistic Driving Study, Phase II—Results of the 100-Car Field Experiment, U.S. Department of Transportation, National Highway Traffic Safety Administration. Report DOT HS 810 593.

FESTA Handbook Version7 (2017). D5.4 Updated Version of the FESTA Handbook, FOT-NetData, European Commission, Available at http://wiki.fot-net.eu/index.php?title=FESTA_handbook.

FOT-NET Data, Available at http://fot-net.eu/.

Hallmark, S., D. McGehee, K.M. Bauer, J.M. Hutton, G.A. Davis, J. Hourdos, I. Chatterjee, T. Victor, J. Bärgman, M. Dozza, H. Rootzén, J.D. Lee, C. Ahlström, I.O. Bagdad, J. Engström, D. Zholud and M. Ljung-Aust. (2013). Initial analyses from the SHRP2 naturalistic driving study: Addressing driver performance and behavior in traffic safety. Transportation Research Board, SHRP2 Safety Project S08.

Helander, M. and B. Hagvall. (1976). An instrumented vehicle for studies of driver behaviour. Accident Analysis and Prevention, 8(4): 271–277.

Ishida, T. and N. Kanda. (1999). An analysis of human factors in traffic accidents using the variation tree method. JSAE Review, 20(2): 229–236.

Kato, K. and S. Kitazaki. (2008). Improvement of ease of viewing images on an in-vehicle display and reduction of carsickness. SAE Paper 2008-01-0565.

Katz, S., P. Green and J. Fleming. (1995). Calibration and baseline driving data for the UMTRI driver interface research vehicle. Technical Report UMTRI-95-2.

Kemeny, A. (2009). Driving simulation for virtual testing and perception studies. Proceedings of the Driving Simulation Conference Europe, Monte-Carlo, 15–23.

Kennedy, R.S., N.E. Lane, K.S. Berbaum and M.G. Lilienthal. (1993). Simulator sickness questionnaire: An enhanced method for quantifying simulator sickness. International Journal of Aviation Psychology, 3(3): 203–220.

Lahlou, S., S. Le Bellu and S. Boesen-Mariani. (2015). Subjective evidence based ethnography: Method and applications. Integrative Psychological and Behavioral Science, 49(2): 216–238.

Lappi, O. and E. Lehtonen. (2013). Eye-movements in real curve driving: Pursuit-like optokines is in vehicle frame of reference, stability in an allo-centric reference coordinate system. Journal of Eye Movement Research, 6(1): 1–13.

Matsunami, H., A. Kubosumi and K. Matsumoto. (2016). Human behavior observation for service science. pp. 49–59. In: S. Kwan, J. Spohrer, Y. Sawatani (eds.). Global Perspectives on Service Science, Japan. Service Science: Research and Innovations in the Service Economy, Springer, New York, USA.

Michon, J.A. and G.A. Koutstaal. (1969). An instrumented car for the study of driver behavior. The American Psychologist, 24(3): 297–300.

Nagai, M. (2006). Research on incident analysis using drive recorder Part 2: Toward active safety assessment. Proceedings of FISTA World Congress.

Parks, A. (2013). The essential realism of driving simulators for research and training. pp. 133–154. In: N. Gkikas (ed.). Automotive Ergonomics: Driver-Vehicle Interaction. CRC Press (Taylor & Francis Group), New York, USA.

Pentland, A. and A. Liu. (1999). Modeling and prediction of human behavior. Neural Computation, 11(1): 229–242.

Rudolf, B., J. Schmidt, M. Grimm, F.J. Kalze, T. Weber, C. Plattfaut and F. Panerai. (2004). Integration of AFS-functionality into driving simulators. Proceedings of the Driving Simulation Conference Europe, 55–165.

Sato, T., M. Akamatsu, N. Imacho, T. Sato, Y. Munehiro, N. Yoneyama and N. Tashiro. (2008). Driving simulator study for driver behavior assessment while driving on down grades with different gradients. Proceedings of Driving Simulation Conference Asia-Pacific, 67–175.

Uchida, N., T. Tagawa and K. Sato. (2017). Development of an augmented reality vehicle for driver performance evaluation. IEEE Intelligent Transportation Systems Magazine, 9(1): 35–41.

Warner, W.H., M. Ljung Aust, J. Sandin, E. Johansson and G. Björklund. (2008). Manual for DREAM 3.0, Driving Reliability and Error Analysis Method, Deliverable D5.6 of the EU FP6 project SafetyNet.TREN-04-FP6TR-SI2.395465/506723, Chalmers University of Technology, Gothenburg, Sweden.

Yasuno, Y., E. Kitahara, T. Takeuchi, M. Tsushima, H. Saitou, M. Imamura and E. Ueno. (2014). Nissan's new high performance driving simulator for vehicle dynamics performance and man-machine interface studies. Proceedings of the Driving Simulation Conference Europe, 4–5.

Zeep, E. (2010). Daimler's new driving simulator—technical overview. Proceedings of the Driving Simulation Conference Europe, 2010, 157–165.

3
Comfort and Quality

3.1 Occupant Comfort During Vehicle Run

3.1.1 Vibration and Comfort

The riding comfort, which is a sense of movement caused by the vibration of the vehicle, is one of the most important functions for the comfort of automobiles. Figure 3.1 shows the relation between the frequency of vibration and the main riding comfort phenomenon and a group of factors. It is possible to evaluate each of these riding comfort phenomena separately because they are felt by different parts and in different manners. For instance, the feeling of bobbing is the sense that the whole trunk of the body is going up and down over the seat; the feeling of choppy ride is the sense of a relative motion of tissues of the femur, the calf, etc., as a whole in relation to the bones; the feeling of buzziness is the sense that part of the skin tissue touching the seat is vibrating, etc.

When the vehicle is moving, the seat, floor, steering and other parts of the vehicle make the body oscillate in different directions, and these vibrations are perceived as a

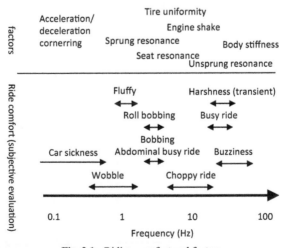

Fig. 3.1: Riding comfort and factors.

psychological phenomenon known as vibration riding comfort. This section presents an overview of the measurement methods based on the ISO, etc., the comprehensive comfort evaluation methods, and discuss the issues and applications in the evaluation of each riding comfort phenomenon.

3.1.1.1 Basic Vibration Measurement and Evaluation Methods

The measurement and evaluation of vibration riding comfort may be conducted through the test of actual vehicles driving on evaluation paths of the proven grounds for riding comfort and through bench tests using vibration tests. Figure 3.2 shows a vibration test system used for riding comfort evaluation. When shaking humans on vibration test machines, it is necessary to guarantee the safety of the participants through output limiters (to prevent excessive acceleration (ISO 2631-1, 1997), safety belts, emergency stop buttons, etc. The basic measurement and evaluation methods for vibration comfort in vehicles are specified in ISOs (2631-1, 1997) (8014, 2005) (10326-1, 1992) (5349-1, 2001) (5349-2, 2001).

The weighting curve of frequencies and the correlation between the evaluation part/direction and the weighting coefficient of frequencies/direction magnification are determined in ISO 063 and ISO 5349, because the frequency properties of the sense or sensation of vibration differs according to the part of the human body, or the direction of the oscillation. Note that in evaluation methods for hand-transmitted vibration, the range of evaluation goes up from 8 to 1,000 Hz, and in methods for the rest of the whole of the body rises from 0.5 to 80 Hz. As to the weighting coefficient of hand-transmitted vibration (W_h), although the direction magnification between each direction and the correction coefficient for vibration of the entire body is not determined, a study that shows that it should be from 0.12 to 0.26 in relation to the surface of the seat (Morioka, 2007).

As is specified on the ISO 10326-1, etc., the vibration of the seat in vehicles is measured by the acceleration of the vibration after installing a disk accelerometer on the surface and the backrest of the seat and an accelerometer for measuring the vibration of the floor under the feet on the floor interface. In ISO 2631-1, the effective

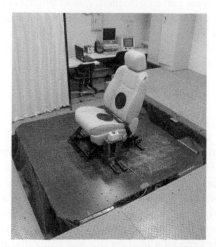

Fig. 3.2: Six degrees of freedom-vibrator base for riding comfort evaluation.

weighted acceleration is used as an index for vibration evaluation when the only vibration of a specific part in a specific direction exercises influence or the combined value of vibration of each point, which is the root-sum of squares of the effective weighted acceleration of each axis when multiple directions have an influence. When multiple parts exert influence, the total combined value, which is the r.s.s. of combined vibration of each point, is used.

3.1.1.2 Riding Comfort Evaluation by Phenomenon

The comfort evaluation of the ISOs evaluates the broad frequency bands and multidirectional vibrations using only the 'comfort/discomfort' parameter. In the automobile field, a multidimensional evaluation, that uses different parameters for each riding comfort phenomenon, is generally utilized. It is possible to reduce the vibration according to the phenomenon if the components or parts correlated to each phenomenon are already known. Regarding each of these phenomena, although companies are developing individual quantitative evaluation indices, they are basically mapping the correlation between sensory evaluation results and vibration data of the selected frequency band by a phenomenon using the principal component and multiple regression analyses (Takei and Ishiguro, 1995). Furthermore, in the harshness at the moment, when the vehicle passes over a bump, the influence of the sound on the level of discomfort is particularly strong; therefore, the quantification is conducted by combining the vibration and the sound.

However, the parts and directions of vibration measurement, the weighting curve of frequency, and the direction magnification used in these quantitative evaluations do not necessarily meet the above-mentioned standards. These standards are based on sensory measurements using sine wave vibrations with a relatively high level of acceleration and are conducted on a flat, rigid chair with a horizontal seat and a vertical backrest. In actual car seats, on the other hand, not only the angles of seats and backrests (Kato et al., 1999), the part in contact with the human body, the surface, the pressure (Morioka and Griffin, 2010), the distribution of vibration (Ittianuwat et al., 2014) are different, but also the ISO sensitivity curve of the sense of vibration, which is the base for the weighting curve of frequency, is amplitude dependent (Morioka and Griffin, 2006). Perhaps, they do not meet the standards because, among other things, one type of weighting factor is not enough for weighting the sensation. Therefore, there is still a need for taking these influences into consideration and developing a generic evaluation index with a higher accuracy.

3.1.1.3 Method for Estimating the Vibration of the Seat when an Occupant is Sitting

As mentioned above, data on the vibration of the seat when a person is sitting is necessary for conducting a quantitative evaluation of the vibration riding comfort. However, when people are actually sitting the vibration data changes significantly according to differences in each person, in the sitting posture, etc. The human body itself is a complex vibration system and the actual vibration of a person sitting cannot be obtained by using rigid body mass model. Some methods are proposed to estimate the vibration of the seat with a person is sitting in a stable manner, as by using a mechanical dummy that places a load on the seat (Mozaffarin and Pankoke, 2008)

(Fig. 3.3), a CAE model of the seat and the human body (Pankoke and Siefert, 2007) (Fig. 3.4), a combination of the actual characteristics of seats and a dynamic mass model for the body (Kato et al., 2004). All of these methods estimate the vibration on the interface between the seat and the body by combining a human body dummy with the same dynamic response (dynamic mass) to vibration with the actual seat or a model of the seat.

Fig. 3.3: Human body dummy for seats (Reprint from Mozaffarin, A., S. Pankoke and H.-P. Wölfel. (2008). MEMOSIK V: An active dummy for determining three-directional transfer functions of vehicle seats and vibration exposure ratings for the seated occupant. International Journal of Industrial Ergonomics, 38: 471–482, with permission from Elsevier).

Fig. 3.4: CAE model for human body vibration (Reprint with permission of SAE International from Pankoke, S. and A. Siefert. (2007). Virtual simulation of static and dynamic seating comfort in the development process of automobiles and automotive seats—application of finite-element-occupant-model CASIMIR. SAE Technical Paper 2007-01-2459, Permission Conveyed through Copyright Clearance Center, Inc.).

3.1.2 Comfort of the Seat

In order to discuss the comfort of the seat it is necessary to consider functions of the seat in relation to the person using it. They are: absorbing the vibration of the vehicle moving not to inflict discomfort on the user, suppressing the unnecessary body movement of the user caused by the vehicle movements, assisting the body movement of voluntary actions to operate control devices to drive, and supporting the user who tries to maintain the position of the body during acceleration/deceleration and cornering.

3.1.2.1 Seat Structure and Vibration Absorption Properties

3.1.2.1.1 Transmission of Vibration through the Seat

The band of frequencies are divided into two bands from the viewpoint of the functions—one is the band ranging from 0 to around 1 Hz; the other ranges from around 1 to 50 Hz. Although the separation is made at 1 Hz, there is no clear border demarcation between them. The upper limit of 50 Hz is the maximum vibration that a person can perceive.

The vibration between 0 and 1 Hz ranges in the zone of movement of the vehicle (i.e., to run, to make a curve or to stop). The goal is to transmit the proper vibrations to the driver. When transmitted, the vibrations from nearby 1Hz up to 50 Hz cause discomfort. In short, the former (0 to 1 Hz) is an N-type characteristic (nominal best characteristic: there is a target value to be pursued), while the latter (1 to 50 Hz) is an S-type characteristic (the smaller, the better). Taking into consideration these characteristics makes it easier to understand the phenomenon under study and the meaning of the characteristics of related components of the actual structure of the seat.

3.1.2.1.2 Issues on the Measurement of the Vibration of the Seat

An established method that is standardized as SAE J 2896 to measure the characteristics of the seat uses a hard indenter in the shape of a person's hips and back (SAE J 2896, 2012). Although this method is comparatively simple and easy to understand having been used for a long time, it contains the following problems:

* It is not possible to obtain a high correlation with the data obtained from experiments using human subjects.
* Because the measured characteristics fully account for the vibration phenomenon of persons, they cannot be the key characteristics of the seat unit to be decomposed into the characteristics of the parts.

3.1.2.1.3 Seat Structure and Specific Characteristics of Vibration

In the case of automobiles, most of seats have a metallic frame fixed to the body. There are also a metal spring, a pad of urethane foam and a slab of cloth; synthetic leather and leather cover complete its structure. Figure 3.5 shows the examples. It is necessary to understand what influence each of the components has on which phenomenon.

The seat plays an important role in phenomena related to vibration problems, as the feelings of bobbing, choppy ride, busy ride, buzziness and shock, which relate to the riding comfort during the transitional vibration. Here, we need to think what component influences which phenomenon.

Table 3.1 depiction of how each phenomenon is influenced by each characteristic of each part, for example, frame characteristics have no influence on the bobbing phenomenon and the resonance frequency of the cushion does have influence.

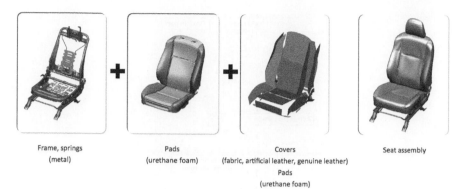

Frame, springs   Pads   Covers   Seat assembly
(metal)   (urethane foam)   (fabric, artificial leather, genuine leather)
Pads
(urethane foam)

Fig. 3.5: Structure of the seat.

Table 3.1: Sensory Evaluation of Vibration and Related Parts and Characteristics.

Parts / Ride	Frame	Cushion		Back rest	R²
Bobbing		Resonance frequency	Decay rate	Deformation	0.96
Choppy ride	1ˢᵗ resonance frequency	Static spring rate at 392N		Deformation at 245N	0.76
Busy ride	Static stiffness (for aft) at 98N	Resilience		Static spring rate at 196N	0.73
Buzziness	Static stiffness (for aft) at 98N	Resilience		Static spring rate at 245N	0.69
Shock	Static stiffness (for aft) at 98N	Deformation at 490N		Deformation at 147N	0.91

The damping factor of the cushion also has a great impact because the pad of the cushion has a strong damping characteristic. The strength of the frame affects the choppiness. Also, the rebound rate of the cushion has influence. The spring and the pad affect the rebound rate of the cushion. As noted above, the strength of the influence of characteristics of the frame, the spring, the pad, etc., differs according to the phenomenon. Consequently, it is necessary to select the characteristic of each component by measuring alternative characteristics that are defined based on the correlation with each sensory evaluation item.

3.1.2.1.4 Vibration Characteristics of the Parts of Seat

During analysis of alternative substitute characteristics of each part, the pad is to be treated carefully. The characteristic value of a certain fixed point of the characteristic is not enough because of the non-linear characteristics. The following equation defined as the elasticity rate of the backrest defines the seat unit characteristics under the assumption of non-linear characteristics (Higuchi et al., 2001).

$$DR_b = D_{490}/D_{245}$$

DR_b: Backrest deflection rate, D_{490}: Amount of deformation of the backrest 490 N, D_{245}: Amount of deformation of the backrest 245 N.

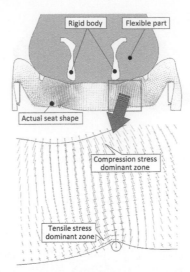

Fig. 3.6: Situation of the pad's internal stress when a person is sitting.

In addition to the foam structure, the partial compression condition when seated is also a cause of non-linear characteristics of the pad. Both compression property and tensile property of the pad are to be considered. Figure 3.6 shows one example of the urethane internal stress (Kawano et al., 2013). It is important to simplify the measure (in this case replacing it only by the compression property), although it is a material property for obtaining a correlation between the characteristic of each part and the characteristics of the seat.

3.1.2.1.5 Changes in the Characteristics of Vibrations on People

It is necessary to examine the changes in the vibrations when the user is seated. Changes in the angle of the upper body and the angle of the thigh cause changes in the center of gravity and each part of the body has its balance affected. This causes a change to level of vibration of the body. This may be verified even in relatively simplified vibration models. Since the human body is a spring with elastic properties, the vibration is coupled with the vibration characteristics of the seat. Another thing that needs to be taken into consideration is that the elastic property of the human body changes according to which part of the body is being supported by the seat. Taking the hips as an example, the human body becomes rigid closer to the properties of the bones and does not present damping characteristics when the pressure concentrates on the ischium. On the other hand, when the pressure is released from the ischium and the load is allocated to the surrounding muscles, the human body becomes flexible closer to the property of muscles and presents strong damping characteristics.

3.1.2.2 Body Movements Caused by Acceleration

With regard to the vibration zone from 0 to 1 Hz that is to be properly transmitted to the driver, it is necessary to take into consideration a variety of inputs that cause significant body movements.

By observing large movements of the body of a sitting person, we may notice that cornering causes the largest movement, which produces a lateral acceleration. When starting to steer from a straight-ahead driving, a relative displacement of the body in relation to the vehicle (seat) is produced due to the inertia caused by the weight of the body itself. In this moment, the influence of the seat is limited to friction. Furthermore, in a relatively short amount of time it passes from transient state to a state where the body weight is moved to the side support. This is a free running time that causes a great discomfort to the occupants of the vehicle. The most simple solution for this free running time is to narrow the distance between the side supports, thus shortening the distance from the user body. However, this design approach is not compatible with different human body sizes. Therefore, we may say that a seat's shape that causes large lateral movements is unavoidable.

With regard to the backward movement of the body caused by the acceleration of the vehicle, only a partial movement of the seat occurs, thanks to the backrest. On the other hand, when it comes to the forward movement of the hip (sliding over the seat) during braking, the shape of the seat contributes only to a certain extent. In the case of a sudden deceleration, the movement has to be stopped by the feet. This is a burden on the muscles and may make the control of the pedals unstable. There is also no support for the upper body when it is moving forward. Therefore, it has to be supported by the body itself (muscular force) by using the steering wheel or the door handles. Furthermore, the characteristics of the seat have a strong influence on vertical movements. In fact, complex movements with forward and backward, vertical, lateral and rotational components occur while the vehicle is running. In this case, the seat may directly affect the movement of the body on vertical directions, while sliding the lower part of the body forward and the support of the body after it touches the side supports. The rest of the supporting function is performed by the body muscle force to maintain a comfortable position using contact points of the human body with the seat.

3.1.2.3 Support Performance of the Seat

3.1.2.3.1 Lateral Movements

For the reasons mentioned in 3.1.2.1.5, there is no appropriate definition for the substitute characteristics value of the vertical movement characteristics of vibration from 0 to near 1 Hz. With regard to lateral movements, it is necessary to take into consideration situations both before and after the body touches the side supports. Figure 3.7 shows how to think about the situation after the touch. In the case of sport seats, where the time until the contact is short, it is effective to support the body shape

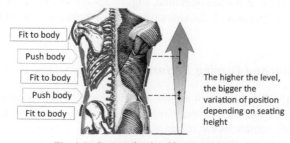

Fig. 3.7: Support by the side support part.

in a well-balanced manner (Senba and Kawano, 2013). When the time of contact is long, we need to take into consideration the additional movement that the human body makes using muscles for keeping the balance (stability) in response to a feeling of discomfort (lack of stable or secure feeling) caused by the movement. This is particularly important for the seats of occupants that are not driving the vehicle and who are exposed to unexpected vehicle movement (Yokoyama et al., 2008).

3.1.2.3.2 Movements of the Head

Figure 3.8 shows time series data of the yaw angle, the steering angle and the roll angle of the head when the vehicle passes through pylons. We must focus on the difference of time $(t_{d1} \sim t_{d4})$ between the changing point of the head yaw angle and changing point of the steering angle of the head (Buma et al., 2009). For the driver, the movement of the head is an attempt to look ahead. After the steering, it turns into a passive roll motion caused by the lateral acceleration of the vehicle. Although it is not a part that provides direct support, as long as the movement of the head does not use the headrest, the head movement as a secondary result of the support by the seat is important because it is responsible for many of visual information used when driving (Hada et al., 2012).

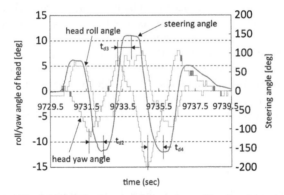

Fig. 3.8: Relation between the steering and head movements.

3.1.2.3.3 Support by the Seat during Driving

The operations of the vehicle are also affected by the movements of the body. Studies have been made to find if the counterforce produced by the seat to a certain part of the body is necessary for a driving operation. Figure 3.9 shows the position of the body during the driving operation. It discusses what part of the scapula should be supported (Hada et al., 2012).

3.1.3 Vibration and Driving Performance

Humans perceive the movement of the vehicle through the vision or the somato sensations and operate the steering wheel and the pedals based on the information obtained. Somato sensations are obtained from a variety of sensory organs as semi-circular canals, otoliths, muscle spindles, tendon spindles and cutaneous receptors.

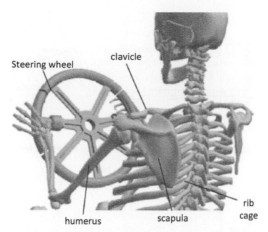

Fig. 3.9: Parts that must be supported during driving operations.

Each of these organs works as sensors for detecting frequencies with different characteristics and the human body perceives the movement of the vehicle or the vibration through the combination of the information provided by them. Therefore, the perception of vibration depends on the frequency. These frequency characteristics are closely related to the resonance characteristics of each part of the body and by amplifying the vibration of the body through the resonance, not only the sensitivity of frequency bands becomes higher, but also the vibration becomes unpleasant. It is already known that vertical vibrations with a frequency band between 4 and 8 Hz cause a strong discomfort to humans as they are highly sensitive to them. For this reason, in order to guarantee the ride comfort, it is necessary to control the vibration within this band in the automotive suspension design.

Figure 3.10 shows the results of the sensory threshold of a sine wave amplitude when a constant frequency of linear and yaw movements is applied to the participant sitting on a rigid seat (Miwa, 1967a; Miwa, 1967b; Chen and Robertson, 1972; Parsons and Griffin, 1988; Irwin, 1981; Yasuda and Doi, 1999). When discussing the steering stability of vehicles, the perception properties of movements of the car within 0–1 Hz in response the steering wheel operations are important, particularly the threshold of perceiving the movement and discriminative sensations as if the difference in the movement was felt or not. It is also important to investigate perceptual properties of visual information in detail when discussing the driving stability (Fig. 3.10 depicts the results of only bodily sensations when visual information is blocked).

Figure 3.11 indicates a six-axis motion base (Muragishi et al., 2007) simulating the movement of the vehicle in response to very small steering operation to analyze the perceptual properties. It adopts a compact parallel link structure for a wide motion range (translation ±0.5 m; rotation ±30 deg.) with a high responsiveness (~ 13 Hz). For the purpose of investigating properties in high-speed driving where the gaze point becomes distant, a distance from the eye-point to the screen is long (6.5 m) with a wide viewing angle (lat. 54 deg., ver. 18 deg.) and a high spatial frequency resolution (28 cpd). By using this device, it is possible to study perceptual properties separating the vision from bodily sensations. Figure 3.12 shows the results of the amplitude threshold of a sine wave when a constant frequency (1Hz) is applied to a

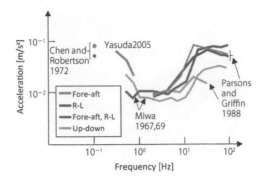

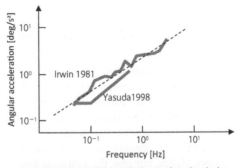

Fig. 3.10: Characteristics of the threshold in relation to the stimulation of bodily sensations.

Fig. 3.11: Six axis motion base.

seated participant (Muragishi et al., 2007). The threshold of the vision is smaller than the threshold of the somato sensations. It means that small movements are perceived mainly by the vision in rotation movements of yaw and pitch.

For improving the comfort of driving as the feeling of steering, etc., it is important to clarify 'how comfortable humans feel with which movement of the car'. There are some studies that aim to replace the indices obtained through a subjective sensory evaluation by expert drivers with objective physical quantities. One of the examples is

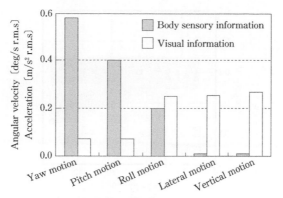

Fig. 3.12: Threshold results.

the one where the structure of the driver's subjective evaluation is analyzed by using the repertory grid method. The results show that the change in the yaw response is the evaluation point (the smaller the change, the better) for the 'sense of lightness (agility)', which is one of the evaluation items of the driving stability. On the basis of Fig. 3.12, it is possible to say that the yaw movement is mainly perceived by the vision. So they investigated the appropriate sensory value of the yaw gain (i.e., the steady gain from the steering angle to the yaw angle speed) by reproducing only the sense of vision through a driving simulator.

Figure 3.13 indicates the results of a simulation experiment and an experiment with real vehicles of the appropriate sensory value of the yaw gain in relation to the speed of the vehicle. The appropriate yaw gain value obtained in only visual information condition was a constant value that does not change with speed, and in the experiment with real vehicles, it was found that it tends to decrease with the speed in high-speed regions. It can be interpreted as follows, based on the properties of the human perception of movements.

Figure 3.12 shows that the vision has a lower threshold for yaw movements and somato sensations have a lower threshold for lateral movements. Each threshold is denoted by ε_r and ε_{dGy} respectively. Figure 3.14 indicates the relation of the vehicle speed and the amplitude of the yaw angle speed at the rotational movement in which the yaw angle speed and the lateral jerk reach the threshold for a constant frequency below 1 Hz$_y$. This is obtained under the assumption that $\omega = \pi$ [rad/s] and the relation between the yaw angle speed r [rad/s], the lateral acceleration G_y [m/s²] and the vehicle speed v [m/s] is supposed to be:

$$G_y = r \cdot v$$

Note that this equation means that the generation of slip angle of the vehicle body is being ignored. Then, the amplitude of the yaw angle speed, where the threshold of the lateral jerk becomes ε_{dGy}, is:

$$r_{\varepsilon dGy} = \frac{\varepsilon_{dGy}}{v\omega}$$

On the other hand, the yaw angle speed is perceived by ε_r regardless of the speed of the vehicle. This means that at low speeds, the yaw angle speed is more perceived

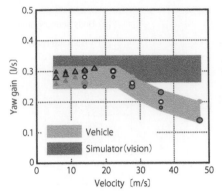

Fig. 3.13: Subjective evaluation of the yaw gain.

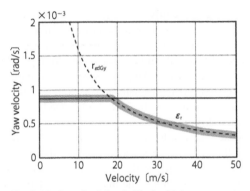

Fig. 3.14: Perceived threshold of vehicle movements.

in smaller movements and at high speeds where $r_{edGy} < \varepsilon_r$, the lateral jerk is perceived more in smaller movements. Since the shape of the threshold corresponds to the appropriate yaw gain value of actual vehicles as shown in Fig. 3.13, this means that in the appropriate yaw gain value for actual vehicles, the gain in the rotational movement felt by the driver is found to be constant with the vehicle speed. Then, it is also understood why the yaw gain becomes constant corresponding to the threshold of the yaw angle speed in simulators, giving visual information without somato sensations.

The results of subjective evaluations, as shown in Fig. 3.13, are obtained from smooth test courses; therefore, the sensation of the vehicle movement will be different under road surface conditions where vibrations with a variety of frequency bands caused by disturbances on the surface of the road are superimposed. Based on human sensory properties, it is considered that the vibrations of superimposed frequency bands become background noise and the movement of the car is not perceived until it is higher than this background noise exceeding the Weber ratio. Therefore, it is important to clarify perception characteristics when high frequency vibrations are superimposed because this delay in perception may influence the driving stability, specially during very small steering operations.

3.2 Acoustic Comfort

3.2.1 Design of the Engine Sound

3.2.1.1 Acoustic Characteristics that Influence Sound Design

The order composition and the linearity are the main acoustic characteristics on which we have to focus when creating sounds (Hirate et al., 2009; Aoki et al., 1987). Figure 3.15 indicates an example of how to create sounds for engines. Regarding the linearity, it shows that the sound pressure changes linearly in relation to changes in the rotational speed of the engine. Sometimes, the balance of the frequency of the engine sound is also taken into consideration. The order composition, which is considered to be the most influential factor in creating sounds for acceleration, is described in this section.

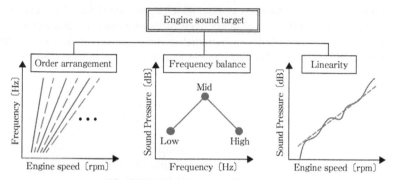

Fig. 3.15: How to create engine sounds.

3.2.1.2 Order Composition of Sounds

The order components of engine sound, suction sound and exhaust sounds are often used in the creation of sounds. A component synchronized with the rotation and combustion of the engine is produced together with higher harmonic waves. The characteristics of these sources of noise are affected by the structure of the chassis and other noise transmission systems emphasizing certain components or changing the balance of components. The impression from a sound changes according to the order balance. Mainly components of integer orders are produced by these sources of noises, but many non-integer components, the so-called 'half-order' components are also produced due to the influence of transfer characteristics. At the end, the balance of these factors determines the impression of the sound.

3.2.1.2.1 Orders and Generation Mechanism

Following are the details of the mechanism of generation of components of integer and non-integer orders according to each component.

(1) *Engine sound*: In 4-cylinders the basic integer order is two times per rotation, i.e., it is a rotation second order component. This is the same as the combustion cycle 1st

order component, which is 1 time per combustion. For example, in 3,000 rpm and 4,500 rpm the rotation secondary component is respectively 100 Hz and 150 Hz. In 6-cylinder engines, the basic order is a rotation third order component. Additionally, higher harmonic components that are double or triple periods are, in general, generated together with basic orders, and in the four-cylinder case, they are rotation 4th, 6th, 8th or higher orders. On the other hand, non-integer orders are generated by influence of the transmission system. For example, when we see the vibration mode of the cylinder block, the response to forces as combustion changes according to the cylinder because usually the structure of the body of the engine is not symmetrical with respect to the position of each cylinder. As a result, non-integer order components are generated, i.e., the engine rotation half-order, because the vibration is repeated in two rotations by period. Also in this case, the sound usually has high-order characteristics with higher harmonics.

(2) *Suction sound*: In suction systems, a pulse of air is generated by the movement of the suction valve in the head of the engine, which opens and closes. This force passes through the group of parts, such as intake manifold, surge tank, etc., connected to each cylinder, propagates through the suction duct moving against the flow of air, and is emitted through the intake. The propagation distances of the pulses of air generated by each cylinder differ because when the intake manifold that is connected to each cylinder is gathered into one suction duct by the surge tank, the distance between this gathering point and the place of the surge tank varies according to the cylinder. Due to this difference in the path length, the phenomenon occurs once the engine rotates twice, and non-integer order components are generated. In many vehicles, suction sounds containing non-integer order components are generated because there are cases in which the path of each cylinder is different due to limitation of space in the engine room. Note, however, that the suction sound will consist only of integer orders if these paths have perfectly equal length. Figure 3.16 shows an example of the structure of intake manifold. The ducts closer to the upstream, the resonator, the air filter, etc., also exert acoustic influence on the suction pulsation, making the balance of frequency change.

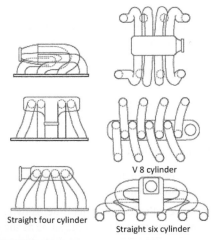

V 8 cylinder

Straight four cylinder

Straight six cylinder

Fig. 3.16: Example of structure of intake manifold.

(3) *Exhaust sound*: The order components in the exhaust system are generated in the same way as in the suction system. The exhaust pulsation created by the movement of the exhaust valve and the influence of the difference in the length of exhaust manifolds of each cylinder on the non-integer orders change the balance of frequency. Additionally, the exhaust sound is affected by damping acoustic characteristics of silencers, etc., that determine the characteristics of the frequency of an exhaust sound. Figure 3.17 indicates the structure of exhaust manifold.

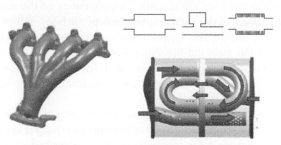

Fig. 3.17: Example of structure of exhaust manifold.

3.2.1.2.2 Relationship of the Order Composition and the Impression of the Sound

When there are mainly integer order components, the impression of the sound is basically clear and smooth. However, the balance in the frequency also affects it because, among other things, the sensory impression may deteriorate if lower-order components are predominant, making the sounds more muffled. Generally, the impression of the sound tends to deteriorate when there are various non-integer order components because the smoothness of the sound decreases. Sometimes when there is an increase in non-integer components at high-rotation speeds of the engine, the impression of the sound becomes extremely rough. Notwithstanding, particularly with respect to sports cars, if there are only integer orders, it gets excessively organized and may be criticized for not giving a sense of speed or power. This means that sometimes a moderate number of non-integer components are necessary for giving a sense of sportiness to the sound. On the other hand, vehicles in which the order components that pass through a certain frequency band according to the acceleration and deceleration of the vehicle are amplified by the resonance characteristics of the propagation system may give a sense of uniqueness to the sound. Therefore, we may say that the order composition and its balance are important factors that might significantly change the impression of the sound.

3.2.1.3 Control of the Sound

Order components are heavily affected by the transmission system until they reach the position of the ear in the cabin, changing how they are heard. It is possible, to a certain extent, to intentionally control these characteristics. The two main methods to control sounds are the one that uses parts that are already equipped in the vehicle and the other that uses additional devices for creating sounds in a more proactive manner.

3.2.1.3.1 Method that Uses Components of the Vehicle

We use resonance characteristics of the propagation system, effects of vibration absorption, acoustic insulation and absorption for the purpose of changing the order composition by making a frequency band more or less emphasized. For example, there is a method in which the spring characteristics of the mount rubber and the resonance characteristics of the plackets change due to the transmission of the vibration of the engine through the mount to the body of the vehicle. For suction sounds, there is a method in which we tune the acoustic characteristics of the resonator and the resonance characteristics of the mounting position of the body. For exhaust systems, it is also possible to achieve the desired resonance characteristics by changing the internal structure of the main muffler. Note, however, that we need to verify if the noise in the exterior of the vehicle complies with the level of noise determined by noise regulations of vehicles when making exhaust/suction sounds louder by tuning exhaust/suction systems.

The methods above change all components whether they are integer or non-integer orders, but there is also a method for controlling only non-integer order components. The sources of non-integer order components are as described above and we may change the order balance by putting them into use, by intentionally tuning them. For suction systems, non-integer order components are suppressed if we match the path lengths of the intake manifold, and increase by using different lengths. In the same way, for exhaust systems, it is possible to control the number of non-integer order components by matching the length from each exhaust port to the gathering point, etc. Usually, there is a difference in the lengths as a result of mounting requirements of the engine compartment or the part surrounding the dash tunnel, but there are also cases where they are intentionally designed with equal lengths. Additionally, we need to take into consideration the impact of the sound tuning on the performance of the engine.

3.2.1.3.2 Method that Uses Devices for Creating Sounds

There is a method in which a branch port that plays the role of a device for generating sound, i.e., a speaker, is attached to the suction duct (Ota et al., 2011). The purpose of this is to force the sound into the interior of the car by installing it at places through which sounds enter the vehicle, as next to the dashboard. If it is installed at a place that is highly sensitive to sounds, in the interior of the vehicle and through which sounds leak from the vehicle, it is possible to minimize the impact on the noise generated by the vehicle.

Furthermore, there is the active sound control technology that may be used as a method for creating sounds more directly. In this method, in order to create the desired sound, the order composition is controlled by installing a speaker in the interior of the car and monitoring the sound field at the position of the reference microphone, sending anti-phase signals to counteract unnecessary order components or sending same-phase signals for stressing components.

Figures 3.18 and 3.19 indicate examples of application of the advance control system and how to make booming noises disappear (Inoue, 2008). In theory it is possible to create any type of sound field and there are even some cases where it has been commercialized. However, we need to consider factors like the size of the space that can be controlled, the stability of the system and the cost.

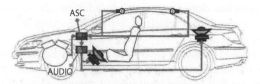

Fig. 3.18: Active control system (Inoue, 2008).

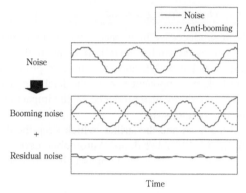

Fig. 3.19: How to make booming noises disappear using active control (Inoue, 2008).

3.2.1.4 Sound Evaluation Methods

In order to determine our target sound and embody it into the vehicle, we need to make a sensory evaluation of the sound and set objective indices for quantitative measurement. We need to find correlated parameters among these, and use them as a base for determining quantitative targets. The quantitative evaluation measures must be used for determining the characteristics of sound sources, propagation systems, devices, etc., and designing the structure. That is how we actually create the desired sounds.

With respect to sound evaluation measures, in addition to traditional physical quantities, like frequency and sound pressure, we must also use psychoacoustic parameters as loudness, roughness, sharpness, etc. There are also cases in which unique evaluation measures that combine these parameters or new parameters are suggested (Biermayer et al., 2001).

3.2.2 Sound of the Door Closing

3.2.2.1 Need for Research on Door Sounds

Although it is complicated to determine if a sound is good or bad because senses and preferences towards sounds vary from person to person, but people tend to prefer quiet sounds that do not contain a signal or that people have no interest in. For the users of the vehicle, doors that make a loud noise are often found to be unpleasant; therefore quiet sounds are found more suitable for doors. However, for the person operating the door, if the door does not make a sound, it is difficult to know for sure if the door is closed or open. As a result, we need a door sound to work as a signal for letting the user know if the door has closed or opened. In addition, the sound of the door may also affect the impression of the vehicle as to its being sporty or luxury (Petniunas et al., 1999; Filippou et al., 2003).

Therefore, it is good to make a sound that is nice to the ears and informs the user of the operation at the same time. As it is particularly important to know for sure if the door closed or not, we will explain what type of sounds are nice for the person who closes the door and not unpleasant for the people around the vehicle.

3.2.2.2 Mechanism of Door Closing Sounds

Door closing sounds are produced by the impact of the door hitting the body of the car. The doors are kept closed by the latch (generally installed in the door) being engaged to the striker (mounted on the body), and the sound is produced by the collision that occurs when the fork, a component part of the latch, engages with the striker (Fig. 3.20). In addition, sounds are also produced by the impact that occurs when the door sash, etc., touch the weather strip.

The functions of the latch and the striker, which keep the door of automobiles closed is different from those of house doors. Automobile doors need to have a two-stage latch mechanism—one in which the door is half-closed and one in which the door is fully closed (hereinafter 'half-closed' and 'fully-closed'). Therefore, door closing sounds of vehicles consist of the collision sound from the door half-closing, the collision sound from the door fully-closing and the collision sound from the door sash, etc., touching the weather strip (Fig. 3.21).

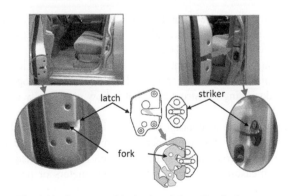

Fig. 3.20: Structure of the latch and the striker in door cars.

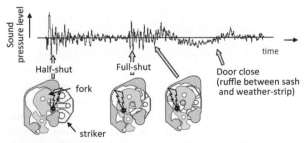

Fig. 3.21: Generation of door closing sounds.

3.2.2.3 Conditions for Good Door Closing Sound

The two conditions for a good door closing sound that gives a sense of luxury, that is not unpleasant and that which assures the user that the door is really closed are as follows:

3.2.2.3.1 Arranging the Distribution of Frequency

We need to produce sounds of low frequency and attenuate it as fast as possible the silencing sounds of high frequency. Note, however, that we should avoid fluctuation of sound pressure at extremely low frequencies that are over 20 Pa and give a sense of pressure on the ears.

Frequencies under 200 Hz, between 50 and 150 Hz and between 50 and 400 Hz are suggested for creating good door closing sounds (Kidachi and Koike, 2006; Kidachi et al., 2006; Maeda et al., 2012). Then it is recommended that we use frequencies between 50 and 200 Hz for the sound to be emitted and frequencies over 500 Hz for the sound to be attenuated.

For example, as part of the study of Uchida et al., Fig. 3.22 indicates the results of a comparison between a solid door closing sound (a) and a door closing sound that lacks solidness (b) (Uchida et al., 2002). The horizontal axis is the time, the vertical axis is the frequency, and the dashed line in the middle of the graph indicates 200 Hz. The part with a dark color at the center of the lower part of the figure indicates areas with high levels of sound pressure, and we see that sounds under 200 Hz in graph (a) are higher than in graph (b). We may achieve a 'deep door closing sound' by increasing components of frequencies under 200 Hz.

3.2.2.3.2 Adding Reverberation Effects: It is Effective to Give Two Sounds with the Same Frequency Components

Successive two sounds with an interval of 25 ms give the impression that the sound is reverberating, a sense of luxury and also a sense of security as to the door being closed or not.

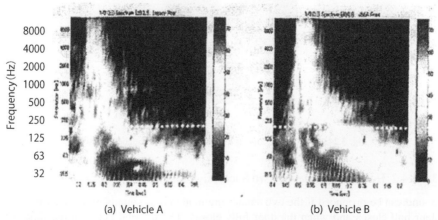

(a) Vehicle A   (b) Vehicle B

Fig. 3.22: Comparison of the distribution of frequency of door closing sounds that give a sense of solidness.

3.2.2.4 How to Realize It

3.2.2.4.1 Method of Producing Sounds of Low Frequency

There are many reports in which low frequency sounds were produced by the mode control of the outer panel of the door. Door sounds are produced by the propagation of the vibration of the latch engaging with the striker, and the outer panel that is large in size produces the louder sound. It is possible to change the sound produced by the door into a low frequency sound by turning the lower-order mode of the outer panel of the door down to 30 ~ 40 Hz through laser holography or acoustic holography (Kidachi et al., 2006) (Fig. 3.23). Also, since the oscillation produced by the engagement with the latch is the vibration source of the sound, it is possible to suppress the oscillation of high-frequency components by coating the striker with resin, thus adding a buffer structure to the place of engagement (Fig. 3.24).

Fig. 3.23: Verification of the low-frequency vibration mode of the door panel (50 to 150 Hz).

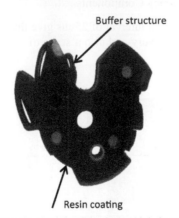

Fig. 3.24: Example of coating for reducing high-frequency sounds.

3.2.2.4.2 How to Produce the Two Successive Sounds

Sound can be produced if the two sounds are made up of sounds produced when the door half closes and when the door fully closes. The interval between the times of emission of these sounds may be determined by the fork part of the latch matched to the speed of the door closing. The interval can be controlled by matching the space between the position of the fork when the door is half-closed and its position when the door is fully-closed to the speed of the door closing. The speed of the door closing is around 0.8 m/s, and because it does not significantly change with the force used to close the door, we can control the interval between the two sounds by matching the position of the two engagement stages.

3.2.2.5 Other Considerations

The results of some studies show that the color of the body of the vehicle and the design of the interior of the car have influence on the evaluation of the door sound; therefore, we need to consider the color of the body and the interior design when discussing the door sound of vehicles (Arimitsu and Toi, 2013).

3.3 Cabin Air Quality

3.3.1 Smells in the Interior of the Vehicle

The smells inside the car originate in the materials used to finish the interior, from the air where the vehicle is traveling or from the air conditioner. It should also be considered that the smell originates in the user, such as tobacco and pet-related odors (Sato, 2004). In evaluating the smells, two methods can be used: sensory evaluation and instrumental analysis. Also, for detecting odorous substances with in-vehicle equipment, a smell sensor (gas sensor) can be applied.

3.3.1.1 Sensory Evaluation

In sensory evaluation, we evaluate the intensity of the smell (it is called 'odor'), the level of comfort/discomfort and the quality of smell. In Japan, we use a scale of six levels (Table 3.2) for evaluation of the intensity of the smell and a scale of nine levels for the evaluation of comfort/discomfort (Table 3.3). The nature of the smell is expressed through terms commonly used for expressing feelings, as sweat, salty, burnt smell, and sometimes through material names, such as flowery, juicy, oily and leathery (Sakakibara et al., 1999).

On the other hand, for the odor concentration, a method of measurement of the intensity of smell uses the dilution rate when the smell cannot be felt anymore after diluting the smell with pure/clean air. The odor concentration can be measured by the triangle odor bag method (Nagata and Takeuchi, 2003). In this method, three bags are

Table 3.2: Six Level-scale for Smell Intensity.

0	Not detectable
1	Barely detectable
2	Weak
3	Easily detectable
4	Strong
5	Very strong

Table 3.3: Nine Level Comfort/Discomfort Scale.

−4	Extremely unpleasant
−3	Very unpleasant
−2	Unpleasant
−1	Somewhat unpleasant
0	Neither pleasant nor unpleasant
1	Somewhat pleasant
2	Pleasant
3	Very pleasant
4	Extremely pleasant

presented to a panel (one with a diluted odor and the other two with odorless air) and the subject is asked to answer the number of the bag he/she thinks has the smell. This process is repeated several times and each time the dilution rate is raised three times. The evaluation ends when we obtain the concentration rate in which the subject is unable to identify the bag containing the smell. The index of odor is calculated using the odor concentration value as follows and is a value representing the actual sense of the smell better than the odor concentration itself.

$N = 10 \times \log S$

N: Index of odor, S: Odor concentration

3.3.1.2 Instrumental Analysis

For instrumental analysis of smell components, Gas Chromatography (GC) is widely used (Brattoli et al., 2013). In order to collect weak odorous substances from the air and concentrate them until the concentration level of the equipment is detected (collecting method), there are the adsorption method, which uses various adsorption substances and the refrigeration method that uses liquefied carbon dioxide and liquefied nitrogen in a cold trap. For sampling low-boiling components that are difficult to collect using adsorbents at normal temperature, canisters (sample collection containers with valves) are used. In case of samples obtained by the adsorption or the canister methods, low-temperature condensation is often used during the sampling to the GC for controlling the width of the peak. It is possible to make a quantitative/qualitative analysis by using a Mass Spectrometer (MS) for components that were separated through the column of the GC. In order to identify the smell component in the interior of a vehicle, we may analyze it by sampling the air (Fig. 3.25) or the gas generated by materials in the interior of the vehicle.

In the analysis of the air in the interior of new cars, Sakakibara reports that more than 200 volatile components where found (Sakakibara et al., 1999). Usually what exists inside cars is a complex odor made up of multiple components. However, this does not mean that the higher the concentration of components gets, the stronger the smell (or discomfort) becomes. That occurs because depending on the components at the lowest level (i.e., threshold) where smells are perceived differently, even small quantities of components with a low threshold affect the odor. The value of the concentration of components divided by the threshold is called Odor Unit, and

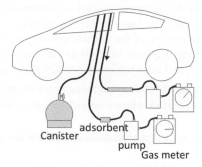

Fig. 3.25: Methods for sampling air components in the interior of the vehicle.

the higher its value gets, more impact it has on the smell. If the threshold of smell component is already known, Odor Unit serves as an index of the impact of each component.

Odor unit = Concentration of smell components/threshold of smell components

With respect to each component of complex smells, the GC-O (Gas Chromatography Olfactometry or GC-Sniffing) method, where the subject smells the odor of its components that have been separated from each other by the GC, is utilized as a means of obtaining the quality and intensity of the smell. As a method using the same system to infer the impact that each component has on the intensity of the whole smell, the AEDA (Aroma Extract Dilution Analysis) seeks the dilution rate where the smell cannot be felt any more by repeatedly diluting the sample smell and submitting it to GC-O. There is also the OSME (Osmegram) that measures the intensity of smell through a scale of strength (Brattoli et al., 2013). When the components cannot be separated by the usual GC analysis (first column), the multi-dimensional GC-MS method, where part of the separated components are trapped by adsorption or refrigeration and introduced into the second column for running another separation, is used (Brattoli et al., 2013). Fukunaga et al. identified 10 components (hexanal, ethylbenzene, dodecane, octanal, tetradecane, longifolene, menthol, calamenene, 4-methyphenol) in the smell of the ceiling material by applying the GC-MS-sniffing method, a two-dimensional GC-MS method, etc., to gases produced by the material used in the interior of vehicles (Fukunaga et al., 2014).

3.3.1.3 *Odor Sensors*

Semiconductor gas sensors are widely used as odor sensors. Ventilation control sensors for automobiles are used for preventing exhaust gases from other cars or smells along the road flowing into the vehicle. With the purpose of quantifying the intensity and quality of smells, systems for evaluating odors by combining multiple sensors have already been commercialized (Shimazu Factory FF-2020, Alpha M.O.S. FOX, etc.).

3.3.1.4 *Odor Control*

There are two aspects to controlling odors: the monitoring of the odor source, by controlling the generation of odor as much as possible, and the reduction of the odor, when the odor has already spread in the room. Examples of monitoring the source in the interior of vehicles are the above-mentioned ventilation control using gas sensors, air conditioner cleaning, etc. The basic reduction counter measure when the odor has already spread in the interior of the vehicle is changing it for outside fresh air, or air fresheners/deodorants may also be used as supplementary means. Aromatics, deodorants, odor removers and odor inhibitors are defined as follows:

- Aromatic: a product that diffuses a fragrance in the room.
- Deodorant: a product that eliminates or reduces the smell through chemical or sensory reactions.
- Odor removers: a product that eliminates or reduces the smell through physical changes.
- Odor inhibitors: a product that prevents the spread or generation of the smell by being added to other substances.

There are four principles for deodorizing or removing odors: the physical method, the chemical method, the biological method and the sensory method. The physical method uses porous materials like activated charcoal and solvents for adsorption, absorption, coating, etc. The chemical method uses chemical reactions as neutralization, oxidation-reduction, etc. The biological method uses microorganisms for decomposing organic matter, chemical preservatives or antiseptics for preventing decomposition by bacteria, etc. The sensory method uses aromatic effects, masking effects, neutralization effects, etc., for reducing or mitigating the smell at a sensory level. Air cleaner products that are developed for residential houses where the ions that mitigate the adherence of smells by releasing ions, are also introduced into the interior of the vehicle (Harada et al., 2011).

3.3.2 Effects of Fragrance

Fragrances are thought to have emotional and physiological effects and have been used in our lives since a long time ago. Fragrances have a wide range of effects—from sedating to stimulating. For example, generally lavender has a relaxing effect; lemon has a refreshing effect.

3.3.2.1 Perception Mechanism of Smells

The mechanism of the human perception of smells is shown in Fig. 3.27. Fragrances in the air are perceived via electrical signals from olfactory cells to the limbic system when substances in the air blend in the mucous membrane of the olfactory epithelium inside the nasal cavity. The intensity of the smell is perceived in proportion to the logarithm of the concentration of smell, following the Weber-Fechner law, indicated in the following equation:

$R = K \log S$

S: concentration of the smell, R: perceive intensity of the smell, K: constant.

It is said that olfaction is not important for human information processing, as compared to other senses (Seo et al., 2010). However, its emotional and physiological impact cannot be ignored. The olfaction is the only sense among the five senses where the electric signal does not pass through the thalamus and is directly sent to

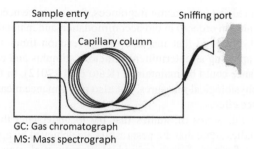

GC: Gas chromatograph
MS: Mass spectrograph

Fig. 3.26: Mechanism of human perception of smells.

the peripheral system. Since the olfaction works directly on the limbic system, which includes the amydala and hippocampus and has a close relation to emotional reactions, it has a considerable impact on primitive emotions or instincts (Lorig and Schwartz, 1988). The effects of fragrances happen when passing from the olfactory system to the cerebrum, and is considered to stop quickly after the use is suspended (Alaoui-Ismaïli et al., 1997).

In automobiles, fragrances have been traditionally used as counter measures against bad smells. Recently, a more positive approach is being taken for adding value, making a good use of the effects on the body and the mind, etc.

3.3.2.2 Emotional and Physiological Effects of Fragrances

Some of the main effects of fragrances are sedating effect, stimulating effect, fatigue mitigation effect, sleep improvement effect, menopause symptoms mitigation effect, dietary effect, immuno-stimulant effect, etc., that work on the autonomic nervous system. There are studies that use a variety of measures as questionnaires, EEGs, blood pressure, skin resistance, heartbeats, biochemicals in the saliva for testing these effects. Influence on the sedating and stimulating effects among the fragrance effects is expected for driving safety. There have been studies on the effects of fragrances with the purpose of mitigating the driving fatigue and maintaining the attention level for driving.

A number of studies have already reported the effects of fragrances in laboratories. Kawamoto et al. reported that lemon fragrance mitigated the exhaustion and maintained vigor for the simple calculation task (Kawamoto et al., 2005). Hiramatsu et al. demonstrated that peppermint has a stimulating effect on drowsiness in an experiment using a cabin mockup (Hiramatsu et al., 1995). It was shown that it is possible to maintain the stimulating effect by applying the fragrance in an early stage when the arousal level is not too low and applying it together with the sound of a buzzer.

Suzuki et al. reported that inhalation of alpha-Pinene mitigated fatigue by analyzing the power level of beta waves of EEG in an experiment using a driving simulator. They also observed that it was possible to reduce the psychological stress and maintain an appropriate headway distance by using a fragrance with alpha-Pinene components (Suzuki et al., 2006). Raudenbush et al. also used a driving simulator to prove that peppermint and cinnamon have stimulating and fatigue mitigating effects on drivers (Raudenbush et al., 2009).

It should be noted that when using fragrances, the fragrance needs to be applied in an intermittent manner in order to avoid accommodation and maintain these fragrance effects. Kato et al. observed that increase of the reaction times of go/no-go tasks was prevented by applying an intermittent olfactory stimulus and concluded that the cognitive performance could be maintained (Kato et al., 2012). In this way, not only psychological or physiological measures, but also performance measures are used for examining fragrance effects.

In contrast to this series of studies that try to understand the effects of each fragrance, some studies argue that the best results are achieved by applying fragrances according to the preference of each individual. Suzuki et al. stated that "fragrance effect is a combination of the influences of both the impression (preference) and the influence of an aromatic component" and the preference may control the effect itself (Suzuki et al., 2006). Nagai et al. analyzed the physiological and psychological effects of fragrances after research participants selected the fragrance they liked the most (Nagai et al., 2000). The results showed that fragrances selected according to the preference of each individual had the effect of enhancing comfort and reducing stress and anxiety. There were also effects on psychological and physical activity with improvement in efficiency of aerobic exercises due to increased vascular activity during muscular movements and maintenance of efficiency in choice serial reaction-time tasks.

3.3.2.3 Future of Vehicles and Smells

Olfaction is largely influenced by the environment and varies from person to person. Previous studies point out sex, age and personality as factors that influence the sense of smell (Mackay-Sim et al., 2006; Larsson et al., 2000). In addition to the individual differences in the sensitivity of smell, there are also large individual differences in preference. For applying the scents in automobiles, we need to work with driver-sensing systems to deal with differences in the sensitivity of smell of each individual or differences in preferences of smell.

There are some interesting uses of fragrances in automobiles. For instance, a fragrance device that allows occupants to enjoy the fragrance selected based on their taste was installed in some French cars. Renault also provides a window washer perfume fluid as a genuine product. This device works by introducing the fragrance into the vehicle via outdoor air inlets when the washer fluid is used. The use of the fragrances in the interior of vehicles will probably increase with the tendency to raise the level of comfort in vehicles.

3.4 Visual Environment of Vehicle Interior

3.4.1 Function and Design of Vehicle Interior Lighting

3.4.1.1 Types of Lighting

Vehicle interior lighting has two purposes: provide safety and convenience to drivers/passengers and create a pleasant atmosphere. Some lighting types can fulfill both the purposes. Figure 3.27 illustrates the main types of vehicle-interior lighting. Besides these types, new types of lighting that are aimed at creating a pleasing atmosphere are

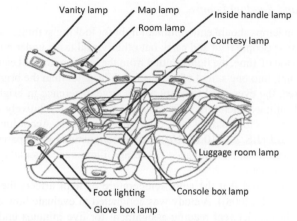

Fig. 3.27: Major types of vehicle interior lighting (Reprint from Watanabe, C. (2015). Concept and possibility of LED light in vehicle compartment space, Toyota Gosei Technical Report, 57: 11–14, with permission).

becoming popular. This indicates that more drivers are interested in lighting that is easy to use and is comforting.

Traditionally, incandescent bulbs were used for vehicle-interior lighting. In more recent years, light emitting diodes (LEDs) are becoming progressively popular as white LEDs are cheaper. Due to their superior characteristics, such as space-saving, energy-efficiency (low-power consumption), cool light, and quick lighting up, LEDs offer convenience from the viewpoint of vehicle-interior design. In addition, because LEDs are easy to manipulate in terms of color selection, brightness, and color tone, they can be used to provide various visual effects.

3.4.1.2 Requirements for Functional Lighting Design and a Study Example

Functional lighting assists drivers and passengers to safely enter/exit a vehicle and provides them with sufficient illumination for in-vehicle activities. The illumine level should be high enough to facilitate in-vehicle activities while it should not be so high as to enable people outside to have a full view of the vehicle interior. Table 3.4 lists the required illumine levels for various interior parts. Individual car manufacturers have set different minimum and target illumine levels based on the activities of drivers/passengers (checking situations and presence/absence of certain things, and control devices) and how they react to the brightness. It should be noted that the level of brightness people prefer sometimes differs among countries. Also, it is necessary to design luminous intensity distribution to meet people's preferences in terms of irradiation range and illumine distribution. In addition, the color of the light (color temperature) needs to be determined, based on people's demands in terms of how they perceive it.

Table 3.4: General Illumine Levels for Various Interior Parts.

Part	Steering wheel	Center console	Shifter	Seat	Foot
luminance [lx]	5 ～ 10	7 ～ 20	5 ～ 10	5 ～ 10	3 ～ 5

3.4.1.3 Map and Reading Lamps

Map and reading lamps should emit optimum light for looking at things and reading in a vehicle. Because people read and carry out other visual tasks under a reading lamp for a certain period of time, the light emitted from the lamp should not cause eyestrain. If the light coming into one's field of view is much brighter than the brightness of the surrounding area, the eyes need to adapt to the large difference in brightness else it leads to visual fatigue. Therefore, it is desirable to maintain relatively constant and mild brightness across the fields of view of drivers/passengers. According to Japanese Industrial Standard (JIS Z 9110 (2010)), it is recommended that spatial uniformity of illumination (minimum illuminance ÷ average illuminance) suitable for reading should be 0.7 or higher.

It is also conceived that the color temperature of light affects the visibility of text (Yamagishi et al. (2008)). A study was conducted to evaluate how easily people can read after they underwent reading assignment for five minutes under a specific type of LED lighting (400 lx). The study found that older people had significantly greater difficulty in seeing letters under low color temperature light than younger people. These results are presumably attributed to the fact that crystalline lenses of aged people have turned yellow. It may be effective to select light of different color temperatures for certain people.

3.4.1.4 Vanity Lamps

A vanity lamp (Fig. 3.28) is installed by a mirror and illumines a user's face. The specifications of the lamp affect the usability of the mirror because complexion and the appearance of makeup vary, depending on lighting conditions. Accordingly, it is important to design the lamp taking into account the brightness (luminance), the color of light (color temperature) and lighting direction.

In a study on vanity lamps (Watanabe, 2015), 15 women were asked to observe and evaluate the appearance of their own faces illuminated by an LED lamp at a color temperature of 2,800 K, 3,300 K, 4,000 K, or 5,000 K in a vanity mirror. There were eight observations in the evaluation: the skin looked beautiful, the color of makeup looked accurate, the appearance looked healthy, the skin had a transparent appearance, the appearance looked natural, the view was bright, it was easy to put on makeup, and the appearance was appreciated. Evaluation used a 3,000 K bulb as a reference and subjects evaluated by scoring out of 100 points.

The scores collected were analyzed, using multiple regressions to identify criteria important for the subjects and to determine how they liked their own appearances in the vanity mirror (Table 3.5). A most important criterion was whether the color of the makeup looked accurate, followed by whether their appearances looked natural. These results indicated that the subjects preferred mirrors that facilitate makeup over mirrors that show their skin beautifully. Figure 3.29 shows the subjects' evaluations of their facial appearances in relation to color temperature. As to whether the color of makeup looked accurate, the most important criterion for the subjects' positive responses was associated with a color temperature ranging between 4,000 and 5,000 K while negative responses were associated with low color temperatures. Based on these results and from an ergonomic viewpoint, optimum color temperature is a most important factor when designing high-quality vanity lamps.

Fig. 3.28: Example of vanity mirror and vanity lamp (Reprint from Watanabe, C. (2015). Concept and possibility of LED light in vehicle compartment space, Toyota Gosei Technical Report, 57: 11–14, with permission).

Table 3.5: Factors Influencing People's Preference Regarding their Appearance in Vanity Mirror (Multiple Regression Analysis) (Watanabe, 2015).

	Partial regression coefficient	Standard error	t value	F value	Standardized partial regression coefficient
The appearance looked natural	0.419	0.0918	4.57	2.78E-05	0.419
It was easy to put on makeup	0.425	0.0752	5.65	5.59E-07	0.425
A transparent appearance	0.197	0.0797	2.47	0.0164	0.197
Constant	5.07E-17	0.0580	8.75E-16	1	

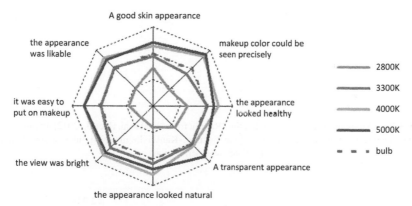

Fig. 3.29: Evaluation of their own appearance in a vanity mirror in relation to color temperature of illumination (Reprint from Watanabe, C. (2015). Concept and possibility of LED light in vehicle compartment space, Toyota Gosei Technical Report, 57: 11–14, with permission).

3.4.2 Comfort Provided by Vehicle Interior Lighting

3.4.2.1 Effect of Shape and Brightness of Light Source on People's Impression of Vehicle Comfort and Spaciousness

In order to investigate the effects of design factors in lighting on the impression of vehicle interior, experiments were performed to evaluate the impression of the vehicle interior in relation to various shapes, arrangements, and luminance of lighting using a simulated vehicle's interior space and assuming that the vehicle is running

(Takahashi et al., 2012). Shapes and arrangements of lighting on the ceiling are shown in Fig. 3.30.

Average comfort scores for individual lighting shapes/arrangements were obtained under the simulated condition in which the vehicle was assumed to be traveling in an urban area (brightness on the road was set at 2.5 cd/m^2). The results are shown in Fig. 3.31. For all shapes/arrangements of vehicle interior lighting, the comfort scores were always higher at luminance of 2.0 or 0.5 lx, compared to when no lighting was used. Moreover, the comfort scores were always higher at luminance of 2.0 lx than 0.5 lx under all conditions studied. The condition in which luminance was at 2.0 lx and the lighting shape/arrangement was A-d, received the highest comfort score. When luminance was at 0.5 lx, lighting shapes/arrangements of B-a, B-b, B-c (lighting source installed on the doors) and A-b were awarded high comfort scores.

Sometimes people have to work or live in a small space. Under such circumstances, they may feel more comfortable in the space if they can make it feel more spacious and open. Concerning the ratio of vertical luminance E_v to horizontal luminance E_h (E_v/E_h), an increase of E_v relative to E_h makes the value of E_v/E_h increase. The relationship of E_v/E_h to 'sense of spaciousness' scores and the relationship of E_v/E_h to 'sense of brightness' scores are shown in Fig. 3.32 (Iwai, 2002). The ratio E_v/E_h and the sense of

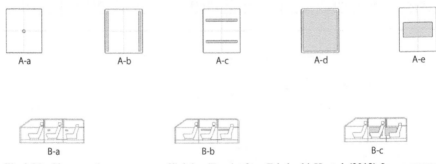

Fig. 3.30: Shapes and arrangements of lighting (Reprint from Takahashi, H. et al. (2012). Improvement in automobile interior comfort by modifying lighting. Kansei Engineering International Journal, 11: 59–65, with permission).

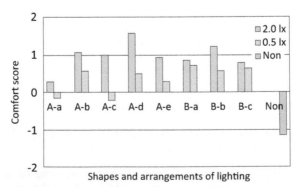

Fig. 3.31: Comfort evaluation scores (Reprint from Takahashi, H. et al. (2012). Improvement in automobile interior comfort by modifying lighting. Kansei Engineering International Journal, 11: 59–65, with permission).

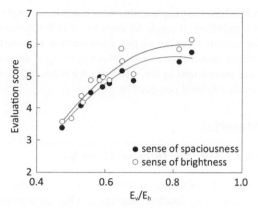

Fig. 3.32: Relationships between E_v/E_h and the 'sense of spaciousness' evaluation scores, and between E_v/E_h and the 'sense of brightness' evaluation scores (Reprint from Iwai, W. (2002). Ceiling mounted luminaire with new type of distribution of luminance intensity developed by the study of 'sensation of room brightness'. Journal of the Illuminating Engineering Institute of Japan, 86: 782–786, with permission).

spaciousness are positively correlated. In other words, as the ratio E_v/E_h increases, the sense of spaciousness also increases. Similarly, as the ratio E_v/E_h increases, the sense of brightness also increases.

3.4.2.2 Poor Visibility of Vehicle Interior from Outside

Drivers/passengers can comfortably sit in their vehicles at night assuming that the inside of the vehicles is virtually invisible to people outside. Relationships among observer-to-vehicle distance, vertical luminance, and how clearly the observers saw faces of the people in the vehicle are summarized in Table 3.6 (The Illuminating Engineering Institute of Japan (1995)). The observers were able to see the eyes, mouths, and noses of the people in the vehicle from 4 m away at vertical luminance of 1.0 lx and from 10 m away at vertical luminance of 2.1 lx. In addition, the observers were able to recognize who the people were in the vehicle from 4 m away at vertical luminance of 1.8 lx, and from 10 m away at vertical luminance of 5.0 lx. These results were obtained under the condition in which vehicle windows had no glass. The presence of a window reduces visibility inside the vehicle as scattering light on the glass surfaces decreases luminance contrast.

Table 3.6: Relationships among Observer-to-Vehicle Distance, Vertical Luminance, and How Clearly the Observers See Faces of the People in the Vehicle (The illuminating Engineering Institute of Japan (1995).

how clearly the observers saw faces	distance (m)	Vertical luminance on face (lx)
Recognize positions of eyes, mouth and nose	4	1.0
	10	2.1
Recognize who is the people	4	1.8
	10	5.0

The condition in which the people in the vehicle are unrecognizable from outside is considered to be satisfactory. It would be even better if the eyes, mouths, and noses of the people in the vehicle were not visible from outside to secure privacy. Taking account of typical distances between two vehicles travelling in opposite directions and between vehicles and pedestrians in the sidewalk, we think the measurements made from 4 m away provide a helpful perspective.

3.5 Interior Materials

3.5.1 Evaluation Criteria for Interior Material

Interior materials need to be evaluated in terms of their appearance, texture, etc., that affect human comfort and other functions unrelated to human comfort. They also need to be evaluated for potential discomfort factors. It should be noted that an interior material may both be advantageous and disadvantageous. For example, a shiny material may be visually appealing, but at the same time, it might hinder the driver's vision if the surface of the interior is reflected on the windows or if it acts as a strong sunlight reflector. This involves not only the material's characteristics but also the shape and layout of the interior parts, which are beyond the scope of this section. We focus on tactile functions in this section.

3.5.2 Gripping Functions

3.5.2.1 Functions of Vehicle Operation System

Various vehicle interior parts have a gripping function, which is especially important for the steering wheel and pedals. With respect to pedals, the physical function is the most important characteristic. On the other hand, in regard to the steering wheel, its tactile function is also an essential characteristic. Relationship between the tactile evaluation and physical characteristics of several steering wheel materials in terms of deflection and compressional resilience are shown in Fig. 3.33. A certain amount of low-level compressional resilience enhances the touch comfort. In addition, increased deflection gives drivers an impression that the steering wheel is luxurious, while

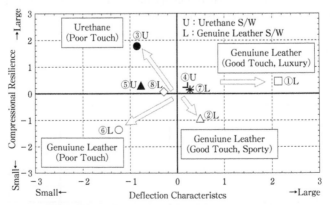

Fig. 3.33: Tactile evaluation of steering wheel texture.

decreased deflection gives them a sporty feeling (Tanaka et al., 2005). For sportiness that is relevant to high steering-control performance, steering wheels require the right amount of deflection.

3.5.2.2 Grips that Support Drivers/Passengers with Physical Stability

There are grips installed on the inner surface of the doors, on the ceiling, and in the area close to the opening part of the doors, to support drivers/passengers getting into and out of the vehicle. These grips clearly serve the purpose of providing them with physical stability when they are seated and getting into/out of the vehicle. The armrests installed in the second-row seats in a van are also designed to serve a gripping function for passengers while getting into/out of the vehicle. From the consideration of safety, adequate materials are selected for these grips so that they can serve their primary functions. It is also important to select appropriate materials to make the grips feel solid and secure. Lastly, it is desirable to make the grips pleasant to the touch. When grips with different surface materials are evaluated in terms of how they feel on holding (having a good grip, easy to hold steadily, etc.), grips with high friction coefficient values usually receive high ratings. The evaluation of surface asperity of grip material differs, depending on participants' age group. Young and middle-aged people tended to give low scores to coarsely surfaces due to pain they felt while handling the grip. In contrast, elderly people tended to give high scores to the same surface as it gave them a good grip. These results are presumably linked to the fact that senior people have lower tactile sensitivity and weakened grip strength.

3.5.2.3 Gripping Functions of Non-grip Parts

Lastly, we discuss the case where people use non-grip parts as a handhold when they accidentally touch these parts. While these parts are not designed to serve as a handhold, people might feel inconvenient if these parts have low friction and thus are slippery. In the architectural engineering, material surfaces are evaluated in terms of their functions to prevent people from falling when they place their hands on the surfaces. A test participant supported their standing posture by placing one of their hands on the test surfaces, which was set either horizontally or vertically, and 750 mm away from them. Then, the surfaces were evaluated for their slipperiness in terms of the subjects' efforts to maintain the standing posture while they added either no further force or stronger force to the surface. As shown in Fig. 3.34, clear relationships were observed between scores of subjects' efforts to maintain standing posture. The study also found that subjects' efforts to maintain standing posture is correlated with the slipperiness of the materials tested (Niimi et al. 2008).

3.5.3 Effect of Sweat

Katz proposed, in 1925, that the tactile feeling of texture consists of four types of sensations: hard/soft, dry/wet, rough/smooth, and warm/cold. However, detailed study revealed that dry/wet sensation is less correlated to physical characteristics as compared to the three other sensations (Sakane and Tachibana, 2007). It is inferred that dry/wet sensation is a mixture of moist sensation and a slimy sensation that are not clearly differentiated. The former gives a positive impression while the latter provides

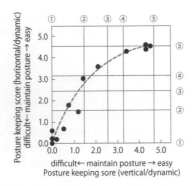 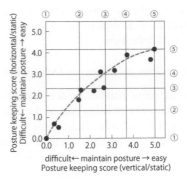

Fig. 3.34: Relationship between indices for posture keeping (Reprint from Niimi et al. (2008) with permission).

a negative impression. Furthermore, the sensation of stickiness, that is also related to dry/wet, might give an even more negative impression. In addition to being a physical characteristic of some materials, stickiness can also be caused by human sweat, so the causes can be complex.

3.5.4 Difference in Skin Structure Among Body Parts

Texture evaluation for vehicle interior materials is often performed using the palm of the hand. However, seat texture is evaluated by people sitting on them thus using their thighs, buttocks, and backs, which are in contact with the seat. Similarly, the texture of armrests is evaluated using arms.

The palm of the hand is covered by glabrous skin. The only other body parts without hair are the lips and soles of the feet. As glabrous skin and non-glabrous skin have different skin structure. The tactile evaluation should be performed carefully as skin structure differences may affect the evaluation results.

3.5.5 Stickiness

As stickiness is perceived when touching a material, stickiness evaluations should be performed by using the palm of the hand. However, there are only a few of this kind of studies available in which high levels of stickiness were studied in the automobile industry. Thus, alternatively, we show a study in the architectural engineering on the relationship between the sole of the foot, another hairless body part, and the floor (Nakagawa et al., 1995).

The sensation of stickiness is said to be closely related to the amount of moisture between the skin and a material (Temming, 1992). In addition, the sensation of stickiness is affected by the relationship between the rate of moisture evaporating from the skin (the sole in this case) and the rate of moisture absorbed through the material surface. In case of a material with a very low moisture absorption rate, the maximum static friction coefficient on its surface affects the sensation of stickiness. Combining these relationships, the stickiness coefficient can be defined by the following equation:

$$C_s = \frac{\mu}{(E + W)/E}$$

where C_s = stickiness coefficient, μ = maximum static friction coefficient of the flooring material, W = rate of the moisture absorbed by the flooring material surface (g/m^{2}h) and E = rate of moisture evaporating from the sole (g/m^{2}h).

The relationship between the sensation of stickiness and the maximum static friction coefficient, and that between the sensation of stickiness and the stickiness coefficient, are presented in Fig. 3.35. For the stickiness coefficient, the relationship at room temperatures of 20 and 30°C is shown (Fig. 3.35 right); the rate of moisture evaporation from individual soles was obtained by separate measurements (36.3 g/m^{2}h at 20°C and 47.4 g/m^{2}h at 30°C). The sensation of stickiness did not differ between the two room temperatures when the stickiness coefficient was up to about 0.03, but it differed when the coefficient was greater than 0.03 (Nakagawa et al., 1995). It is considered that similar relationships would be observed when the palm of the hand is used instead of the sole. The material-related factors affecting the sensation of stickiness are the maximum static friction coefficient and the surface absorption rate. Both the factors are closely related to the sense of touch. The study of the sense of touch should take into account the room temperature variations and the negative effect of sweating.

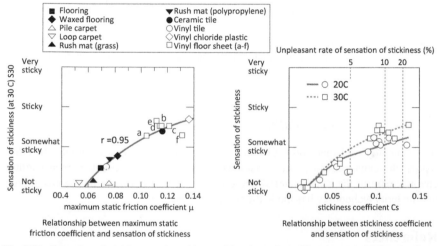

Fig. 3.35: Sensation of stickiness evaluated by touching the surface of interior material (Reprint from Nakagawa et al. (1995) with permission).

3.5.6 Thermal Sensation

People sense warmth and coldness through their entire body surfaces. Even though people wear clothing, the vehicle seats, which have close contact with human bodies, have a large impact on people's temperature sensation, It is more efficient, both in energy and time, to control seat-surface temperature than air temperature in the vehicle, and thus, it is important to consider warming/cooling functions of the seats.

A seat is heated usually by copper, carbon, or other kinds of wires distributed just beneath the seat surface. The thermal conductivity of a seat surface material has a relatively minor influence on the heating function as the heat source is positioned close to a seated person. More importantly, there are concerns about sweating, which was also mentioned in relation to the sensation of stickiness and concerns about low-temperature burns, which may occur in hairy body parts where thermo receptors are sparse.

A seat is cooled by blowing air from the inside of the seat to the surface. Also, it is common for cool air to be circulated around the human body surface through air suction. In some cases, air temperature is controlled by using the Peltier device as a heat source, but in the majority of cases, room air cooled by an air conditioner is circulated into the seat. Accordingly, seat cover materials should be breathable, allowing a seated person to be exposed to air blown from the seat interior.

3.5.7 Breathable Seat Materials and Structures

The breathability of seat materials has a major influence on air conditioning through the seat. Even in a seat without air-conditioning function, the breathability of materials has a huge impact on the dampness of the seat surface.

The breathability of fabric is influenced by yarn density, yarn thickness, yarn twist count, and material thickness. A complex interaction among these factors determines a material's breathability. In addition, a backing material, which structurally fortifies a fabric material from behind, also affects breathability. Backing materials used to reinforce moquette fabrics are membranous and are hardly breathable. Accordingly, their use requires caution. Technologies to maintaining the breathability of a fabric material have been proposed.

When genuine and synthetic leathers are used as seat cover materials, their breathability is enhanced by making numerous holes. This treatment also affects the texture of the material.

3.5.8 Texture and Durability

It is generally said that good textured clothing materials are not durable. In other words, the enhancement of one characteristic of a material might worsen another characteristic of the same material, although this may not always be the case given that material technology has made dramatic progress.

References

Arimitsu, T. and T. Toi. (2013). Influence of automobile body color on the impression of door closing sound evaluated from the visual and auditory matching. Transaction of Society of Automotive Engineers of Japan, 44(6): 403–1408 (in Japanese).

Alaoui-Ismaïli, O., E. Vernet-Maury, A. Dittmar, G. Delhomme and J. Chanel. (1997). Odor hedonics: Connection with emotional response estimated by autonomic parameters. Chemical Senses, 22(3): 237–248.

Aoki, H., M. Ishihama and A. Kinoshita. (1987). Effects of power plant vibration on sound quality in the passenger compartment during acceleration. SAE Technical Paper 870955.

Biermayer, W., S. Thomann and F. Brandl. (2001). A Software tool for noise quality and brand sound development. SAE Paper 2001-01-1573.

Brattoli, M., E. Cisternino, P.R. Dambruoso, G. de Gennaro, P. Giungato, A. Mazzone, J. Palmisani and M. Tutino. (2013). Gas chromatography analysis with olfacto-metric detection (GC-O) as a useful methodology for chemical characterization of odorous compounds. Sensors, 13(12): 16759–16800.

Buma, S., H. Kajino, J. Cho, T. Takahashi and S. Doi. (2009). Analysis and consideration of the driver motion according to the rolling slalom running. Review of Automotive Engineering, 30(1): 69–76.

Chen, P.W. and L.E. Robertson. (1972). Human perception thresholds of horizontal motion. Journal of the Structural Division. Proceedings of the American Society of Civil Engineers, 98(2): 1681–1695.

Filippou, T.G., H. Fastl, S. Kuwano, S. Namba, S. Nakamura and H. Uchida. (2003). Door sound and image of cars. Fortschritte der Akustik DAGA'03, Deutsche Gesellschaft für Akustik, Oldenburg.

Fukunaga, Y., A. Sawada, S. Ogawa and K. Tatsu. (2014). Consideration for analysis method, used to assess odor in car. Proceedings of JSAE Fall Conference, 154-14: 9–12 (in Japanese).

Hada, M., T. Yamaguchi, H. Goto and K. Kawano. (2012). An analysis on ease of steer by supporting shoulder girdle. Proceedings of JSAE Spring Conference, 6-12: 15–20 (in Japanese. English abstract is available as SAE Technical Paper 08-0028).

Harada, S., S. Yukawa, M. Kitahira, I. Tokai and N. Temporin. (2011). Air quality improvement technology for car interior. Journal of Automotive Engineers of Japan, 65(12): 35–41 (in Japanese).

Higuchi, Y., S. Iwasaki, M. Shingo, S. Kitazaki, K. Kato and H. Takezawa. (2001). Development of a method of predicting vibration absorbent on seat unit. Proceedings of JSAE Autumn Conference, 95-01: 5–8 (English abstract is available as SAE Technical Paper 08-0289).

Hiramatsu, M., J. Kasai and M. Taguchi. (1995). Study on the effects of odor on relieving drowsiness. Proceedings of JSAE Fall Conference, 26-2: 88–93 (in Japanese. English abstract is available as SAE Technical Paper 928201).

Hirate, N., M. Komada, T. Yoshioka, S. Thomann and F. Brandle. (2009). V6-SUV Engine Sound Development. SAE Technical Paper 2009-01-2177.

Inoue, T. (2008). Development of active sound control technology. JSAE Symposium Text, No. 06-08: 59–64 (in Japanese).

Irwin, A.W. (1981). Perception, comfort and performance criteria for human beings exposed to whole body pure yaw vibration and vibration containing yaw and translational components. Journal of Sound and Vibration, 76(4): 481–497.

ISO 10326-1 (1992). Mechanical vibration and shock—Laboratory method for evaluating vehicle seat vibration—Part 1: Basic requirements.

ISO 2631-1 (1997). Mechanical vibration and shock—Evaluation of human exposure to whole-body vibration—Part 1: General requirements. Amendment 1(2010).

ISO 5349-1 (2001). Mechanical vibration and shock—Laboratory method of human exposure to hand-transmitted vibration—Part 1: General requirements.

ISO 5349-2 (2001). Mechanical vibration and shock—Laboratory method of human exposure to hand-transmitted vibration—Part 2: Practical guidance for measurement at the workplace.

ISO 8041 (2005). Human response to vibration hock—Evaluation of human exposure.

Ittianuwat, R., M. Fard and K. Kato. (2014). The transmission of vibration at various locations of vehicle seat to seated occupant body. Proceedings of the Inter Noise, 293: 1–12.

Iwai, W. (2002). Ceiling mounted luminaire with new type of distribution of luminance intensity developed by the study of 'sensation of room brightness'. Journal of the Illuminating Engineering Institute of Japan, 86(10): 782–786.

Kato, K. and T. Hanai. (1999). Measurement of the equivalent comfort contours for vibration in a sitting posture. Proceedings of the Inter Noise, 99: 955–960.

Kato, K., S. Kitazaki and H. Tobata. (2004). Prediction of seat vibration with a seated human subject using a substructure synthesis method. SAE Technical Paper 2004-01-0371: 1–10.

Kato, Y., H. Endo, T. Kobayakawa, K. Kato and S. Kitazaki. (2012). Effects of intermittent odors on cognitive-motor performance and brain functioning during mental fatigue. Ergonomics, 55(1): 1–11.

Kawamoto, R., C. Murase, I. Ishihara, M. Ikushima, J. Nakatani, M. Haraga and J. Shimizu. (2005). The effect of lemon fragrance on simple mental performance and psycho-physiological parameters during task performance. Journal of UOEH, 27(4): 305–313.

Kawano, K., Y. Nishigaki, K. Yokota, H. Nakamura and H. Yamada. (2013). Analysis of relationship between seat's ride comfort and urethane foam's inner stress. Transactions of Society of

Automotive Engineers of Japan, 44(2): 543–548 (English abstract is available as SAE Technical Paper 08-0622).

Kidachi, J. and M. Koike. (2006). Study of door-closing sound mechanism. Proceedings of JSAE Spring Conference, 37-06-170: 1–4 (in Japanese. English abstract is available as SAE Technical Paper 08-0161).

Kidachi, J., T. Tomita and T. Takasu. (2006). Measurement technology for door-closing sound analysis. Proceedings of JSAE Spring Conference, 37-06-170: 5–8 (in Japanese). English abstract is available as SAE Technical Paper 08-0162).

Larsson, M., D. Finkel and N.L. Pedersen. (2000). Odor identification: influences of age, gender, cognition and personality. The Journals of Gerontology Series B: Psychological Sciences and Social Sciences, 55(5): 304–310.

Lorig, T.S. and G.E. Schwartz. (1988). Brain and odor: I. Alternation of human EEG by odor administration. Psychobiology, 16(3): 281–284.

Mackay-Sim, A., A.N. Johnston, C. Owen and T.H. Burne. (2006). Olfactory ability in the healthy population: Reassessing presbyosmia. Chemical Senses, 31(8): 763–771.

Maeda, T., S. Okuya, T. Murakami and Y. Nakahara. (2012). Study of improvement technology of door closing sound focused on generation mechanism of low frequency sound (First Report). Proceedings of JSAE Spring Conference, 22-12-109: 17–22 (in Japanese. English abstract is available as SAE Technical Paper 08-0105).

Miwa, T. (1967a). Evaluation methods for vibration effect, Part 1. Measurements of threshold and equal sensation contours of whole body for vertical and horizontal vibrations. Industrial Health, 5: 183–205.

Miwa, T. (1967b). Evaluation methods for vibration effect. Part 2. Measurement of equal sensation level for whole body between vertical and horizontal sinusoidal vibrations. Industrial Health, 5: 206–212.

Morioka, M. and M.J. Griffin. (2006). Magnitude-dependence of equivalent comfort contours for fore-aft, lateral and vertical whole-body vibration. Journal of Sound and Vibration, 298(3): 755–772.

Morioka, M. (2007). Equivalence of sensation for vertical vibration between the hands, seat and feet. Proceedings of the 42nd United Kingdom Conference on Human Responses to Vibration, 125–133.

Morioka, M. and M.J. Griffin. (2010). Frequency weightings for fore-and-aft vibration at the back—Effect of contact location. Contact area and body posture. Industrial Health, 48(5): 538–549.

Mozaffarin, A., S. Pankoke and H.-P. Wölfel. (2008). MEMOSIK V—An active dummy for determing three-directional transfer functions of vehicle seats and vibration exposure ratings for the seat occupant. International journal of Industrial Ergonomics, 38: 471.

Muragishi, Y., K. Fukui, E. Ono, T. Kodaira, Y. Yamamoto and H. Sakai. (2007). Improvement of vehicle dynamics based on human sensitivity (First Report). Development of Human Sensitivity Evaluation System. SAE Paper 2007-01-0448.

Nagai, M., M. Wada, N. Usui, A. Tanaka and Y. Hasebe. (2000). Pleasant odors attenuate the blood pressure increase during rhythmic handgrip in humans. Neuroscience Letters, 11 August, 289(3): 227–229.

Nagata, Y. and N. Takeuchi. (2003). Measurement of odor threshold by triangle odor bag method. Odor Measurement Review, 118: 118–127.

Nakagawa, T., I. Matui and N. Yuasa. (1995). A study on sticky sensation of floor coverings. Journal of Finishings Technology, 3(1): 11–19 (in Japanese).

Niimi, K., K. Goto, R. Kudo, Y. Yokoyama and H. Ono. (2008). Investigation on evaluation method of slip resistance of building elements and member with hands from the view point of posture keeping, study on the evaluation method of contact resistance of building element and members with hands from view point of posture keeping for avoidance of falling down (Part 1). Transactions of AIJ. Journal of Structural and Construction Engineering, 73(631): 1489–1494 (in Japanese with English abstract).

Ota, K., M. Maeda, H. Tanaka and K. Suzuki. (2010). Research on effectiveness of supplying alpha-pinene to driver for preventive safety. Denso Technical Review, 15: 45–51 (in Japanese).

Ota, K., I. Fukumoto, J. Yoshida, Y. Kato, H. Akamatsu and H. Andou. (2011). Development of sound creator, the device for sound quality (Second Report). Proceedings of JSAE, Spring meeting, No. 57–11 (in Japanese. English abstract is available as SAE Technical Paper 08-0245).

Pankoke, S. and A. Siefert. (2007). Virtual simulation of static and dynamic seating comfort in the development process of automobiles and automotive seats—Application of finite-element-occupant-model CASIMIR. SAE Technical Paper, 01-2459: 1–11.

Parsons, K.C. and M.J. Griffin. (1988). Whole-body vibration perception thresholds. Journal of Sound and Vibration, 121(2): 237–258.

Petniunas, A., N.C. Otto, S. Amman and R. Simpson. (1999). Door system design for improved closure sound quality. SAE Technical Paper 01-1681.

Raudenbush, B., R. Grayhem, T. Sears and I. Wilson. (2009). Effects of peppermint and cinnamon odor administration on simulated driving alertness, mood and workload. North America Journal of Psychology, 11(2): 245–256.

SAE J 2896 (2012). Motor Vehicle Seat Comfort Performance Measures.

Sakakibara, K., K. Kaitani, C. Hamada, S. Sato and M. Matsuo. (1999). Analysis of odor in car cabin. JSAE Review, 20(2): 237–241.

Sakane, T. and M. Tachibana. (2007). Sense of touch of epidermal (Genuine leather/synthetic leather/artificial leather). Proceedings of JSAE Fall Conference, 99-07: 25–28 (in Japanese; English abstract is available as SAE Technical Paper 08-0448).

Sato, S. (2004). Air quality in auto-cabin. R&D Review of Toyota CRDL, 39(1): 36–43.

Senba, R. and K. Kawano. (2013). Car seat shape and sitting comfort. Journal of Society of Automotive Engineers of Japan, 67(5): 65–69 (in Japanese).

Seo, H.S., M. Guarneros, R. Hudson, H. Distel, B.C. Min, J.K. Kang, I. Croy, J. Vodicka and T. Hummel. (2010). Attitudes toward olfaction. Chemical Senses, 36(2): 177–187.

Suzuki, K., M. Yasuda, Y. Sassa and S. Harada. (2006). Effects of aroma of trees on active safety during car driving. Transactions of Japan Society of Mechanical Engineers, Part C, 72(11): 3584–3592.

Takahashi, H., H. Asakura, H. Terashima, T. Irikura, S. Ii and D. Wakita. (2012). Improvement in automobile interior comfort by modifying lighting. Kansei Engineering International Journal, 11(2): 59–65.

Takei, K. and M. Ishiguro. (1995). Evaluation of ride comfort on the basis of subjective judgement. R&D Review of Toyota CRDL, 30(3): 47–56 (in Japanese).

Tanaka, C., K. Makino, Y. Miyahama and M. Kyogoku. (2005). Improvement of tactile feel of genuine leather steering wheel. Proceedings of JSAE fall Conference, 130-05: 21–24. (in Japanese; English abstract is available as SAE Technical Paper 08-0628).

Temming, J. (1992). Assessment of the summer suitability of car seats. 2nd International Conference of Vehicle Comfort, Ergonomics, Vibrational Noise and Thermal Aspects, pp. 349–358.

The Illuminating Engineering Institute of Japan, Kansai Branch (1995). Survey analysis on improvement of street lighting (in Japanese).

Uchida, H., S. Nakamura, K. Yamada, K. Nagano, N. Kojima and M. Mikami. (2002). Development of solid door closing sound. Proceedings of JSAE Autumn Conference, 95-02: 17–20 (in Japanese with English abstract).

Watanabe, C. (2015). Concet and possibility of LED light in vehicle compartment space. Toyota Gosei Technical Report, 57: 11–14 (in Japanese).

Yamagishi, M., K. Yamaba, C. Kubo, K. Nokura and M. Nagata. (2008). Effects of LED lighting characteristics on visual performance of elderly people. Gerontechnology, 7(2): 243.

Yasuda, E. and S. Doi. (1999). Analysis of human vibration sensitivity by vibration components. Proceedings of JSME, 99-59: 123–126 (in Japanese with English abstract).

Yokoyama, S., K. Tanida and K. Hashimoto. (2008). Trial method of evaluating muscle strain while maintaining sitting posture during cornering and strain reduction by seat improvement. Transaction of Society of Automotive Engineers of Japan, 39(4): 177–182 (English abstract is available as SAE Technical Paper 08-0025).

4

Driver State

4.1 Driving Fatigue, Workload, and Stress

4.1.1 Stress and Strain

The terms 'stress' and 'train' originally came from mechanics. Stress is 'an external force' and strain is 'the measure of deformation of the object to which that force is applied'. By directly applying these definitions to automobile driving, stress becomes the driving environment while strain becomes the internal condition of the driver caused by such an environment. In muscle fatigue, a heavy object is stress and the muscle activity caused by carrying that object is strain. The result from continuous muscle activity is fatigue. Muscle fatigue is caused by accumulation of lactic acid produced in muscles and exhibits reduced force capacity. As a result, in order to continue generating the same amount of force, more muscle fiber activity is required, and the effort to increase the physiological activity is felt as fatigue.

In 1971, Holger Luczak showed the relation between stress, strain, and workload schematically (Luczak, 1971). Luczak defined external stimuli as stress, the resulting changes caused in a human as strain, and fatigue is the result of the relationship between strain and individual endurance limit. He showed the relation between stress and strain by the concept of mental capacity, incorporating the idea of the mental workload that was introduced for flight task in 1960s (Fig. 4.1).

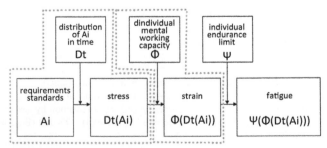

Fig. 4.1: Luczak's fatigue model (Reprint from Luczak, H. (1971). The use of simulators for testing individual mental working capacity. Ergonomics, 14(5): 651–660, www.tandfonline.com, with permission).

4.1.2 Driver Fatigue

In the case of stress on muscles, the relation between stress and muscle activity is relatively simple, though there may be some issues, like how multiple muscles are used. As a result of continuous muscle activity, the force capacity of muscles declines. Because of this, fatigue results and is sometimes defined as a decline in muscle performance caused by continuous stress. However, the relation between stress, strain, and fatigue is not so simple in the case of a mental task, such as automobile driving. It may seem that fatigue is caused by long hours of driving, but a clear decline in driving performance is often not observed in a real road environment. Discussions have been ongoing since the 1950s about what driver fatigue is (Crawford, 1961). Now the problem that arise is how to measure fatigue. If fatigue is defined as a decline in driving performance, it would imply that fatigue cannot be measured unless a decline in driving performance caused by fatigue is measured.

Fatigue can be subjectively defined as 'a condition of feeling tired', but there is no guarantee that the feeling of tiredness is the same for all people. Meanwhile, in studies using physiological indices that are objective, electroencephalogram (EEG), electrocardiogram (ECG; heart rate and its variability), GSR, electro-oculogram (EOG), eye blink, eyes-closed duration, and pupil diameter, etc are observed. As for driving performance, increased corrective steering movements and increased speed variations are observed and vigilance is found to decline with fatigue. With regard to performance other than driving, a delay in reaction time and a decline in scores of psychological tests (such as a crossing-out test, time estimation, or arithmetic task) are observed. Lal and Craig concluded that EEG measures might be promising in their critical review (Lal and Craig, 2001), but none of these are stable indices. Moreover, when studying the relation between objective indices and fatigue, an analysis must inevitably be made by comparing the indices with the subjective feeling of tiredness.

Since driver fatigue may occur in various modes depending on various causes, it is necessary to classify the types of driver fatigue. Jennifer F. May classified driver fatigue into sleep-related fatigue and task-related fatigue (Fig. 4.2) (May and Baldwin, 2009). Sleep-related fatigue is caused by sleep deprivation (sleep debt), prolonged

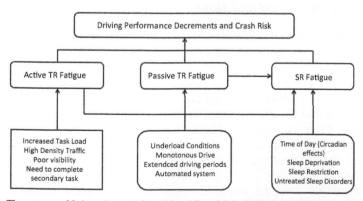

Fig. 4.2: Three types of fatigue (Reprint from May, J.F. and C.L. Baldwin. (2009). Driver fatigue: The importance of identifying causal factors of fatigue when considering detection and countermeasure technologies. Transportation Research Part F, 12: 218–224, Copyright (2009), with permission from Elsevier).

wakefulness, or circadian rhythms. While sleepiness tends to occur between 2:00 and 6:00 a.m. and around 2:00 p.m., a simulator test showed that speed variability tended to be greater during these hours as compared to other times of the day (Lenne et al., 1997). Task-related fatigue can be further divided into active task-related fatigue and passive task-related fatigue. The former fatigue results from driving in high-density traffic or poor visibility (a high task- demand situation) and the latter results from long hours of monotonous driving (a low task-demand situation). Passive task-related fatigue tends to affect sleep-related fatigue, and driving on a monotonous road induces sleepiness. Such interaction is one of the causes that make the phenomenon of driver fatigue complicated. Further, the driver's emotional state, such as anxiety and personality or character also affects fatigue.

Since fatigue involves various factors and causes, diverse physiological responses and performance changes, it is difficult to operationally define fatigue with these variables. Ross O. Phillips proposed the following definition of fatigue, incorporating all of these (Phillips, 2015): "Fatigue is a suboptimal psychophysiological condition caused by exertion. The degree and dimensional character of the condition depends on the form, dynamics, and context of exertion... The fatigue condition results in changes in strategies or resource use such that original levels of mental processing or physical activity are maintained or reduced." In other words, fatigue is described as the expression of changes in the resource input (effort) for maintaining the driving performance corresponding to the high or low task demand or corresponding to changes in the arousal level (see 6.1.2.2) (Fuller, 2005), how the performance is maintained or how the resource is input is expected to differ by the driver's personality and coping capacity. This model explains the diversity of the fatigue phenomenon.

Desmond and Mathews proposed a hypothesis that the range of adapting driving performance to driving demand becomes narrow when fatigued. This may occur when the demand is low because, in a high-demand condition, a driver makes effort to perform to the demand, to avoid increase in risk. He showed deteriorated performance in a low-demand situation when fatigued using driving simulator (Desmond and Matthews, 1997).

4.1.3 Mental Workload and Tasks

Research on mental workload increased since the 1970s, prompted by research on aircraft pilots. In physical work, stress can be clearly defined as, for example, the load of an object to be lifted up. In mental work, however, the specific works to be performed for a task to be achieved (such as win the air battle) cannot be clearly defined. Since work stress (workload) cannot be defined by work itself, the concept of mental workload became necessary. In the process of developing this concept, the NASA Task Load Index (NASA-TLX) for subjective assessment was established (see 5.1.1.2), while heart rate variability (HRV) drew attention as a physiological index (see 4.4.4.2), and the subsidiary (secondary) task method was established (see 5.1.2).

The concept of mental workload consists of task demand, mental resources required for performing the demanded task, and the resulting performance. They are difficulties in a given task, the amount of resources input and the effort made for performing the task, and the level of performance obtained (Fig. 4.3, Schlegel, 1993).

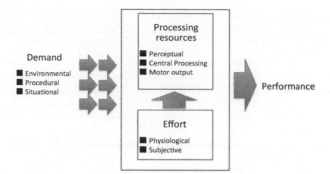

Fig. 4.3: Schematic of Schlegel's mental workload (modified from Schlegel, 1993). Driver mental workload. pp. 359–382. *In*: B. Peacock and W. Karwowski (eds.). Automotive Ergonomics, with permission.

In the NASA-TLX, a task comprises physical demands, mental demands, and temporal demands (see 5.1.1.2), and the operator performs the activity according to subjective assessment of the levels of the demands. Subjective workload also involves the results of subjective evaluation of the operator's task performing (own performance, effort, and frustration level).

4.1.4 Mental Workload Described in ISO 10075

ISO 10075-1: 'Ergonomic principles related to mental workload—Part 1: General issues and concepts, terms and definitions' defines terms relating to mental workload (ISO 10075-1, 2017). While this ISO standard uses the term 'mental workload' in its title, it defines such terms as 'mental stress' and 'mental strain'. 'Mental stress' is defined as influences impinging upon a human being from external sources and affecting it mentally, and 'mental strain' as the immediate effect of mental stress within the individual, being in line with the concept of Luczak's model. The ISO standard clearly indicates that mental strain differs from person to person, stating that it depends on a person's individual habitual and actual preconditions, including individual coping styles. In that respect, the differences in the individual input efforts are being taken into consideration.

This ISO standard mainly focuses on influences on humans from mental workload, and divides such influences into facilitating effects and impairing effects (Fig. 4.4). Facilitating effects are classified into effects with short-term potential and those with long-term potential. Short-term effects mean short duration of the exposure to mental stress; effects that begin immediately after the exposure and effects appearing transitory. Long-term effects mean long duration of the exposure, delayed effects and cronic effects. Following are the examples.

Facilitatory effects with short-term potential
- Warming-up effect: Reduction of the effort required to perform the task after frequent exposure to the stress
- Activation: Internal state with increased mental and physical functional efficiency
- Learning: Process based on experiences which lead to enduring changes

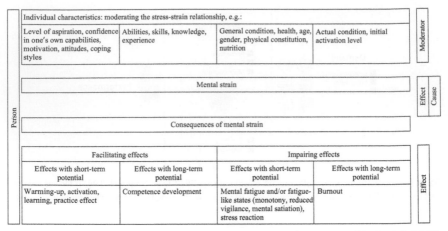

Fig. 4.4: Framework of mental workload described in ISO10075 (Reprint from ISO 10075-1. (2017), a part of Figure A.1).

- Practice effect: Enduring change in individual performance associated with learning process

Facilitating effects with long-term potential

- Competence development: New acquisition, consolidation, enhancement of cognitive, emotional and motor skills and abilities

Impairing effects with short-term potential

- Mental fatigue: Temporary impairment of mental and physical functional efficiency
- Fatigue-like state: Sates within individual as a result of the effects of mental strain which quickly disappears after change in the task and/or environment

Impairing effects with long-term potential

- Burnout: State characterized by perceived mental, emotional and/or physical exhaustion and reduced performance capacity, resulting from prolonged exposure to mental stress leading to impairing short-term effects.

The ISO standard assumes mental work to be work in a labor environment where a certain fixed task is conducted continuously. Compared to physical work, mental work can be implemented with more freedom in terms of style. Since automobile driving involves larger changes in demand as compared to mental work in labor, there is more freedom in the tasks to be performed. Accordingly, care must be taken when using these terms for automobile driving, for example, when comparing monotony in typing data with driving on an expressway, the latter is considered to involve larger changes in the required task due to changes in the road traffic environment. Also, whereas mental fatigue is defined as a temporary impairment of mental and physical functional efficiency, such impairment appears as delay in work time in the case of laborious work, but in automobile driving, as described above, it is difficult to observe the impairment of efficiency in the driving performance.

4.1.5 Task Demand, Mental Resource and Fatigue

Introduction of the idea of demand and resource allocation makes it easier to understand the structure of fatigue. Fatigue was basically regarded as decline in efficiency resulting from continuation of the same task. In automobile driving, however, the performance is maintained even in fatigued condition, which seems contradictory. This can be understood as follows. In a driving task, the risk of accidents rises unless the driving performance falls short of the demand (see 6.1.2.2, Fuller, 2005), so resources are supplied in order to maintain such a performance. If the performance efficiency declines due to continuous engagement in a task, an increased amount of resources is allocated to compensate for the efficiency decline and to keep the risk level under check. Since the supply of additional resources requires mental effort, such effort is subjectively perceived as a feeling of strain.

The decline in efficiency is affected not only by the continuation of the active or passive task, but also by circadian rhythms and arousal level. Meanwhile, because the human body has a recovery mechanism to remedy the efficiency decline, the efficiency may recover not only during rest, but also during a task. Physiological fatigue is likely to be caused by the supply of resources as well as by the physiological mechanism that operates for the recovery. Such compensation mechanism and recovery mechanism are considered as causes for a variety of the fatigue phenomena.

4.1.6 Difference Between the Concept of Mental Workload and the Concept of Stress/Strain

In the concept of mental workload, the task demand corresponds to stress in the sense that a task is to be performed and the amount of resources or effort allocated to a task correspond to the strain. However, the concept of stress/strain is a concept of fixed or mandatory tasks whereas the concept of demand and effort on how to perform a task depends on or is left to the person who carries out the task.

In physical work, stress is the load of an object to be lifted up, and muscle activity required for the lifting takes place. In contrast, in mental work, the task to be performed cannot be explicitly defined. Mental work is normally complicated, and a person determines the accuracy and speed of a task (task demand) in the light of a given goal (what task should be completed by when). How effort is to be supplied to perform the task is varied, and the person chooses the effort according to his or her own abilities, motivation and personality (see 6.1.2.2).

In the case of automobile driving, 'moving from location *A* to location *B* without causing an accident' is the given task, but how safety is to be ensured is left to each driver. The driving task entails selection of the route, setting the time required for reaching the destination, selecting the speed and the headway distance, and determining which objects require visual attention and what amount of resources should be allocated to each task. The resources that a driver uses in driving changes, depending on the driver's driving skill, the ability of driving-related information processing, the driver's physical and mental conditions at the time, and the surrounding traffic conditions. This means that the human functions that contribute to the driving task are not always the same and this because of the variety of physiological phenomena.

4.1.7 Driver's Stress

The terms 'stress' and 'stressor' for biological organisms have come into wide use since Hans Selye coined these terms for his stress theory of general adaptation syndrome in the mid-20th century (Selye, 1976). Selye's *stress* is a non-specific (not dependent on the type of stimuli) adaptation syndrome that occurs in response to noxious external stimuli. It is a general response of the endocrine system, including the adrenal gland and autonomic nerves occurring to facilitate the organism to adapt to a stimulus (an electric shock or cold-water stimulation in the case of animal studies). Selye indicated that the response has three stages: the alarm reaction, the stage of resistance, and the stage of exhaustion. When the organism's capability of adapting to the stimulus gets exhausted, irreversible tissue damages appear. Here, *stress* is a response, and stimuli that cause a *stress* response are called *stressors*. It should be noted that the terms 'stress' and 'strain' had already been used in mechanics, and meanings of Selye's '*stress*' and '*stressor*' differ from those of stress and strain in mechanics.

Selye's works are studies on animals with noxious stimuli; thus it was physiological *stress*. Studies on psychological *stress* were inspired by Selye's theory. The stress experienced by humans is almost always accompanied with a subjective experience of distress. Richard S. Lazarus proposed a cognitive appraisal model whereby a psychological *stress* is relevant to an appraisal by a person as to whether he/she can cope with the environment (Lazarus and Folkman, 1984). The person appraises whether a particular event in his/her daily life is taxing and endangers his/her well-being and whether the person can cope with the event.

In the 1970s, occupational *stress* came to be actively studied. Robert Karasek presented that the relationship between job demand and decision latitude is relevant to the risk of illness (Karasek and Theorell, 1990). Here, the job demand is decided in relation to the person's skill. Karasek indicated that a person engaged in a job with high-job demand and high-decision latitude faces a low risk of illness, but a person engaged in a job with a high job demand and low decision latitude faces a high risk of illness. If a person has discretionary control over a task, he/she is likely to feel less *stress* and face a lower risk of falling ill.

Although *stress* theories related to the well-being may not be directly associated with automobile driving, stimuli from the driving environment affect the emotional state of a driver. Negative emotions may arise by the threatening behavior of other traffic users or an unexpected traffic jam, for example. Another important viewpoint is coping. There is no way a driver can solve a traffic jam. A driver must find other coping methods, else he/she will be *stressed*. Gerald Matthews worked on driver's *stress* from this point of view. Matthews et al. proposed a model for driver's *stress* that distinguished three types of stress. First is short-term stress, such as anxiety; second is cognitive stress, such as appraisal of the difficulty to cope with, and third is the personality trait that is vulnerability to driver stress. They developed the Driver Behaviour Inventory (DBI) as a tool to the driver's *stress* (Gulian et al., 1989). In the process of developing DBI, they identified three major factors for *stress* vulnerability—'driving aggression', 'dislike driving' and 'driving alertness'. The aggression items included driver's reaction of irritation, impatience, and behavioral aggression. The dislike items were feelings of anxiety, unhappiness, and lack of

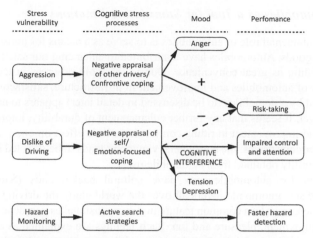

Fig. 4.5: A transactional model of driver stress traits and their effects on cognitive processing, mood state and behavior (Matthews, G. et al. (1998). Human Factors, 40: 136–149, Copyright© 1998 by Reprinted by permission of SAGE Publications, Inc.).

confidence. The alertness items monitored difficult conditions and potential hazards. Among the three, dislike was strongly related to emotional stress symptoms, such as a tense and depressed mood. Therefore, they focused on DIS (Dislike of Driving) score obtained from DBI (Matthews et al., 1996). Example items for DIS were 'driving usually does not make me happy', 'I am more anxious than usual in heavy traffic'. Matthews et al. proposed a hypothetical transaction the model of driver's *stress* as shown in Fig. 4.5 (Matthews et al., 1998). It indicated how aggression, dislike of driving and hazard monitoring (corresponding 'alertness') affected the cognitive processes, mood and performance. In their subsequent studies, they proposed five dimensions of driver's *stress*—vulnerability, aggression, dislike of driving, hazard monitoring, thrill seeking and fatigue process. The five dimensions of coping with driving *stress* are confrontive coping, task-focus coping, emotion-focus coping, reappraisal and avoidance (Matthews, 2002).

4.2 Enjoyment Generated by Automobiles

4.2.1 Utility of Automobile Use

In this section, we discuss the utility of automobiles. In general, it is divided into two categories: experienced utility and memorized utility. While the former looks at enjoyment in driving, while the latter looks at feelings of well-being on remembering the feeling convenience in transportation. Both categories of utility are important for improving the human well-being. In the first part of the section, we discussed automobiles as a tool for stimulating positive emotions, i.e., experienced utility, referring to Csikszentmihalyi's *flow theory*. We then discussed the subjective well-being and satisfaction with life, i.e., memorized utility and the role of automobiles in human well-being based on its subordinate concepts.

4.2.2 Automobiles as a Tool for Stimulating Emotions

The most fundamental role of automobiles is to serve as a means for transportation of humans and goods. Automobiles have contributed to enhancing our satisfaction with life by providing us great convenience. Due to the rapid progress in technological development of automobiles and improvement in infrastructure, satisfaction with life brought about by automobiles (to be discussed in detail later) appears to have risen to the prime level. It seems that any further enhancement of durability, improvement of reliability, and improvement in riding comfort have little effect on one's satisfaction. Under such circumstances, the consumers' viewpoint tends to be directed at values of experienced utility obtained from automobile usage.

Affection for automobile may have cultural backgrounds (Sachs, 1984). However, we see automobile lovers all over the world enjoy the driving itself. This means that there must be a common feature in driving that makes people feel good and experience enjoyment, pleasure and fun. Such feelings are explained as experienced utility and must be one of the values that automobiles can create and further enhance. The automobile industry is now aware that there are market needs for enjoyment and other related feelings, i.e., experienced utility, as an additional value of automobiles.

A large part of automobile driving is enjoyable, but what generates the enjoyment of driving remains unknown. Thus, it is difficult to decide on the direction for developing enjoyable automobiles. If a theory on the mechanism of generating enjoyment is developed, proper direction can be made at the initial stage of product development. So it may become possible to develop enjoyable products more efficiently. In the following sections, we explain the theory of generating enjoyment by referring to the flow theory developed by Mihaly Csikszentmihalyi, a prominent psychologist who pioneered positive psychology.

4.2.3 Flow Theory of Csikszentmihalyi

We sometimes feel enjoyment while driving an automobile. People who see automobiles as a tool for enjoyment have skills that help in free control of automobiles and enjoyment in conversing with automobiles. As the engine gives sweet sounds and the automobile accelerates smoothly, the driver feels as if there is a flowing conversation between the automobile and him/her. It is difficult to express this feeling in words, but if dared to do so, it would be as if the driver becomes fully involved in the flow of events, and feels at one with the automobile. The automobile's response to the driver's action arouses feeling of unity with the automobile and that he/she can drive well without thinking about the action. The driver enjoys driving, feels confident that he/she can drive perfectly, and becomes immersed in driving. Interestingly, this flowing feeling is known to commonly occur not only in automobile driving but also in various activities such as rock climbing, dancing, sewing, chess and even performing surgery (Csikszentmihalyi, 1975).

In social psychology, studies have been made by naming this feeling flow (Csikszentmihalyi, 1975; Csikszentmihalyi, 1990). According to Csikszentmihalyi's flow theory, flow is the positive effect that wells up when a person is voluntarily engaged in an activity of optimal difficulty by fully demonstrating his/her skills. It is considered that there are different levels of intensity of flow, ranging from micro flow, which occurs through casual activities in everyday life, to rather deep flow, which occurs when a highly skilled person engages in a specialized activity (Csikszentmihalyi, 1990). People who experience the flow more often in their daily lives are known to have better psychological well-being (Asakawa, 2010), because the feeling flow with increase in time for doing things is same feeling enjoyment with increase in time spent. Thus, flow is considered an index in improving subjective well-being (Salanova et al., 2006).

According to the flow theory, it is important that the difficulty of driving tasks and the driver's skills are appropriately balanced (Fig. 4.6). In order for the flow (enjoyment) to occur in automobile driving, the operation of the automobile needs to be moderately difficult and the driver's driving skills need to be moderately high. While the concept of flow is easy to understand, it is difficult to quantify how 'moderately' each of the two factors should be. This is because the level of difficulty and the diver's skills are assessed based on the driver's subjectivity (senses). While the level of difficulty (e.g., the steering efficiency) and performance (e.g., steering response) can be calculated as objective numerical values, only the driver can determine whether they are appropriate for him/her. How an appropriate level of difficulty can be created is a challenge in automobile development.

Flow is thought to be more likely to occur for a task that immediately feeds back the results of one's act (Csikszentmihalyi, 1990). If there is immediate feedback, a person can intuitively determine whether or not his/her act was successful. It is, for example, a task where pressure on the accelerator pedal is immediately fed back as acceleration of the automobile. The ability to obtain immediate feedback from an act means that the person is able to feel himself/herself to be the actor (i.e., the sense of agency). That feeling is one of the causes of flow.

An automobile with which flow is obtainable would be therefore an enjoyable automobile. Flow, however, accompanies a sense of immersion. Therefore, flow in automobile driving tends to be regarded as being likely to cause risky driving. When

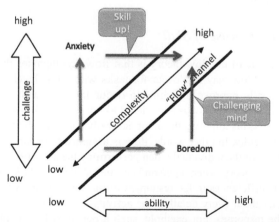

Fig. 4.6: Diagram of flow. Flow is considered to occur when the level of difficulty of a task and an individual's skills are balanced. Skills increase in the process of moving from state 1 to state 4. In state 2, the individual's skills are higher than the level of difficulty of the task, giving rise to boredom. In state 3, the level of difficulty of the task is higher than the individual's skills, giving rise to anxiety. Flow occurs when moving from state 2 or 3 to state 4.

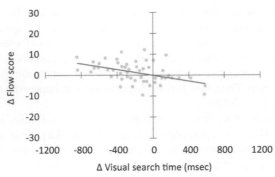

Fig. 4.7: Correlation between flow and improvement of performance (visual search time) (modified from Kaida, K. et al. (2012). As the flow score increases, the time required for finding a specific stimulus from the many stimuli (visual search time) shortens. Ind. Health, 50(3), with permission).

thinking about applying the flow theory to automobiles, there is a need to examine the relationship between flow and the risk of accidents. Kaida showed that, in a preliminary experiment in the laboratory, the visual search skills improved along with an increase in the sense of immersion accompanying the flow (Fig. 4.7, Kaida et al., 2012). It is still unknown whether this result can be applied to actual driving. Some think that the application of flow in driving carries a risk, but it may be incorrect. Rather, a driver in a flow state may actually be able to understand the surrounding situation more precisely than usual and thus drive smoothly in a flowing manner.

According to another study, flow is less likely to occur when the arousal level is low (Kaida et al., 2012). The cognitive-energy theory of Dov Zohar et al. (Zohar et al., 2005) states that attentional resources are preferentially allocated to negative effect when the arousal level is low. This makes it less likely for a positive effect to occur. Since flow is one type of positive effects, the arousal level would need to be maintained in order to cause the flow.

4.2.4 Flow and Increase of Skills

Another characteristic of the flow theory is that flow accompanies the improvement of skills (Fig. 4.6). Flow may first occur for tasks with a low level of difficulty, but as the skills gradually improve, it begins to occur for tasks with a higher level of difficulty. Since the stronger flow is considered to occur for tasks with a high level of difficulty, when the skills improve, the actor naturally comes to challenge tasks with a higher level of difficulty. In other words, flow is assumed to occur concurrently with an increase in skills of the individual when performing the activity.

Under the flow theory, when a person becomes skilled at a task, he/she is able to distinguish subtle differences in the responses to his/her activity. A skilled actor will become conscious about every detail of the activity, and come to enjoy the slightest differences in the responses. For example, such a person will feel subtle changes in the brake pedal feel and the tyre grip, depending on the air temperature and humidity. If a person becomes skilled at tasks, he/she will be able to find infinite changes in any kind of task. Csikszentmihalyi interprets that this phenomenon results from increased complexity of the person's consciousness that occurs with the flow (Csikszentmihalyi, 1990).

If flow occurs in the process of responding to subtle changes of tasks (process of developing internal complexity), a tool suitable for flow may be a tool that subtly changes according to the environment. This is because flow occurs in the process of developing the skills to gauge such subtle changes. We sometimes find it more amusing to drive old automobiles with manual transmission. Before automatic control technologies using computers developed, automobiles were equipped with many devices, such as the choke valve and manual transmission, and their responses were inclined to change according to air temperature and humidity. This means that responses can vary almost infinitely through different combinations of devices. Given the characteristics of flow, flow may likely occur in the process of developing (or adjusting) the skills to feel the changes of such varying status of devices, and this may be the reason that people find it more enjoyable to drive classic or manual-transmission automobiles.

Automobiles in recent years are equipped with automatic control devices, and their safety and comfort have remarkably improved as compared to old automobiles. Automobiles came to operate in a stable manner, irrespective of the changes in the external environment. However, the discussions so far suggest that the development of such automatic control may have caused automobiles to lose their enjoyment aspect. Even if that is the case, we cannot take cutting-edge automobiles back to the state that existed several decades ago. A future challenge may be to revive the feeling of old automobiles by using the latest automatic control technology.

Meanwhile, flow is sometimes regarded negatively as a factor for becoming completely absorbed in such activities, as playing computer games or gambling. Flow that impedes social adaptation without promoting growth is called *junk flow* in the sense that it does not contribute to a healthy increase in skills. Flow in a socially unadaptable case should be treated separately from *creative flow*, which the earlier part in this section has explained and which is often mentioned as flow.

4.2.5 Flow and the Zone

Flow is sometimes referred to as 'the zone' among racing drivers. The zone seems to be a term originating from Yuri L. Hanin's IZOF model (Individual Zones of Optimal Functioning model) (Hanin, 1978; Hanin, 2000). According to Hanin, the zone is a specific emotional state which is controlled for producing optimal performance in sport (Hanin, 2000). The zone theory is characterized by the approach of focusing on the positive aspect of anxiety and managing it. The theory is interesting as it considers that both positive affect (including joy) and negative affect (including anxiety) contribute to enhancing the driver's performance.

The zone is considered to be a similar concept as flow, but while the flow theory regards flow to be a positive effect resulting from human behavior, the zone theory regards the zone to be an effect which the individual creates by him/herself in advance in order to demonstrate his/her skills. The two theories fundamentally differ as to whether the effect is evoked naturally or by preparation. An article by Ruiz provides an extensive review on previous studies on the emotional states leading up to the zone (Ruiz et al., 2015).

4.2.6 Effects of Feelings of Enjoyment

Does improvement of automobile driving skills lead to improvement of other skills in daily life? For example, does improvement of skill to predict risks at intersections in automobile driving help prevent careless injuries when cooking with a knife? If so, it would be wonderful. However, psychological tests usually fail to demonstrate the spillovers that a skill acquired in one task can be transferred to other tasks.

Nevertheless, such generalizations may occur indirectly through positive emotions. For example, people who have more opportunities to experience flow in their daily lives are known to have higher subjective well-being (Salanova et al., 2006). When subjective well-being increases, the person becomes positive toward matters that involve him/her and becomes motivated to take action on other tasks as well. If a person engages in matters in a motivated manner, flow (including micro flow) is likely to occur in all kinds of tasks, and as a result, the task performance improves. At the same time, the skills for these tasks upgrade. Accordingly, positive affect can be considered to have a favorable influence on other tasks as well.

Therefore, a chain of flow is possible and it may be triggered by automobiles. Flow experienced during automobile driving is expected to raise the person's motivation for other activities and have an overall influence. If this is regarded as a generalization of skills to cover other tasks through positive emotions, the enjoyment of automobiles can be a starting point for enhancing subjective well-being.

The effect whereby positive emotion arising from a task spreads to other tasks is called 'spillover effect'. According to a survey conducted by Kaida in Stockholm County, Sweden, citizens who shifted their travel mode from automobiles to public transportation in response to the introduction of congestion charging (which was aimed at controlling car traffic volume within the city by charging a certain amount of Swedish kroners to car entries to the city) subsequently performed pro-environmental behaviors more frequently that are not directly related to automobiles, such as saving electricity and water (Kaida and Kaida, 2015).

4.2.7 Subjective Well-being and Automobiles

According to Diener, a prominent psychologist and pioneer in subjective well-being research, people's subjective well-being can be defined by three aspects: satisfaction with life, effects, and virtue (Diener, 2009). These three aspects are distinct in that they are subjectively reported (based on self-awareness) and are different in nature from the quantified objective measures of well-being in the field of economics.

The satisfaction with life aspect of subjective well-being is determined by the level of achievement in an individual's life and goals. For example, a person has a high level of satisfaction with life if his/her desires, such as being materially satisfied or being able to physically move freely, or having social connections with others, and being in harmony with the social environment, are satisfied. Satisfaction with life includes an objective aspect, such as being able to walk freely, but it also largely covers subjective aspects, including memorized experience. Satisfaction with life is considered to be the first well-being factor that is fulfilled by automobiles. This is because the fundamental role of automobiles is to give an individual the freedom of mobility, thus allowing the individual to go to a place where he/she desires, to meet a person who he/she desires, and to deliver an object that he/she desires to give to another person. All these satisfy the individual's fundamental desires. Also, the sense of psychological freedom of being able to go anywhere one desires by automobiles could be one of the pillars that supports the enhancement of satisfaction with life.

The effects aspect of subjective well-being refers to the various emotions in daily life, including flow. Positive effect has the characteristic of occurring in the process of pursuing one's interest or interacting with others and plays the role of searching for new possibilities. On the other hand, negative effect occurs in a situation where an urgent action is required and plays an important role for survival, such as running away or fighting. In order to maintain a high level of subjective well-being, it is necessary to maintain an appropriate balance between positive effect and negative effect. Fredrickson states that the optimal ratio of positive to negative effect for sustainable maintenance of subjective well-being is 3:1 (Fredrickson, 2009). The rate of positive effect is higher than that of negative effect because the negative influence that negative effect has on subjective well-being is stronger than the positive influence that positive affect has (Matthews, 2002). If automobiles function as a tool for evoking positive effect, they may be able to be used as a tool for turning the emotional balance in daily life towards the positive direction.

Lastly, we consider the virtue aspect of subjective well-being. Virtue is the code of conduct that is socially regarded as good and is acquired gradually throughout one's lifetime. For example, in modern society where a lifestyle with low environmental load (low CO_2 emissions) is considered to be desirable, environmentally friendly thoughts and behavior, such as energy-saving behavior, are known to have an effect of enhancing subjective well-being (Kaida and Kaida, 2016a; Kaida and Kaida, 2016b). In this context, the desire of wanting to drive an environmental-friendly automobile may be based on the instinct of wanting to enhance subjective well-being by accumulating virtues.

Unlike other tools, automobiles are a special tool that can contribute to all three aspects (satisfaction with life, effects, and virtue) of subjective well-being. The idea of an ideal automobile which people seek in automobiles may be a vehicle that stimulates all the aspects of satisfaction with life, effects (or flow), and virtue.

4.3 Arousal Level

4.3.1 Arousal Level and Sleepiness

Several factors are involved in the occurrence of sleepiness: the process of recovering from lack of sleep (process of maintaining sleep homeostasis), biological rhythms, such as the circadian rhythm and circasemidian rhythm; the arousal-enhancing process through motivation, posture (such as lying down or sitting), external stimulation (such as light or sound), or physical activity; sleep inertia, which occurs immediately after awakening, and instability of sleep and arousal caused by a lack of orexin. Sleepiness occurs due to several factors and their interaction. It is a complex phenomenon whose nature has not yet been sufficiently clarified. There has not been a consensus about a definition for theoretical treatment for sleepiness. Accordingly, sleepiness is currently being defined operationally and quantified by focusing on certain phenomena caused by sleepiness.

Of late sleep propensity has often been used as an operational definition of sleepiness. Sleep propensity refers to the amount of time it takes to fall asleep. Another operational definition of sleepiness used before the concept of sleep propensity is somnolence (or drowsiness). Somnolence is a drowsy state that occurs when shifting from arousal to sleep. It causes changes in recognition, behavior, performance, subjective experience of sleepiness, and the autonomic nervous system and central nervous system. However, it does not necessary accompany the subjective experience of sleepiness. Although somnolence is correlated with sleep propensity, these two types of sleepiness should be distinguished (Johns, 2010). Most of the currently-developed sleepiness-evaluation methods measure either sleep propensity or somnolence.

Terms associated with sleepiness include arousal, vigilance, and alertness (Oken et al., 2006). Arousal refers to non-specific activation of the central nervous system within the range of sleep to excitement. Arousal can be measured by operationally defining it as low amplitude and fast wave of surface EEG. When the arousal level declines, somnolence or sleep occurs. On the other hand, an excessively high arousal level leads to a hyperarousal state. The highest performance can be achieved when the arousal is at a medium level (Yerkes and Dodson, 1908). Such arousal level is called the 'optimum arousal level'.

Vigilance means a state where the subject continues to pay attention to a specific event over a certain period, and is used synonymously with sustained attention. Sometimes alertness is divided into 'phasic alertness' and 'tonic alertness', the former being used synonymously with orienting response, and the latter synonymously with vigilance or sustained attention. Vigilance and tonic alertness respond sensitively to sleep deprivation and changes in biological rhythms (Lim and Dinges, 2008), and decline in association with somnolence (Doran et al., 2001). However, vigilance/alertness is also affected by factors other than those that cause sleepiness, such as progress of the time on a task (Lim and Dinges, 2008) or hyperarousal (Edinger et al., 2013).

One of the similar concepts to sleepiness is fatigue. The Japanese Society of Fatigue Science defines fatigue as a decline in the ability and efficiency of mental and/or physical activities caused by excessive mental or physical activity or disease, and is often accompanied by a peculiar sense of discomfort and a desire to rest. With regard to the relationship between sleepiness and fatigue, sleepiness is considered one symptom of fatigue and sleepiness and fatigue as separate symptoms but with some parts in common. The former idea defines sleepiness as a physiological desire to sleep (Dement and Carskadon, 1982). If fatigue is defined as a physiological desire to rest (Williamson et al., 2011), sleepiness becomes one symptom of fatigue since sleep is included in rest. Under this concept, the physiological desire to sleep (sleepiness) is caused by the sleep control mechanism, while the arousal control mechanism is considered as a sleepiness-masking factor. Under the latter idea, sleepiness is considered to be caused not by the sleep control mechanism alone, but by the sleep control mechanism and the arousal control mechanism as well as their interactions (Johns, 2010). Since the operation of the arousal control mechanism and the interaction between the arousal control mechanism and the sleep control mechanism do not necessarily relate to the physiological desire to rest, sleepiness may change, irrespective of fatigue.

4.3.2 Sleepiness Measurement Methods

Sleepiness associated with the low level of arousal causes traffic accidents. Therefore, the aptitude test of drivers who exhibit excessive sleepiness and real-time measurement of sleepiness during driving are important for automobile technology. Sleepiness evaluation methods can be roughly divided by what is measured (sleep propensity or somnolence) and whether it is an objective measurement method or a subjective measurement method (Johns, 2010). Table 4.1 shows how they are classified. Each evaluation method only identifies a certain aspect of sleep propensity or somnolence. As for which evaluation method should be used, this could be decided depending on the conditions under which the experiment is conducted (laboratory or field) or by the aim of the research. Next, the major measurement methods for sleep propensity, vigilance, and subjective sleepiness are described. Indices for real-time measurement of sleepiness during driving are discussed in the section on arousal level measurement methods (see 4.3.3).

Table 4.1: Sleepiness Evaluation Methods.

Sleepiness	Objective/subjective	Typical measures
Sleep propensity	Objective	Multiple sleep latency test
		Maintenance of wakefulness test
		Behaviroal maintenance of wakefulness test
	Subjective	Epworth sleepiness scale
Somnolence, Drowsiness	Objective	Kalorinska Drowsiness Test
		Alpha wave attenuation test
		Pupillographic Sleepiness
		Psychomotor Vigilance test
		PERCLOS (percent time that the eyes are more than 80 percent closed)
		Johns Drowsiness Scale
	Subjective	Stanford Sleepiness Scale
		Karolinska Sleepiness Scake
		Visual Analogue scale

4.3.2.1 Sleep Propensity

Methods to objectively evaluate sleep propensity during daytime include the multiple sleep latency test (MSLT) and the maintenance of wakefulness test (MWT) (Littner et al., 2005). The MSLT measures the amount of time taken to fall asleep when sleeping is allowed, whereas the MWT measures the ability to remain awake when being instructed to maintain wakefulness (ability to stay awake). Since these two texts are not highly correlated with each other, they are considered to evaluate separate processes (Arand et al., 2005).

In the MSLT, a participant lies down on a bed installed in a dark room with eyes closed, and the time it takes for the participant to fall asleep is measured. A test that takes about 20 minutes each is conducted four or five times a day. In the MWT, the participant lies down on a reclining chair installed in a dark room with eyes open, and the time it takes for him/her to fall asleep is measured. A test that takes 20 minutes or 40 minutes each is conducted four times a day. In the MSLT, the participant is told to sleep, while in the MWT, the participant is told not to sleep. Both tests measure the amount of time taken for falling asleep (sleep latency) by using sleep polygraph recording. When evaluating sleepiness through the MSLT or MWT, experiments should be carefully designed because sleep latency is sometimes extended by arousing stimuli, such as exercise or motivation.

Figure 4.8 shows the sleep latency measured by the MSLT and MWT (Arand et al., 2005). Sleep latency is longer for the MWT than for MSLT because in the MWT, the participant is told to stay awake. Also, for both tests, the sleep latency becomes longer with age.

The MWT focuses on the ability to stay awake (Arand et al., 2005), which is an important ability in the duties of an occupational driver. The MWT is known as a predicator of the performance in a driving simulation task (Sagaspe et al., 2007). Therefore, discussions are under way as to whether the MWT is applicable to assessment of the risk for accidents caused by sleepiness.

Since implementation of the MSLT and MWT requires the facility and equipment for sleep polygraph recording and a skilled tester, who can measure and interpret sleep EEG, simpler methods are required. As a simple method, the Oxford sleep resistance test (OSLER test), which does not require measurement or interpretation of test EEG, has been developed (Bennett et al., 1997). In the OSLER test, a participant sits on a comfortable chair installed in a dark room, and pushes a button in response to a targeted stimulus (red light-emitting diode) that is presented for one second every three seconds over a period of 40 minutes per test. The OSLER test judges the participant as fallen asleep when no response is made seven consecutive times (21 seconds). The sleep latency highly coincides between the OSLER test and the MWT (intra-class correlation: ICC = 0.94) (Krieger et al., 2004).

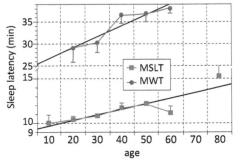

Fig. 4.8: Comparison of the sleep latency measured by the MSLT and MWT (Reprint from Arand, D. et al. (2005). The clinical use of the MSLT and MWT. Sleep, 28(1): 123–144. By permission of Oxford University Press).

4.3.2.2 Vigilance

According to a meta-analysis to examine various cognitive/memory functions of the brain that are more likely to be affected by sleep deprivation, a lower level of arousal was found likely to appear, particularly as a deterioration in vigilance (Lim and Dinges, 2010). With a low arousal level, omission errors (lack of response to a targeted stimulus) as well as commission errors (response made when no stimulus has appeared or response to a non-targeted stimulus) increase. Since omission errors become more likely to arise when the arousal level has lowered, the subject makes an intentional or unintentional effort to compensate for the errors (compensatory effort) (Doran et al., 2001). Compensatory effort sometimes causes commission errors, but may succeed in making a normal response even when the arousal level has lowered. However, compensatory effort under a low arousal level does not last long. Accordingly, when the arousal level is lowered, normal response and deteriorated performance appear in turns. This is called 'wake state instability' (Doran et al., 2001).

The psychomotor vigilance test (PVT) is an excellent index for evaluating deterioration in vigilance based on wake-state instability (Basner and Dinges, 2011). In the PVT, the participant presses a button immediately when a counter starts to run once every 2 to 10 seconds, and the test takes 10 minutes to finish. The performance indices include the response time, response delays of 500 ms or more (omission errors), and the number of unrequired responses (commission errors). A strictly-controlled laboratory test has revealed that PVT results are highly sensitive to sleep deprivation (Basner and Dinges, 2011).

The PVT satisfies various requirements for a work aptitude test, such as high sensitivity and specificity, reliability, no learning effect, short-time measurement, instant feedback of results after measurement, and a function to identify any attempts to falsify a response. Therefore, its practical application as a sleepiness-related work aptitude test is expected. However, in order to conduct the PVT as a work aptitude test for automobile driving, etc., there is an operational difficulty that makes the test take 10 minutes to complete. To improve the usability of the test in the occupational field, it is necessary to shorten the test duration of the PVT.

As solutions to this problem, two shorter versions of the PVT—the PVT-A (Adaptive-Duration Version of the PVT) (Basner and Dinges, 2012) and the PVT-B (Brief PVT) (Basner et al., 2011) were developed. The PVT-A shortens the test duration by way of stopping the test once enough information has been obtained while conducting the 10-minute PVT. Since enough information can be obtained quickly when there is hardly any response delay or there are frequent response delays, the test duration becomes shorter. The vigilance level (classified as high, medium, and low) obtained by the PVT-A highly coincides with that obtained by the 10-minute PVT (sensitivity: 93.7 per cent; specificity: 96.8 per cent), and the test duration has been shortened to less than 6.5 minutes on an average (Basner and Dinges, 2012). When determining the work aptitude by classifying the vigilance level into three levels, the PVT-A is more useful than the 10-minute PVT.

The PVT-B adopts a method to shorten the test duration from 10 minutes to 3 minutes. However, it has changed the definition of response delay from 500 ms to 355 ms and the inter-stimulus interval from 2–10 seconds to 1–4 seconds. The PVT-B's accuracy of differentiating sleep deprivation (total sleep deprivation, partial sleep deprivation) is 22.7 per cent less than the 10-minute PVT. However, as the test duration is shortened by 70 per cent, there is sufficient merit in conducting the PVT-B in the occupational field (Basner et al., 2011). The PVT-B performance measured during sleep deprivation of 34 hours indicated changes during the day similar to the performance on a task simulating luggage-screening for security check by an airport screener, suggesting that the PVT-B has the ability to predict work performance in the occupational field (Basner and Rubinstein, 2011). Nevertheless, for practical use of the PVT and its shorter versions as work aptitude tests, sufficient studies need to be made for their use in the occupational field.

4.3.2.3 Subjective Sleepiness

Methods to subjectively evaluate the chronological changes in sleepiness include the Stanford sleepiness scale (SSS) (Hoddes et al., 1973), the Karolinska sleepiness scale (KSS) (Akerstedt and Gillberg, 1990), and the visual analog scale (VAS) (Monk, 1987) (while the SSS incorporates the fatigue viewpoint, the KSS only focuses on sleepiness; the two indices are highly correlated with each other, and their results are not so different). These subjective evaluation methods are highly sensitive to variations of sleepiness within an individual. Also, they indicate high reproducibility when having the same participant conduct the evaluation under controlled environmental factors and living conditions. When using the same participant, the KSS also indicates a relatively high correlation with objective sleep indices, such as EEG and response time (Kaida et al., 2006). However, there is sometimes a gap between subjective sleepiness and objective sleepiness even within the same participant.

As shown in Fig. 4.9, those whose sleep periods were consecutively restricted to four hours or six hours per night showed an increase in the number of lapses (response delays, indicating response latency of 500 ms or more) in the psychomotor vigilance test (PVT) as the number of days progressed. By the last day, the number of lapses worsened to the level equivalent to one night of total sleep deprivation for participants who continued to sleep six hours per night. It worsened to the level equivalent to two nights of total sleep deprivation for those who continued to sleep four hours per night. On the other hand, as the number of days progressed, the increase in subjective sleepiness ratings slowed down and the difference in the sleepiness level between those who slept six hours per night and those who slept four hours per night diminished. This suggests that subjective sleepiness tends to be underestimated (Van Dongen et al., 2003). Also, because subjective sleepiness is reduced even by modest arousing stimuli (Kaida et al., 2007), it is necessary to conduct subjective evaluation under controlled environmental factors. Since vigilance could deteriorate while not being aware of sleepiness, objective evaluation rather than subjective evaluation is essential for evaluating a driver's sleepiness.

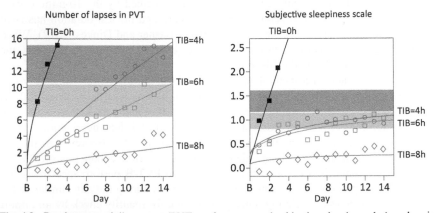

Fig. 4.9: Psychomotor vigilance test (PVT) performance and subjective sleepiness during chronic partial sleep deprivation (Reprint from Arand, D. et al. (2005). The clinical use of the MSLT and MWT. Sleep, 28(1): 123–144. By permission of Oxford University Press).

4.3.3 *Arousal Level Measurement*

Methods for real-time measurement of constantly changing sleepiness while driving can largely be divided into methods for measuring driving performance (driving behavior) and physiological indices related to sleepiness. This section introduces recent studies on methods for measuring the arousal level while driving.

4.3.3.1 *Driving Behavior*

Techniques for measuring driving performance (driving behavior) are indices that are directly linked to the occurrence of accidents and so are useful indices for preventing accidents. A method for estimating a declined level of attention caused by sleepiness based on driving behavior has already been put to practical use (a driver's attention monitoring system, MDAS). This method focuses on the fact that a decline in the attention level causes reduced frequency of driving operations (such as indicator or gear-shift operation) (an increase in the degree of monotonicity), a decreased amount of correctional steering for following the target course, and difficulty in driving straight inside the traffic lane (an increase in the meandering rate), and determines the attention level by using fuzzy inference based on the three data items—degree of monotonicity, steering amount, and meandering rate (Yamamoto, 2001). However, driving behavior is affected not only by a decline in the arousal level, but also by physiological factors including the driver's physical condition and fatigue, psychological factors including rushing mentality and increased stress, and environmental factors including the weather and traffic conditions.

4.3.3.2 *EEG*

Methods to measure vigilance (a state where the subject sustains attention on a specific object over a certain period) by analyzing EEG components related to sleepiness are being studied. EEG has an advantage of high temporal resolution, and when measuring micro sleep (the section where the EEG in the theta band of 4–7 Hz arises for three consecutive seconds), evaluation can be made in three seconds at the shortest. However, there are cases where micro sleep does not arise even if there is a response delay (Priest et al., 2001). It has also been indicated that the EEG band which is highly correlated with the psychomotor vigilance test (PVT) (see 4.3.2.2) is EEG in a lower band (1.0–4.5 Hz) from the front region of the head than the theta band (Chua et al., 2012). EEG, which tends to become unmeasurable due to artifact, has a disadvantage of requiring the attachment of electrodes.

4.3.3.3 *Rating Based on Facial Expressions*

Methods have been developed to evaluate sleepiness based on facial expressions using video images of head-and-chest by several researchers (Wierwille and Ellsworth, 1994; Kitajima et al., 1997). For example, the method proposed by Kitajima (Kitajima et al., 1997), two examiners independently evaluate sleepiness according to the following five categories.

1. Appears not sleepy at all (the gaze movement is rapid and frequent; the blink frequency is stable; and the head movement is active and accompanies body movement).

2. Appears a little sleepy (the eye movement is slow; and the lips are open).
3. Appears definitely sleepy (the blink is slow and frequent; there is mouth movement; the subject reseats himself/herself; and the subject puts his/her hand to his/her face).
4. Appears very sleepy (there is seemingly intentional blinking; there are unnecessary movements of the whole body, such as shaking the head or moving the shoulders up and down; frequent yawns and deep breathing are observed; and both the blink and eye movement are slow).
5. Appears extremely sleepy (the subject shuts the eyelids; the head tilts forward; and the head is thrown back).

The recommended method is to evaluate sleepiness at 5–20-second intervals and to obtain the average for the 60-second period. It has been shown that this technique provides high coincidence between the examiners (0.756) and high correlation with subjective sleepiness (0.795).

4.3.3.4 Pupil Diameter

A low level of arousal causes constriction of the pupils and a low-frequency pupil fluctuation of 0.8 Hz or less. These characteristics can be applied as a method to measure arousal level (Shirakawa et al., 2010). When the pupil diameter is measured and an interview on subjective sleepiness is conducted at the same time, it is found that the subject becomes aware of sleepiness when a pupil fluctuation is observed. A pupil fluctuation is considered to occur when constriction of the pupil, which is caused by hyperactivity of the parasympathetic nervous system associated with a low level of arousal, and dilatation of the pupil, which is caused by hyperactivity of the sympathetic nervous system associated with intentional or unintentional effort to maintain wakefulness, arise repeatedly in turns. However, since the pupil is strongly affected by mental factors other than brightness and sleepiness, this method has many challenges to overcome in order to be put into practical application for measuring sleepiness in an actual vehicle.

4.3.3.5 Eye Movement

4.3.3.5.1 Saccade

It is known that saccade characteristics and frequency serve as indices for the arousal level. With low level of arousal, the peak velocity (PV) of the saccade declines, the duration (D) becomes longer, and the kurtosis of the saccadic velocity waveform (the ratio of the peak velocity to the duration: PV/D) also decreases (Ueno et al., 2007). By presenting a sound stimulus when the PV/D value declines, the PV/D value increases after the stimulus as compared to when no stimulus is presented (Ueno et al., 2007). Accordingly, PV/D can be considered as an index that changes in response to the arousal level.

When the arousal level declines, the frequency of saccades with a PV of less than 40 deg/s or a time interval of less than 0.2 seconds increases. The corrective saccades increase with a decline in the saccade accuracy associated with a decline in the arousal level. Saccades can be a useful index in evaluating changes in the arousal level when having the eyes open.

4.3.3.5.2 Slow Eye Movement

Sometimes, the eyeballs exhibit a slow pendulum movement when shifting from an arousal state to a sleep state. This is called 'slow eye movement' (SEM). In electrooculography, eye movement with a sinusoidal waveform and amplitude of 50 nV or 100 nV or more and a duration of one second or more is defined as a SEM. In a driving simulator (DS) experiment, where the subject was instructed to press the brake pedal instantly when the brake lamps of the leading vehicle are illuminated, it has been demonstrated that the reaction time becomes longer and the accident rate increases when SEM is present, compared to when it is absent (Shin et al., 2011). It was also shown that the response time shortens and the accident rate decreases if a warning sound is given when SEM arises (Shin et al., 2011). However, it is necessary to distinguish between SEM associated with low arousal level and vestibulo-ocular reflex and smooth pursuit eye movement that arise under a high arousal level.

4.3.3.5.3 Vestibulo-ocular Reflex (VOR)

The VOR is a reflex response that results from the eyeballs moving involuntarily in the direction opposite to the head movement in order to compensate for visual instability caused by the head movement (see 4.4.3.1). When the arousal level is high, the ratio of the ideal angular velocity of eye movement for compensating for the head movement (estimated by measured head motion) to the actual angular velocity of eye movement (VOR gain) approaches 1, but as the arousal level lowers, the value decreases and the residual standard deviation of the two angular velocities increases (Hirata et al., 2009). When objective sleepiness detection thresholds are set by using the VOR gain decrease rate and the residual standard deviation increase rate, and the point of time when either of these measures exceed the threshold continuously for more than 40 seconds is regarded as the occurrence of a sign of sleepiness. The sign of sleepiness could be identified in 112.5 seconds (maximum: 260 seconds; minimum: 20 seconds) prior to the occurrence of subjective sleepiness during a monotonous DS driving (Hirata et al., 2009).

4.3.3.6 Eyelid Activity

Various indices related to eyelid activity (such as blinking and opening/closing) change in correspondence to low arousal level. For a healthy subject, the following appear as sleepiness increases—an increase or decrease in the number of blinks; extension of the time required for eyelid opening/closing; a decline in the speed of eyelid opening/ closing; an increase in the eye closure duration; an increase in PERCLOS (per cent of time the eyes are more than 80 per cent closed) (Abe et al., 2011).

4.3.3.6.1 PERCLOS

Among the eye-related measures, PERCLOS is an index that can measure vigilance with high accuracy due to its high degree of coincidence with vigilance deterioration (Chua et al., 2012; Abe et al., 2011; Dinges et al., 1998). Figure 4.10 shows the PERCLOS and PVT performance during chronic sleep restriction. The correlation between PERCLOS and PVT performance was 0.77, and PERCLOS changed with change in the PVT performance (Chua et al., 2012). PERCLOS could also predict

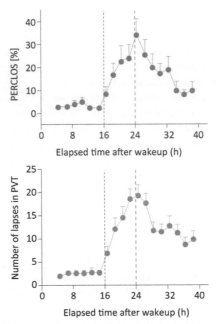

Fig. 4.10: Psychomotor vigilance test (PVT) and PERCLOS during chronic sleep restriction (Reprint from Chua, F.C. et al. (2012). Heart rate variability can be used to estimate sleepiness-related decrements in psychomotor vigilance during total sleep deprivation. Sleep, 35(3): 325–334. By permission of Oxford University Press).

a deterioration in vigilance not only in a vigilance task using visual stimuli, but also in an auditory vigilance task (Ong et al., 2013). This means that the relationship between PERCLOS and a vigilance decline is not only due to the stimulus caused by eye closure and physical blocking of visual information given to the central nervous system, but also by declined activity of the central nervous system. PERCLOS can detect vigilance deterioration with high accuracy, but it also has some problems. When measuring sleepiness using PERCLOS, it is recommended to take measurement for at least one minute (Dinges et al., 1998). However, since vigilance changes at intervals shorter than one minute, PERCLOS cannot detect vigilance that changes moment by moment.

A study on measuring vigilance while the participant is traveling on an actual road was conducted, targeting the occupational truck drivers (Dinges et al., 2005). This study uses indices including PERCLOS, lane deviations, and psychomotor vigilance test to examine the effects of their feedback. Under the condition where feedback was given from the device, driving performance measures during night-time driving improved, and the sleep hours after returning home increased. The vigilance detection system not only plays the role of indicating the vigilance state, but also the role of raising self-awareness about the importance of securing sleep hours.

4.3.3.6.2 Integrated Indices of Eye-related Measures

Indices that evaluate vigilance by using multiple eye-related measures have been developed. One of them is an index called the 'Johns drowsiness scale' (JDS) (Johns et al., 2007). A glass-type device for drowsiness detection using the JDS has already been commercialized. The JDS is a method to estimate sleepiness based on multiple eye-related measures (blink duration, relative speed of eyelid opening/closing, etc.) by using multiple regression analysis, and rates sleepiness by scores in 11 levels from 0 to 10. The JDS score increases as the performance declines in a vigilance test or a DS task (Johns et al., 2007, 2008).

Figure 4.11 shows the values of JDS and an auditory vigilance test during a chronic sleep-restriction period. The correlation between the JDS and auditory vigilance test performance (response time) is 0.87, and the JDS shows changes corresponding to the auditory vigilance task performance during sleep restriction (Ftouni et al., 2013a). Measurement has also been conducted in an actual vehicle, and it was found that the JDS scores of nurses measured while driving before and after a night shift were related to inattention during driving (Ftouni et al., 2013b). However, since there were cases where vigilance deterioration could be detected more accurately using eye-related measures independently than using the JDS (Ftouni et al., 2013b).

A device (driver status monitor) that can measure values that are highly correlated with sleepiness ratings by facial expressions based on multiple indices, such as the degree of eye opening, the average blinking duration, distribution of the continuous eye closing time, PERCLOS, and the number of blinks, has also been developed (Omi, 2016).

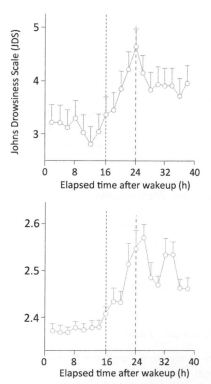

Fig. 4.11: Psychomotor vigilance test (PVT) and Johns Drowsiness Scale during chronic sleep restriction (Reprint from Ftouni, S. et al., (2013). J. Biol. Rhythms, 28(6): 412–424. Copyright© 2013, by permission of SAGE Publications, Inc.).

4.3.3.7 Heart Rate

There have also been studies to measure vigilance by using heart rate variability. Chua et al. clarified that some components of heart rate variability (0.02–0.08 Hz) increase with an increase in delayed responses during a psychomotor vigilance test (Chua et al., 2012). Although heart rate variability was less accurate than PERCLOS, its accuracy was close to PERCLOS as compared to subjective indices and EEG. This component showed changes corresponding to not only the PVT performance during chronic sleep restriction (Chua et al., 2012), but also the PVT performance during repeated partial sleep deprivation (4 h sleep for 5 nights) (Henelius et al., 2014). However, because its accuracy in predicting the PVT performance measured during normal sleep (8 h sleep for 5 nights) is low, in order to measure vigilance by using heart rate variability of 0.02–0.08 Hz, it is necessary to acquire information on sleep hours in advance (Henelius et al., 2014).

It has also been shown that the standard deviation of the RR interval has a higher correlation with the vigilance task performance than the conventionally used low-frequency component (0.04–0.15 Hz) or high-frequency component (0.15–0.50 Hz) of heart rate variability (Kaida et al., 2007). Furthermore, the standard deviation of the RR interval can predict the vigilance task performance one minute ahead, and is robust against individual differences. Although there is a need to collect data for a certain period of time in order to analyze heart rate variability, deviation can be calculated from data for a few heart beats. So the method of evaluating sleepiness by using the deviation of heart rate has advantage in temporal resolution.

4.3.3.8 Summary

For measurement of the arousal level in an actual vehicle, development is essential in the following two aspects: (i) high-accuracy sleepiness indices that are not easily affected by external physical environment (such as humidity or brightness) and by psychological factors other than sleepiness (such as rushed state and increased stress); and (ii) a device that can stably measure the sleepiness indices even where there are external factors, such as light and vibration.

4.3.4 Arousal-enhancing Technology

Sleepiness induces attentional lapses during driving and serves as a cause of traffic accidents. It is known that sleepiness-related accidents represent about 10–20 per cent of all traffic accidents (Connor et al., 2002). In order to take effective counter measures against sleepiness, we need to know the mechanism of how sleepiness actually arises. This section discusses the mechanism of sleepiness and then describes measures for reducing sleepiness.

4.3.4.1 Sleepiness and Arousal Level

The term 'sleepiness' is a subjective expression in which a human being assesses his/her level of physiological condition. The physiological conditions of the body used to assess sleepiness are called the 'arousal level'. When the arousal level declines, the brain function declines and the work performance deteriorates.

The arousal level is determined by interaction between the sleep center and the arousal center in the brain (Fig. 4.12) (Saper et al., 2001). The activation of the sleep center is determined mainly by the waking hours and the biological clock. Therefore, the activation of the sleep center is enhanced when the waking hours become longer or when a certain time arrives (around 4:00 a.m. and around 2:00 p.m.). The activation of the sleep center weakens when the subject sleeps. The activation of the arousal center becomes relatively enhanced when the subject sleeps, and the see-saw balance between the sleep center and the arousal center tilts toward arousal. As a result, the arousal level increases (Fig. 4.12).

The important point is that because the effect on subjective sleepiness and work performance would differ depending on whether to activate the sleep center or the arousal center, when we think about measures for preventing sleepiness, we need to consider which center the prevention measure would work on. Taking an overview of the findings on sleepiness prevention measures to date, measures that activate the sleep center (sleeping) tend to improve both subjective sleepiness and work performance, while those that activate the arousal center tend to improve subjective sleepiness but have less influence on work performance (to be discussed later in this section).

The arousal level is influenced not only by the waking hours and sleep but also by the biological clock. In particular, the influence of the biological clock becomes stronger under sleep deprivation (Van Dongen et al., 2003). Sleepiness typically tends to occur at around 4:00 a.m. and 2:00 p.m. and becomes stronger at the former hours than the latter. Also, the sleepiness around 2:00 p.m. is influenced not only by the biological clock but also by having meals (Reyner et al., 2012). Since accidents and work errors occur at hours when the sleepiness strengthens, many measures against sleepiness target these hours.

The indices used for estimating the arousal level are subjective sleepiness and work performance. In most cases, these indices change in sync with each other. However, the two do not always coincide with each other. For example, it is known that, after a long sleep deprivation (e.g., staying awake all night), work performance deteriorates while subjective sleepiness does not increase in some cases. Also, after exercise, subjective sleepiness decreases but work performance does not improve (Matsumoto et al., 2002).

Such a gap between subjective sleepiness and work performance is called the 'masking effect' of the arousal level (Kaida et al., 2007). It is considered that, when the masking effect of the arousal level arises during automobile driving, work performance is overestimated (the subject does not notice that his/her driving performance has declined) and thus the risk of causing a traffic accident increases (Dorrian et al., 2003).

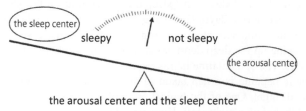

Fig. 4.12: Arousal level determined by the balance between the arousal center and the sleep center in the brain.

In order to prevent traffic accidents, it is important to identify the conditions under which the masking effect of the arousal level tends to occur.

4.3.4.2 Counter Measures against Sleepiness, Napping

The most effective counter measure to raise the arousal level is, naturally, to sleep. This is because, while sleeping, sleep substances (such as adenosine) that have accumulated during arousal are removed and the neural networks that became inefficient during arousal are optimized. 'Fatigue' that has accumulated during arousal is removed by sleeping. Also, sleeping restrains the activation of the sleep center and makes the activation of the arousal center predominant. As a result, the arousal level rises.

Sleeping is a method to directly activate the sleep center, and even a short sleep (15–20 minutes) has an effect to raise the arousal level (Kaida et al., 2013; Hayashi et al., 2003; Takahashi and Arito, 2000). Sleep researchers call sleeping of less than four hours as 'napping'. For convenience, napping is divided into three types by its purpose—appetitive napping, replacement napping, and prophylactic napping (Dinges, 1992).

Appetitive napping is a short sleep that is taken to deal with mild sleepiness. About 15–20 minutes of napping is sufficient, and it has an effect of improving both the mild subjective sleepiness and deterioration of work performance in the afternoon.

The reason for making the nap-time short is to prevent slow-wave sleep (i.e., deep sleep), which appears after sleeping for 20–30 minutes or more (depending on the age). When the subject wakes up from slow-wave sleep, a sleepy state called 'sleep inertia' occurs, and the work performance and sleepiness become worse than before the nap. Since this can be a cause of an accident, the timing of waking up should be carefully decided. One of the methods for preventing sleep inertia is to keep the nap-time short and avoid waking up from slow-wave sleep.

Replacement napping is napping taken in response to sleep deprivation. This napping is needed to make up for the lack of slow-wave sleep or rapid eye movement (REM) sleep. Therefore, the nap time is often set at 30 minutes or more so as to take a slow-wave sleep. In many cases, the nap time is set at about 90 minutes. This is because the cycle of non-rapid eye movement (non-REM) sleep and REM sleep is about 90 minutes. A slow-wave sleep ends after 90 minutes; so there will be less probability of waking up from a slow-wave sleep. If about 90 minutes of napping is secured, it is possible not only to replace the sleep that was deprived on the previous day, but also to restrain the influence of sleep inertia.

However, it needs to be noted that the maximum amount of slow-wave sleep that one can have in one day is fixed at a certain level. One's night-time sleep will be shallower if he/she has had slow-wave sleep during daytime on the same day. If one has slow-wave sleep during daytime, the sleep latency on the day become long, and this can disrupt the biological clock. Accordingly, even in the case of replacement napping for making up for sleep deprivation, it is better to shorten the sleep time as much as possible. The optimal time in this regard is about 90 minutes.

Prophylactic napping is a nap taken in advance when the subject is expected to stay awake all night. One of the typical examples is napping for some length of time immediately before a night shift. However, due to the problem of the biological clock, it is difficult to take a nap at an arbitrary time. In particular, it is hard to sleep at around 8:00 p.m., when night shift workers typically want to sleep in preparation for their work. The time window from around 10:00 a.m. to 8:00 p.m. is called the 'forbidden zone' when the biological clock does not let the subject sleep. Therefore, a prophylactic nap often cannot be taken before a night shift in reality.

It would be more realistic to take an appetitive nap or replacement nap during the work hours in the case of a night shift. For instance, when a long-distance truck driver or a taxi driver works on a late-night shift, it is desirable to control sleep by securing nap time even during the work hours. This would have an effect of preventing late-night traffic accidents. Accordingly, it would be desirable for automobiles traveling a long distance at night to have seats that are suitable for taking a nap (Iizuka and Kaida, 2011).

4.3.4.3 Counter Measure against Sleepiness, Other than Napping

One of the ways to raise the arousal level other than napping is to take caffeine. Caffeine is a substance that restrains the activation of the sleep center by suppressing the activity of the sleep substance (adenosine) accumulated during arousal (Fredholm et al., 1999). It has been reported that caffeine intake reduces vehicle fluctuations during automobile driving, and caffeine is considered to have an effect of preventing accidents at least temporarily (Reyner and Horne, 1997). However, since caffeine does not reduce the actual sleep substance that is the cause of a decline in the arousal level, its effect is limited.

It has also been reported that the effect increases when caffeine is combined with a short nap. Taking into account the characteristics that caffeine intake requires about 30 minutes to exhibit its effect, we can expect synergetic effect of having caffeine and nap if we take a short nap (less than 30 minutes) just after taking caffeine (Hayashi et al., 2003; Reyner and Horne, 1997).

Another measure is bright light (2000 lx or more), in particular, light of the blue wavelength (450–550 nm) is considered to have a strong effect (Cajochen et al., 2005). In fact, exposure to bright light in late afternoon onwards is reported to improve both sleepiness and work performance (Phipps-Nelson et al., 2003). However, this measure has yet to be put to practical use (Taillard et al., 2012), for example, a study that attempted to raise the arousal level by using blue light during automobile driving at night-time found that blue light causes eye strain and thus its practical application is difficult. Also, light has a strong influence on the biological clock (Czeisler et al., 1986), so there are concerns that countermeasures that use light may have a side effect of disrupting the biological clock. Moreover, the mechanism through which bright light directly affects the arousal level remains largely unknown, thereby making its practical application difficult.

Other sleepiness mitigation measures include chewing gum, washing the face with cold water, exposure to cold wind, listening to the radio, receiving vibration stimuli, and engaging in conversation. In particular, engaging in conversation is considered rather more effective to increase arousal level measured by electroencephalography (Kaida et al., 2007). We can easily implement these measures in daily life. However, in most cases, they only tentatively improve arousal level and slightly enhance the work performance.

Considering the above, we can assume that the effect of improving both subjective sleepiness and work performance mainly arises when the activation of the sleep center is restrained. While the stimuli that enhance the arousal center temporarily reduce subjective sleepiness, they have limited effect on work performance and possibly cause a masking effect on the arousal level. Therefore, in order to prevent accidents attributed to sleepiness, it is more desirable to directly activate the sleep center.

4.3.4.4 Summary

Human's sleepiness is an ambiguous psychological concept, as 'fatigue' and 'stress' are (see 4.1). There is no index that can precisely capture the concept of sleepiness itself; so we must rely on subjective evaluation or work performance to estimate sleepiness. When discussing sleepiness mitigation measures, it is important to understand the mechanism of sleepiness and also to understand whether the prevention measures work on the sleep center or the arousal center. This is because measures that work on the sleep center improve not only subjective sleepiness but also work performance, but others, i.e., working on the arousal center, merely improve subjective sleepiness and often have hardly any effect in improving the work performance. Even worse, stimuli on the arousal center could cause a masking effect on the arousal level, which makes vehicle drivers unaware of their reduced actual work performance, and thus could be counter productive in preventing accidents during driving. In order to avoid an erroneous consequence from applying a measure for preventing sleepiness, we must have a good understanding of the activation mechanism of the measure to be introduced. In addition to that, it is crucially important to give sufficient consideration to avoid the masking effect on the arousal level by applying the sleepiness-mitigation measures.

4.4 Techniques for Measuring/Analyzing Physical Conditions

4.4.1 Significance of Introducing Biosignal Measurement

4.4.1.1 Purpose of Biosignal Measurement

In the automobile field, people subject to measurement would be drivers, passengers, and pedestrians. When using biosignal measurement for evaluating the vehicle comfort, driving workload or usability of in-vehicle system, changes in the biological state of the driver are used as measures. Until recently, biosignal measurement was mainly applied in the development phase, such as for deciding on the specifications in the design or evaluating prototypes and identifying points to be improved. It may also be possible to use biosignal during actual driving and feed it back to product development and serve the design in the coming big-data era. Furthermore, biosignal sensing can be a part of the driver support system such that the measured biosignal is used to adapt the system to the driver state, or to detect the driver's intention.

In this manner, biological measurement is expected to be applied in various development phases of vehicles or in-vehicle systems. Before applying, it is necessary to decide on what should be measured and how within the constraints in the measuring environments, estimate what kind of information can be obtained and approximate the certainty of information, including S/N of the measurement and the state estimation accuracy of the algorithm.

4.4.1.2 Activities of an Organism and Biological Systems

An organism consists of systems that have various functions for interacting with the surrounding outer world (environment), adapting to the environment, and maintaining the individual and species. Activities of an organism include behavioral and physiological adaptations to the environment, such as active coping (fight-or-flight response) and passive coping (freezing response) against changes in the external environment, restoration and repair for maintaining the internal environment of the organism, adaptation to external environmental conditions, nutritional intake behavior, and circadian rhythms. For these activities, an organism has systems, such as the nervous system, sensory system, motor system, circulatory system (vascular system and lymphatic system), digestive system, respiratory system, endocrine system, and immune system. All of these systems are directly or indirectly related to the diverse biological activities mentioned above.

These systems are composed of multi-scale subsystems, ranging from the whole-body level to the genetic or molecular level. They are not independent parallel systems, but are systems to realize an efficient and robust system as a whole by interacting and coordinating with and complementing each other. Biosignal measurement measures the state of these biological systems, but since they are related to various activities of the organism, it should be noted that the measurement is affected by various factors relating to such activities. For example, the cardiovascular system which supplies nutrition and oxygen relates to all activities, so the signal from the cardiovascular system is substantially affected by the state of various activities of the organism.

4.4.1.3 Advantages and Disadvantages of Biosignal Measurement

Features of biosignal measurement are that it enables (a) objective measurement, (b) continuous and quantitative measurement, (c) measurement that does not cause excessive workload, (d) identification of changes which the subject is unaware of, and (e) interpretation and use of indices based on physiological mechanisms. Here (a) means that the measurement does not require the subjective judgement of the observer, but reflects the subjective view of the subject measured. In subjective assessment and behavior observation, the ambiguousness of the assessment criteria as well as individual differences in judgement and variations within an individual are disadvantages as a measure. In the case of physiological indices, on the other hand, as long as the method of deriving the indices is fixed, assessment can be made based on the same criteria every time. Nevertheless, physiological indices have a number of aspects that make them difficult to handle, for example, expression of biological conditions in physiological indices differs with individuals, index values differ greatly among individuals, and the values for an individual change depending on the time of the day and are easily affected by diet and exercise.

Now that (b) is dependent on the physiological amount to be measured and the measurement method, quantitativeness is a useful characteristic as a measure. However, the quantitative biosignal index should be interpreted carefully. The relationship between the human state and the obtained index value is never linear (Fig. 4.13). The relationship is sigmoid (Fig. 4.13b) at best; however, since the high-sensitivity range is limited, extrapolation is not applicable. The relation is often not a

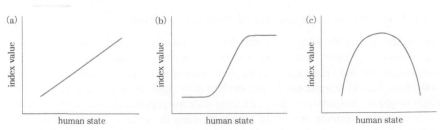

Fig. 4.13: Schematic diagram showing the relationship between the human state and the index value
(a) linear, (b) sigmoid, and (c) inverted U-shape.

monotonic function (such as an inverted U-shape; Fig. 4.13c). In this case, there is a risk of making an opposite interpretation.

Biosignal measurement (c) does not ask for an additional workload on participants as in the case of subjective evaluation or the subsidiary task technique that imposes an additional task. On the other hand, it causes more than a little strain, such as wearing a sensor or restriction on movement or posture; so there is a challenge to how measurement can be made while causing less constraint.

With regard to (d), changes which the subject is unaware of cannot be identified through subjective report and performance can be maintained through effort even if the human state changes. This aspect often causes a gap between changes in physiological indices and results of other indices.

Here (e) is the aim of use of biosignal measurement.

4.4.1.4 Potential of Biosignal Measurement

The potential of biosignal measurement as to what kind of state and characteristics of the driver can be presumed from the respective indices obtained by the measurement are summarized in Table 4.2.

4.4.2 Indices of Central Nervous System Activity

In most of the processes through which a human recognizes the circumstances and behaves appropriately, the cerebral cortex plays the role of a command center. Ever since human electroencephalography was reported in 1929, many matters have been clarified with regard to the correlation between cerebral cortex activity patterns and the mental state or mental process. In particular, the understanding of the brain functions has grown in an accelerated manner from the 1990s onward with the introduction of new non-invasive brain-function-measuring techniques, such as functional magnetic resonance imaging (fMRI) and functional near infrared spectroscopy (fNIRS). In addition, study fields called 'neuroergonomics' and 'neuromarketing' have been cultivated, and there are increasing moves to use the brain function measurement in industrial scenes. In automobile development as well, it may be possible to evaluate changes in the driver state which were not observed through subjective report, behavioral measures and peripheral physiological responses alone, by measuring the activity pattern of the diver's cerebral cortex and interpreting it based on research findings already obtained.

Table 4.2: Potential of Indices Obtained by Biosignal Measurement.

Type of Index	What Internal States can be Identified from the Index (Potential)	Points that Require Caution
Electroencephalogram (EEG) Background EEG Event-related potentials (ERP)	- Arousal level - Degree of relaxation - Habituation - Attention, selective attention - Expectation - Willingness to exercise	- Differences depending on the reference electrode - Even if a certain internal state causes a certain EEG index, the opposite is not necessarily true - Number of trials conducted for measuring the ERP - Relationship between ERPs and background EEG - Photosensitive epilepsy
Gaze/eye movement	- Will to capture visual information (level of skill, vigilance) - Attention span - Interest, concern - Mental tension, anxiety	- Measurement accuracy - Definition of gaze Separation from micro eye movement, gazing area, gaze direction, and point of gaze - Can gazing be regarded as capturing of visual information? - Eye strain
Eye blink	- Arousal level - Concentration of attention (particularly visual task) - Interest in visual stimuli - Frustration, anxiety	- Separation from physiological eye blink - Eye strain
Autonomic nervous system indices	- Surprise, near miss - Tension, excitement - Concentration of attention - Weariness (feeling fed up or jaded)	- Instead of a single index, multiple evaluation is necessary - Multiplicity of autonomic nervous system control (the index value and the internal state do not correspond one-to-one) - Individual differences in reactivity and response patterns
Musculoskeletal system indices	- Voluntary muscle activity: willingness to exercise - Mimetic muscles: emotions - Unnecessary muscle activity: Mental tension - Muscle fatigue	- Electrode positioning - Noise identification

This section focuses on EEG, fMRI, and fNIRS as widely-used non-invasive brain-function-measuring techniques, and outlines their principles and characteristics. Then it also briefly touches on the critical flicker fusion frequency (CFF), which has been used as an index for central nervous system fatigue since long.

4.4.2.1 Electroencephalogram (EEG)

The oldest method for non-invasive observation of human brain-activity patterns is electroencephalography. The electroencephalogram (EEG) amplifies the electrical potential difference of tens of microvolts that occur between two electrodes attached on to the scalp, and records the signals continuously. The cerebral cortex is said to contain more than ten billion nerves cells (neurons) and these cells are connected by synapses to form a network structure. When a neuron receives a signal from another neuron though a synapse, its cell membrane potential changes (post-synaptic potential). The EEG is considered to reflect the post-synaptic potential of the neuron group. The human brain is always active electrically, even during sleep, so potential changes that are measures as EEG are also constantly occurring (stationary EEG). The frequency bands of stationary EEG are known to change, depending on the arousal level.

The frequency bands of EEG are the following, from low to high—the delta band (0.5–4 Hz); the theta band (4–8 Hz); the alpha band (8–13 Hz); the beta band (13–30 Hz); and the gamma band (> 30 Hz). When a human is carrying out work with a sufficient arousal level, activities in the beta band and the gamma band become dominant. However, if the arousal level declines slightly, activities in the alpha band will be observed. If the arousal level declines further to a state of strong sleepiness, activities in the theta band become conspicuous, and from the sleep onset period to the sleep state, activities in the delta band become notable. Many studies have reported that, in automobile driving as well, the amplitude in the alpha, theta, and delta bands increases significantly due to mental fatigue or sleepiness resulting from long hours of driving or sleep restriction (Lal and Craig, 2001). Therefore, stationary EEG is considered an effective index for estimating a driver's arousal state.

In order to evaluate cognitive information processing for events that occur in the external world, it is necessary to obtain the event-related potentials (ERPs). An ERP is a potential change that is evoked when specific cognitive-information processing takes place, and its amplitude is about several µV. Since the signal-to-noise ratio of an ERP is lower than that of stationary EEG, with an amplitude of several tens of µV, which occurs irrelevant to cognitive information processing, there is a need to repeatedly collect EEG data for same events and remove noise by averaging (Fig. 4.14).

ERPs consist of various components depending on their latency (the time from the onset of an event to generation of a potential), polarity, and potential distribution on the scalp. Studies reveal which components reflect what kinds of cognitive information processing. In the case of visual information processing, for example, *P1* and *N1* components, that reflect the initial-stage visual information processing, the *N2* component that reflects cognitive monitoring, and the *P3b* component relating to reorientation of attention for a task and updating of working memory, are evoked. In addition, the *N170* component that arises when detecting a face, mismatches negativity that arises when detecting an unpredicted event, the *N2pc* component that arises when shifting spatial attention, and the *P3a* component that arises when attention is captured exogenously, are also sometimes evoked in visual information processing.

In order to obtain ERPs, it is necessary to identify the onset of an event at an accuracy of milliseconds, but during actual work, such as automobile driving, the onset of an event is often not necessarily clear. Therefore, when estimating a driver's cognitive process, some efforts that differ from measurement within a laboratory would be required. For ERP measurement parameters, etc., see the guidelines published by the Society for Psychophysiological Research (Picton et al., 2000).

Table 4.3 shows a list of EEG components that may be used as indices.

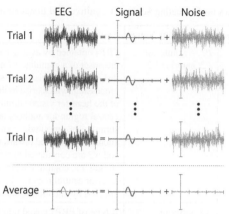

Fig. 4.14: Averaging for ERPs. Since an ERP (signal) has smaller amplitude than stationary EEG (noise), the potential change associated with the processing of an event cannot be evaluated by observing the EEG. ERPs occur synchronously with the onset of an event, whereas stationary EEG is asynchronous; so when EEG data is measured several times for same events and is averaged, the amplitude of ERPs does not change, but that of stationary EEG approaches zero. ERPs are calculated by such averaging and the cognitive process is estimated.

4.4.2.2 Functional Magnetic Resonance Imaging (fMRI)

The method of measuring the electrical activity of neurons as EEG provides higher temporal resolution (the accuracy of identifying 'when' a brain activity occurred) but lower spatial resolution (the accuracy of identifying 'where in the brain was active) in principle. Cerebral cortex is known to have different functions by its area (functional localization). Precise investigation of which area of cerebral cortex was active becomes an important clue in investigating what kind of cognitive process is involved.

One of the methods for estimating the activity area with high spatial resolution, although the temporal resolution is low, is functional magnetic resonance imaging (fMRI), which measures local blood flow changes in the brain. It is known that when neurons in the brain become active, the blood flow increases near the activity area, and the percentage of deoxygenated hemoglobin in blood decreases. The fMRI estimates the activity area in the brain based on changes in magnetic resonance signals, taking advantage of the magnetic property of deoxygenated hemoglobin. The spatial resolution of fMRI is said to be around 1 mm. The fMRI is frequently used in the basic brain science field, but the equipment is large and the head of the person observed needs to be fixed to the equipment. So it is not suitable for applied studies where measurement during actual work is required. Even so, there has been a study report in which a driver's brain activity patterns have been identified by using simple driving simulation (Calhoun et al., 2002), and fMRI is expected to yield results also in the field of applied studies.

4.4.2.3 Functional Near Infrared Spectroscopy (fNIRS)

Functional near infrared spectroscopy (fNIRS) is another method that evaluates blood flow changes in the brain and is similar to fMRI. The fNIRS is a method to emit near-

Table 4.3: EEG Indices for Evaluating Sensory, Cognitive, and Behavioral Processes and States.

	Cognitive Process	Component Name	Explanation
Indices for sensory information processing or attention	Sensory information processing/ attention	*P1*, *N1*, anterior *P2*, anterior *N2*	ERPs obtained by averaging EEGs time-rocked to the onset of a stimulus. *P1* and *N1* are components that reflect the initial-stage sensory information processing, which arise in the posterior region of the head for visual stimuli and in the anterior central region for auditory and somato-sensory stimuli. The amplitude is known to increase when attention is directed toward stimuli. Anterior *P2* and *N2* are considered to reflect the distinction of stimuli and cognitive monitoring, and arise in the anterior central region. These components are also affected by attention to stimuli.
	Shift of spatial attention	*N2* posterior contralateral (*N2pc*)	A type of ERP evoked upon a horizontal shift of attention. When shifting attention to the left (*right*), *N2pc* arises in the posterior temporal region of the right hemisphere (*left* hemisphere).
	Detection of unpredicted events	Mismatch negativity (MMN)	A type of ERP evoked when the predicted event does not match (mismatches) the occurred event where events occur in a series. MMN arises in the posterior temporal region for a visual event and in the anterior central region for an auditory event.
	Attention to visual features	Selection negativity (SN)	A type of ERP evoked when visual stimuli with attended features (such as a specific color or driving direction) appear. It arises in the posterior region.
	Capture of attention	*P3a*	A type of ERP evoked when attention is captured exogenously by notable stimuli. It arises in the centro-parietal region (slightly more anterior than later-mentioned *P3b*).
	Capture of visual information after eye movement	Eye-fixation-related potential (EFRP)	ERPs obtained by averaging based on the time of the end of saccadic eye movement (the start of gaze). It reflects capture of visual information at a new gaze position. The amplitude of its *P1* component is known to increase when attention is directed to visual information.
	Attention to objects	Steady-state visual evoked potential (SSVEP)	When observing a blinking visual stimulus (e.g., 20 Hz), potential changes at the same frequency are observed in the posterior region. Such synchronization is called SSVEP. When attention is directed to a specific visual object, the SSVEP synchronized with the blinking frequency of the object is known to increase.
	Face detection	*N170*	A type of ERP evoked when detecting face stimuli. It arises in the posterior temporal region (predominantly in the right hemisphere). It is also known to be evoked by visual stimuli similar to face stimuli (e.g., an image of a front view of an automobile).

Table 4.3 contd. ...

...Table 4.3 contd.

	Cognitive Process	Component Name	Explanation
Indices for consciousness and degree of concentration	Allocation of attention resources and updating of working memory	*P3b*	A type of ERP associated with allocation of attention resources for a task and updating of working memory. It arises in the parietal region. *P3b* widely reflects higher cognitive processes, and is the most frequently used ERP as an index for investigating the cognitive function. It is sometimes called *P300*.
	Detection/ evaluation of a missing stimulus or response	Missing stimulus potential (MSP)	A type of ERP evoked when the occurrence of an event is incidentally missing where events occur in a series or where events occur by the subject's own behavior. Its most conspicuous component is the positive component that arises in the parietal region. It is considered to reflect the process of detecting or evaluating any missing event.
	Attention to emotional arousal stimuli	Late positive potential (LPP)	A type of ERP obtained by averaging based on the time of the onset of a stimulus that arouses an emotion (particularly an unpleasant stimulus). It arises in the parietal region. It is considered to reflect the changes in the state of attention in line with the processing of emotional stimuli.
	Retention of working memory	Contralateral delayed activity (CDA)	A type of ERP evoked while retaining visual information in the working memory. It is observed as potential changes in the posterior temporal region on the opposite side of the spatial position of the memorized information (in the *left* hemisphere if having memorized information in the *right* visual field).
	State of preparation before taking an action	Contingent negative variation (CNV)	A type of ERP that occurs between a preparatory signal (get set) and a response stimulus (go) where the signal and the stimulus are presented sequentially. It arises in the anterio-central region. Its amplitude changes depending on the arousal level and the state of attention.
	Shift from a resting state to a working state	Alpha-blocking	When mental work is started after a resting state, the EEG amplitude in the alpha band (8–13 Hz) attenuates. This is called alpha-blocking. It is used as an index, indicating a shift from a resting state to a working state.
	Concentrated state	Frontal-midline theta (fm theta)	EEG in the theta band (4–8 Hz) which arises in the anterior region (near the midline) in a state where the subject is concentrating in work. It is used as an index of the degree of concentration, but *fm theta* does not arise for some people even in a concentrated state.

Table 4.3 contd. ...

...Table 4.3 contd.

	Cognitive Process	Component Name	Explanation
Indices for preparation for an action and self-monitoring	Preparation for an action	Lateralized readiness potential (LRP)	A type of ERP obtained by averaging based on the time of an action such as button pressing or the onset of a muscle activity before an action. It is observed as potential changes in the central region on the opposite side of the moving limb (in the left hemisphere in the case of an action using the *right* hand) before the start of the action. It is considered to reflect brain activity relating to preparation for an action, and its latent time serves as an index for the time necessary from the start of preparation for an action to the actual action.
	Imaging of an action	Mu rhythm	The EEG rhythm in the alpha band (8–13 Hz) around the central sulcus is called *mu* rhythm, and its amplitude is known to be attenuated by performing an action or imaging an action. This amplitude attenuation is called event-related desynchronization (ERD), and is sometimes used as a signal for brain-machine interface.
	Detection of an error after an action and correction of behavior	Error-related negativity (ERN), error positivity (*Pe*)	A type of ERP obtained by averaging based on the time of an action such as button pressing or the onset of a muscle activity before an action. ERN arises in the anterior central region immediately after (recognizing about) having made an erroneous reaction. It is considered to reflect detection of an error or reaction interference. *Pe* is a type of ERP evoked following ERN. It arises in the centro-parietal region. It is considered to reflect detailed evaluation of an error and correction of behavior. These components are effective indices for investigating the state of self-monitoring of the subject's own action.
	Recognition of feedback of a negative result	Feedback-related negativity (FRN)	A type of ERP obtained by averaging based on the time of the onset of a stimulus of feedback of a negative result presented after a reaction (feedback indicating that the choice of reaction was erroneous). It is considered to reflect processes similar to those of the above-mentioned ERN.

infrared light from the scalp and measure the reflected light on the scalp (Fig. 4.15). It estimates changes in the concentrations of oxygenated hemoglobin and deoxygenated hemoglobin in the cerebral cortex based on differences in their absorption spectra. Changes in the hemoglobin concentration in blood in the brain are associated with the brain activity areas as mentioned above. In the case of fNIRS, the near-infrared light emitted only reaches a certain depth, so only blood flow changes in the surface layer of the cerebral cortex can be measured.

While the spatial resolution depends on the distance between the emission probe and the detection probe, a distance of about 30 mm (in the case of an adult) is required

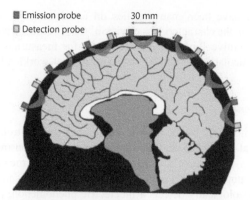

Fig. 4.15: Schematic of brain function measurement by fNIRS. The measuring equipment has multiple emission probes and detection probes. Near-infrared light emitted from an emission probe is absorbed and scattered in the head, including the cerebral cortex, and its reflected light is measured by the detection probe.

for measuring the reflected infrared light that reaches the cerebral cortex. Therefore, compared to fMRI, fNIRS provides lower spatial resolution, but its equipment is compact (some are small enough to be vehicle-mountable) and allows freer movement of the head. These characteristics of fNIRS are extremely advantageous in applied studies where measurement during actual work is required, and there has been a report of a study evaluating the driver state by using a high-fidelity driving simulator (Tsunashima and Yanagisawa, 2009). However, since the measurement is affected by such factors as the probe stability, hair, and scalp blood flow, attention is required when measuring and interpreting the results.

4.4.2.4 Critical Flicker Fusion Frequency (CFF)

While critical flicker frequency (CFF) for flicker stimuli is not a method for directly measuring brain activity, it has been used since 1940s in industrial scenes as an index reflecting central nervous system fatigue (Brozek and Keys, 1944). CFF is obtained by gradually lowering (raising) the blink frequency of visual stimuli and measuring the frequency of the critical point at which the blink becomes identifiable (becomes unidentifiable). The CFF is known to decline due to mental fatigue, and the relationship between the variation pattern of a driver's CFF and occurrence of accidents has been discussed in 1950s–1970s (Yajima et al., 1976). The physiological mechanism of the change in CFF is unclear, but it is considered that CFF reflects the activity status of the cerebral cortex because the fusion threshold of the nerve response to flicker stimuli becomes lower from the retina towards the nerve center. Since CFF has been used in an enormous number of study reports in an industrial setting and is easier than the method of directly measuring brain activity patterns, it is considered an effective index for evaluating the state of the central nervous system.

4.4.3 Indices Relating to the Visual System

Vision is the most important sense for obtaining information for driving, and nearly 90 per cent of the external world information necessary for driving is said to be visual information. There have been many studies on car driving and use of in-vehicle equipments, based on the characteristics of vision. Also, because visual characteristics are affected by cognitive function, they are used for measuring and evaluating a driver's states (particularly the mental state and mental workload), creating a new study field.

4.4.3.1 Eye Movement

Eye movement is measured by the optical method and the electrooculograph (EOG) method. The optical method is classified into corneal reflection method and limbus tracking method. In the case of a device that adopts image measurement, such as video imaging, the pupil diameter can also be measured at the same time. The EOG method electrophysiologically detects a bioelectric phenomenon, and in the case of measuring eye movement, it detects the difference in the electrostatic capacitance of the cornea and that of the retina by sensors attached around the eyes. In particular, the optical method is capable of measuring the point of gaze and the distance and speed of gaze movement with high accuracy. It has wide application possibilities, including evaluation of how a driver is trying to capture visual information and evaluation of a driver's information-processing workload level and attention level (Holmqvist et al., 2011).

Eye movement is mainly classified by whether the movement is voluntary or involuntary. Voluntary eye movement is the binocular eye movement that occurs when intentionally changing the gaze direction. It is called 'conjugate movement' when the two eyes move in the same direction, and 'disjunctive movement' when they move in different directions. Conjugate movement includes saccade (saccadic eye movement) where the movement is high speed, and smooth pursuit where the movement is low speed. Disjunctive movement includes convergence where the two eyes rotate inwards when shifting the point of gaze in the far-near direction while keeping the same gaze direction, and divergence where the two eyes rotate outward.

On the other hand, involuntary eye movement occurs unintentionally, reflexively, or constantly, and includes vestibulo-ocular reflex (VOR), which occurs in response to movement of the head or body, and miniature eye movement, which is eye movement during gaze. The former VOR is a reflex of rotating the eye in the opposite direction from the head rotation when rotating the head. It plays the role of stabilizing gaze within a space, fixing the image on the retina and correctly conveying the visual information to the brain. There is a study that uses this VOR as an index to conduct quantitative evaluation of driver distraction (Obinata et al., 2008) and examine a technique to detect signs of sleepiness (Hirata et al., 2009). Meanwhile, the latter miniature eye movement refers to the constant miniature movement of the eye that is observed when a human is gazing at one point, and it does not affect stereoscopic vision.

Many studies use the voluntary eye movement as an index for mental workload. The following are some of them. A study focused on the state of the driver and indicated that when cognitive workload increases during driving, the shift of the gaze slows down and is detrimental to the visual search behavior (Recarte and Nunes, 2000). Another study measured the eye movement while driving a driving simulator and indicated that the duration of eye fixation differs by cognitive workload (Shinoda, et al., 2001). Further, in a study that measured the eye movement during actual driving under various road conditions (a congested road or expressway), the frequency of shift of gaze was found to increase and the duration of eye fixation and the shift distance became shorter when the road was more congested. The result is explained based on the rule that the amount of a driver's information processing resources is constant and on the relationship between the depth and the width of the processing (Miura, 1992).

Under a condition where the task demand at each point of gaze is large during driving, the amount of attention for processing information increases and deep processing is conducted as a result of requiring a large amount of mental resources. But under a condition where the task demand at each point of gaze is small, the amount of attention becomes little, and shallow processing is conducted with a small amount of mental resources. Also, supposing that the capacity of mental resources is constant under a condition that requires deep processing, the amount of attention that can be allocated to the surroundings decreases and the area around the point of gaze for which the driver can process information becomes smaller. Under a condition that only requires shallow processing, the amount of attention that can be allocated to the surroundings increases, and the area for which the driver can process information becomes larger, which corresponds to the width of the processing. The study indicates that the width and depth of processing are in a trade-off relationship in the same manner (Fig. 4.16). In other words, if the width of processing is the range in which visual information can be collected (visual field), the narrowing of the visual field can be explained based on the rule that the amount of a driver's mental resources is constant and on the relationship between the depth and the width of the processing. Details of the visual field are discussed in the next section.

In these evaluations, saccadic eye movement, which is large and quick rotating of eyes in order to shift the point of gaze, is used as the main index. Saccadic eye movement is generally defined as eye movement with a rotational angular velocity of 30°/sec or more and a travel distance of 3.5° or more. With regard to its dynamic property, the relationship between the peak velocity and amplitude of the saccade is a non-linear relationship as expressed by the following formula (Smit et al., 1987):

$$V(A) = V0 \{1-\exp(-\alpha A)H\}$$

where $V(A)$: peak velocity, A: amplitude, $V0$ and α: constants.

In saccadic eye movement, a slight deviation could occur when placing the point of gaze on a visual object. In such a case, a corrective saccade, defined as eye movement with a rotational angular velocity of 30°/sec or more and travel distance of 1.0–3.5° that occurs within 100–150 ms after the saccade, occurs and corrects the

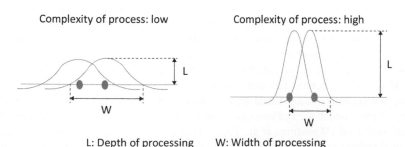

L: Depth of processing   W: Width of processing

Fig. 4.16: Relationship between the width and depth of processing.

deviated point of gaze. In other words, the frequency of occurrence of corrective saccades is also used as an index of mental workload.

4.4.3.2 Visual Field

A human has a visual field of about 200°, but is capable of seeing objects in detail only within the range of the central vision of about 2°. However, the central vision is not the only vision that contributes to cognition. The areas of the peripheral visual field, which are close to the point of gaze and are perceived relatively clearly, also contribute to cognition. Therefore, it is important to know the properties of the peripheral visual field, and many studies have been conducted on the visual field during automobile driving.

The visual field has a number of definitions. They include a collection of visual directions in which the retina is sensitive where the eye is fixed to the primary position (looking straight ahead in a seated position) (definition in the medicine) and the entire space that is in sight at that time (definition in the psychology). In medicine, the spatial range in which a prescribed circular visual target can be identified from a uniform background while gazing at a fixation point with monocular vision and fixing the eyeball direction is defined as a stationary field. In perceptual psychology, this is classified into a kinetic field and a static field based on the measurement method. The range in which visual information can be collected, which is effectively used at work, is called a 'useful field of view' or 'working conspicuity field'. The range of this useful field of view is not absolutely decided, but changes according to the visual task or work conditions.

Previous studies on the useful field of view during automobile driving include a study that measured the range in which the visual target can be perceived when a driver's visual load has increased by way of dividing the view into equal areas and identifying the range in which the perception rate was significantly high. A study that evaluated such a range based on the area in which a driver can react to light spots was presented during actual driving (Ikeda and Takeuchi, 1975). Both these studies indicated that the useful field of view shrank with increase in the visual load, such as road congestion. However, these studies have not necessarily used accurate definitions of the visual field or accurate measurement/evaluation methods, such as assuming the useful field of view to be circular in advance and using the radius from the fixation point as the scale of range.

In contrast, a method has been proposed to quantitatively measure the useful field of view by regarding the distance from the point of gaze to the visual target as the strength of stimuli, estimating the psychometric curve that represents the functional relationship between that distance and the detection probability, and obtaining the point of subjective equality with a detection probability of 50 per cent, defining such point as the boundary of the useful field of view (Yamanaka et al., 2006) (Fig. 4.17).

Moreover, a method has been proposed to estimate the useful field of view during automobile driving based on the saccade amplitude and distribution of the points of gaze, which are eye movement indices that are affected by an increase in the mental workload (Morishima et al., 2015). This method is considered to be a highly practical technique because it can estimate the useful field of view without requiring presentation of and reaction to visual targets, which was necessary in conventional measurement methods.

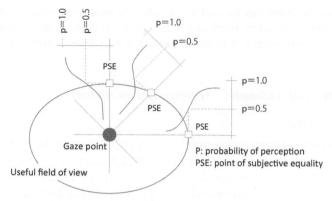

Fig. 4.17: Visual field estimation method using a psychometric curve.

4.4.3.3 Eye Blink

Eye blink can be measured by electrical potential changes in the vertical direction, using the EOG method or by image processing. In studies on a driver's eye blink, the arousal level is generally evaluated and indices, such as the eye-closing time, blink frequency (rate), and blink waveform are proposed. A relationship between the blink waveform and the arousal state is that a small blink of closing the eye from an half-open state or opening the eye after incompletely closing it occurs in the initial state of arousal lowering, and a large blink occurs when the subject realizes the arousal lowering and is struggling against sleepiness. Furthermore, a small and slow blink of the eyelid, moving from a half-open state, arises when the arousal lowering progresses and the subject is unable to struggle against sleepiness.

Eye blink is a useful index for not only evaluation of the arousal state; many studies suggest the possibility that the blink rate and blink waveform characteristics will serve as useful indices, reflecting the cognitive state and the attention state. There have been reports of a study indicating that an increase in the blink rate corresponds to the end of information processing and a study indicating that, for a task involving external attention, the blink rate decreases as the task difficulty increases, and for a task involving internal attention, the blink rate increases as the task difficulty increases (Veltman and Gaillard 1996). Eye blink can also be used for evaluating the cognitive workload on the driver, in combination with findings such that eye blink is restrained when trying to capture visual information from the external world.

4.4.3.4 Pupil

The pupil is the part surrounded by the iris. Humans have a pupil of an oval shape, close to a perfect circle. The light reaction, where the pupil becomes smaller on light entering the eye is called the 'pupillary reaction'. There is also the near reaction where the pupil becomes smaller in line with convergence when the subject looks at a near target. Moreover, since the pupillary sphincter for constriction of the pupil is controlled by parasympathetic nerves and the pupillary dilator is controlled by sympathetic nerves, the pupil is constricted when the subject's mental activity declines due to strong sleepiness or fatigue, and is dilated when the subject becomes excited.

Accordingly, the pupil can be used as an index for evaluating mental activity, but since the change caused by brightness occurs larger than the change caused by the autonomic nervous system, it is difficult to use it for mental activity in real-world settings.

4.4.4 Indices of Autonomic Nervous System Activity

4.4.4.1 Heart Rate

A waveform of an electrocardiogram (ECG) is shown in Fig. 4.18. Interval of the peak of the QRS wave, which reflects contraction of the left ventricle of the heart, is called the RR interval (RRI), based on which the heat rate (HR) per minute is calculated. An ECG is easily obtained by attaching electrodes to two points on both sides of the heart and an area for grounding, while conducting differential amplification. The HR can be measured as long as the electrodes are attached to two areas with phase difference in the propagation of potential from the heart, not only in the chest part or on the four limbs, but also on the shoulder, face, or the head. The HR can also be measured based on heart sounds, ballistocardiogram, or rib cage movement. Studies are under way to carry out measurement that causes less burden by using capacitive coupling-type electrodes, microwaves, and distance sensors.

The HR increases upon activation of the cardiac sympathetic nervous system or activity reduction of the cardiac parasympathetic nervous system; it decreases in the opposite case. Generally, the HR increase is used as an overall index for physical and mental strain while HR decrease is used as an index for a relaxed state and a decline in the arousal level. There is also an attempt to separate HR changes beyond that required for physical activity, as those resulting from psychological factors (Blix, 1974). With regard to mental workload, the degree of HR increase differs, depending on the characteristics of the workload and the coping action. It is said that the HR decreases when the subject intends to capture information by focusing attention on the external world, and increases when the subject focuses attention to inner activity or intends to block information coming from the environment (Lacey and Lacey, 1978). The HR increases as a result of defective reflex (DR), and decreases as a result of orienting reflex (OR) which is the 'what is it?' reflex to novel stimuli.

Even where the stimuli are the same, the response differs, that is, whether the subject feels that his/her life is in danger. For example, a mere difference in the intensity of sound decides whether the response will be DR or OR. Even where DR would occur in an on-road experiment, OR could occur in a laboratory experiment, and thus a difference may occur in the results.

The HR increase has been reported where mental effort is required in laboratory-simulated work or in an actual setting that involves tension, and HR decrease is reported

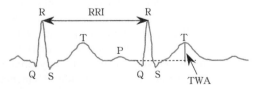

Fig. 4.18: ECG waveform and the obtainable indices.

in the case of fatigue, monotony, and low arousal level. The driver's excitement, anger, and frustration caused by traffic conditions or bad manners of other drivers are considered to show an HR increase. Temporary HR increase is observed immediately following a sudden disturbance that provokes feeling of danger, and sustainable HR increase is observed during difficulty (Kuriyagawa et al., 2009).

4.4.4.2 Heart Rate Variability (HRV) Indices

Variation of instantaneous HR (or RR interval [RRI]) is called heart rate variability (HRV). Frequently used HRV indices are the high-frequency component (HF, about 0.15–0.4 Hz), which are mainly variations associated with respiration, the mid-frequency component (MF, 0.08–0.12 Hz), which is associated with blood pressure Mayer waves, and the low-frequency component (LF, 0.04–0.15 Hz), which contains components of lower frequency than the MF component (Fig. 4.19).

The HF often increases when the subject is relaxed and decreases due to mental workload. The main cause is considered to be reduced baroreceptor reflex sensitivity, but a rise in respiration frequency also has a significant influence. It is referred to as an index of cardiac parasympathetic nerves, but it is better to consider it as reflecting the level of the resting level of the cardiorespiratory system. The LF (MF) also reflects mental effort and decreases due to reduced baroreceptor reflex sensitivity when work involves high mental workload or information processing workload. But when sympathetic nerves are active and an increase of Mayer waves has a larger influence, then the LF (MF) increases. It also increases due to boredom with monotonous work or mental fatigue, and the major cause for this is irregular respiration. The principal cause for LF (MF) increase at the time of speaking, becoming motion sickness, or to struggle against sleepiness is also irregular respiration (an increase in the number of temporary large breathing). The LF or LF/HF is sometimes used as an index for the sympathetic and parasympathetic nervous system balance, but simple interpretation is risky because change factors are diverse. For determining the sympathetic and

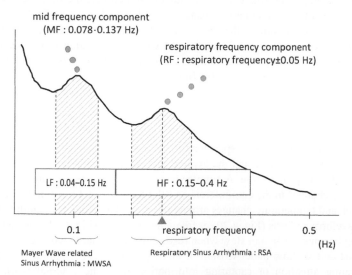

Fig. 4.19: HR spectrum and HRV indices.

parasympathetic nervous system balance, HR changes should also be evaluated first. If the HR does not increase significantly, the LF/HF increase should not be interpreted as the cause for predominance of the sympathetic nervous system, and other factors (such as irregular respiration or speaking), should be considered.

The definitions (frequency bands), and quantifying methods of HRV indices vary among researchers, which makes comparison of research results difficult. Real-world settings are not in a steady state, and involve many work-dependent changes, so how the segment for analysis is selected has a larger influence on the results than differences in the analysis techniques.

4.4.4.3 Blood Pressure and Pulse Waves

Blood pressure is an objective variable of the control system of autonomic nerves and contains important information. The methods for non-invasive, continuous measurement of blood pressure include applanation tonometry and the volume-compensation method, but both pose many challenges in order to be applied to real-world settings. There are studies on indirect measurements using pulse waves to estimate blood pressure.

Pulse waves are measured by attaching a photoelectric sensor to a fingertip or earlobe. The pulse rate can be obtained from pulse waves. The pulse rate is highly correlated with HR, and in many cases, they are interchangeable. However, the pulse rate differs from HR in a strict sense as it reflects the transmission characteristics of blood vessels. Since it has a disadvantage of poor accuracy in cycle calculation, although this depends on the algorithm, caution is required when using the pulse rate in place of HR when obtaining HRV indices. For stable measurement, it is necessary to keep the sensor attaching position and the level of pressure on the skin the same. In the actual environment, the influence of sunlight also causes a problem.

When peripheral sympathetic nerves are activated due to emotion or mental work, the blood vessels become constricted, while the pulse waves become round and small and the pulse amplitude (PLA) becomes lower. Sometimes, the baseline fluctuation increases. In some cases, contrary changes are observed, such as PLA lowering at the fingertip but increasing at the earlobe and the head. In the case of OR, vascular constriction is observed at the fingertips and vascular dilatation in the head. These are considered to be purposeful reactions for securing the necessary cerebral blood flow for information processing by the brain. When startling stimuli are presented, a temporary PLA decrease as well as a baseline decline are observed, indicating DR.

4.4.4.4 Respiration

With regard to respiration, the respiration flow rates are measured by having the subject wear a mask or mouthpiece and the expired gas is analyzed in basic studies, but the burden on the subject is large. A simpler way to measure respiration is to estimate based on the changes in the circumference or the forward-backward movement of the body trunk and temperature changes around the nose. The index is quantified through either waveform analysis (Fig. 4.20) or frequency analysis (Fig. 4.21) for each breath. Breathing becomes deep and slow when resting; deep and large when under physical strain; fast and shallow under tension or anxiety; and could temporarily stop when concentrating attention or capturing information from the external world. When

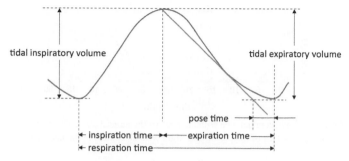

Respiratory driving = tidal inspiratory volume / respiration time
Respiratory timing = inspiration time / respiration time

Fig. 4.20: Respiratory waveform and the obtainable indices.

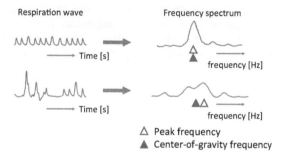

△ Peak frequency
▲ Center-of-gravity frequency

Fig. 4.21: Frequency analysis of respiratory waveform.

boredom with monotonous work, sleepiness, or motion sickness occurs, temporary large breathing gets mixed frequently, and respiration becomes more irregular. While the respiration rate and regularity while resting and patterns of respiratory changes in response to a sudden event or near crash differ greatly among individuals, they are relatively reproducible within the same individual.

4.4.4.5 Electrodermal Activity

The electrodermal activity (EDA) is obtained by capturing electric changes of the skin associated with sweat-gland activity controlled by peripheral sympathetic nerves. It used to be called galvanic skin response (GSR) long ago, but in recent years, it has been classified into skin potential, skin resistance, skin conductance, skin impedance, and skin admittance according to the measurement method. Each has two types of names for level change (level [L]) and temporary change (response [R]). For example, skin conductance (SC) has two names: SCL and SCR. The EDA does not necessarily coincide with the measurement result of a perspiration meter.

The electrodermal level of an emotional sweating area (such as the palm or the sole) reflects the arousal level and is used for evaluating the arousal level of the driver. Various stimuli and emotions cause temporary electrodermal responses. The EDA does not change its direction by DR or OR, as in the case of heart rate, and its reaction is in the direction of increased arousal level both for DR and OR. Figure 4.22 shows an example of changes in skin conductance with changes in the arousal state.

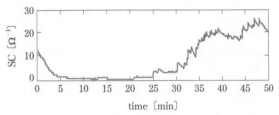

Fig. 4.22: Changes of skin conductance with changes in the arousal state. The arousal level was low in the first half of the period and flight with sleepiness in the last half.

4.4.4.6 Skin Temperature

Skin temperature is measured by having the subject attach a thermistor or thermocouple to the skin surface. Non-contact measurement is also possible by using an infrared camera (thermography). It is necessary to ensure that the measuring part is not exposed to direct sunlight or wind. The skin temperature increases when the skin blood flow increases, but this relationship is not linear. It is used for evaluating thermal comfort and estimating the warm/cold feeling. When peripheral sympathetic nerves are activated due to mental stress or emotion, and the blood vessels are constricted, the blood flow rate decreases with decrease in the peripheral skin temperature. In reverse, the temperature increases when relaxed or when the arousal level declines. A decrease of skin temperature caused by mental workload is observed notably at the nose. In order to reduce the influence of the environment temperature, the temperature difference between the nose and an area that is closely linked to changes in the body trunk, such as the forehead, can be used as an index. OR also causes a decrease of the peripheral skin temperature associated with blood vessel constriction, but it is not suitable to use this phenomenon for identifying the correspondence with events because the time constant is long. Temperature also decreases due to a large breath, such as a sigh, or smoking.

4.4.5 Facial Expression

We often say that anger or fatigue shows on one's face. People have the ability to guess another person's emotions or physical conditions from that person's facial expression. In recent years, research and development have been conducted on systems to detect fatigue or sleepiness from the driver's facial image, which is precisely an approach to estimate the driver's physical condition from his/her facial expression. There is a possibility that various technologies for estimating the driver state from facial expressions will develop in the future. This section reviews conventional findings from studies on facial expressions that would be useful for such technology development.

4.4.5.1 Anatomy of Mimetic Muscles

Anatomical studies have revealed the parts of mimetic muscles that create expressions and their actions (Faigin, 2012). Mimetic muscles, which are also called facial muscles, are characterized by being fixed to the bone or tendon on one end and fixed to the skin or another muscle on the other end. Therefore, when a muscle contracts,

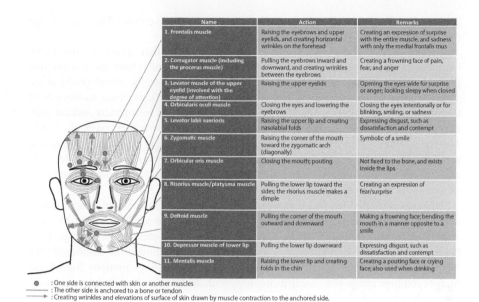

Name	Action	Remarks
1. Frontalis muscle	Raising the eyebrows and upper eyelids, and creating horizontal wrinkles on the forehead	Creating an expression of surprise with the entire muscle, and sadness with only the medial frontalis mus
2. Corrugator muscle (including the procerus muscle)	Pulling the eyebrows inward and downward, and creating wrinkles between the eyebrows	Creating a frowning face of pain, fear, and anger
3. Levator muscle of the upper eyelid (involved with the degree of attention)	Raising the upper eyelids	Opening the eyes wide for surprise or anger; looking sleepy when closed
4. Orbicularis oculi muscle	Closing the eyes and lowering the eyebrows	Closing the eyes intentionally or for blinking, smiling, or sadness
5. Levator labii suerioris	Raising the upper lip and creating nasolabial folds	Expressing disgust, such as dissatisfaction and contempt
6. Zygomatic muscle	Raising the corner of the mouth toward the zygomatic arch (diagonally)	Symbolic of a smile
7. Orbicular oris muscle	Closing the mouth; pouting	Not fixed to the bone, and exists inside the lips
8. Risorius muscle/platysma muscle	Pulling the lower lip toward the sides; the risorius muscle makes a dimple	Creating an expression of fear/surprise
9. Deltoid muscle	Pulling the corner of the mouth outward and downward	Making a frowning face; bending the mouth in a manner opposite to a smile
10. Depressor muscle of lower lip	Pulling the lower lip downward	Expressing disgust, such as dissatisfaction and contempt
11. Mentalis muscle	Raising the lower lip and creating folds in the chin	Creating a pouting face or crying face; also used when drinking

● : One side is connected with skin or another muscles
— : The other side is anchored to a bone or tendon
→ : Creating wrinkles and elevations of surface of skin drawn by muscle contraction to the anchored side.

Fig. 4.23: Eleven mimetic muscles that create six emotions (sadness, anger, happiness, fear, disgust, and surprise) (Faigin, 2012).

the skin is pulled towards the direction where the muscle is fixed, making wrinkles and raised parts, while creating a facial expression. There are over 30 mimetic muscles. Among these, mainly 11 muscles are involved in the generation of emotional facial expressions of sadness, anger, happiness, fear, disgust and surprise (Fig. 4.23). The levator muscle of the upper eyelid for opening the upper eyelid anatomically belongs to, not a mimetic muscle (controlled by facial nerves), but an eye muscle (controlled by oculomotor nerves). But, since the degree of eye opening is said to reflect the subject's degree of attention (Faigin, 2012), the muscle is largely involved in facial expressions.

Facial-electromyogram (f-EMG) is useful for analyzing the activities of the respective mimetic muscles, and is frequently used in studies on emotions (Dimberg, 1990; Aoi et al., 2011). Facial expression is a system that is extremely sensitive to distinction of emotions, and stronger reactions are shown on the f-EMG when the subject actually feels an emotion, rather than on merely imagining the emotion (Brown and Schwartz, 1980). When the subject feels a positive emotion, such as happiness, the activity of zygomatic muscle increases whereas the activity of the corrugator muscle decreases, and when the subject feels a negative emotion, such as sadness or anger, the opposite tendencies are indicated (Cacioppo et al., 1988; Dimberg, 1982; McCanne and Anderson, 1987). There have been findings that activities of mimetic muscles have an influence on emotions.

4.4.5.2 Relationship Between Facial Expression and Emotion

According to the current established theory (Russell and Fernández-Dols, 1997), seven basic emotions (happiness, surprise, fear, anger, disgust, contempt, and sadness) can be read from facial expressions unless the subject hides, restrains, or leaks the emotion.

This is part of an innate signaling system that is irrelevant to culture, language, or age (Ekman and Friesen, 1971; Izard, 1994).

Physiological and psychological findings indicate that many of the facial expressions are innate, that is, genetic (Darwin et al., 1872) and infants show distinctive physiological or behavioral reactions to facial expressions of the rearer (Field et al., 1982). Reactions to facial expressions with negative emotions, such as anxiety or threat are sensitive (Hansen and Hansen, 1988) and the right hemisphere and the right amygdaloid nucleus of the brain are involved in such reactions (Morris et al., 1998). Moreover, when a stimulus of facial expression, such as anger or smile, is perceived, activities of mimetic muscles corresponding to that facial expression are reflexively evoked (300–400 ms) (Dimberg et al., 2000), and the emotion corresponding to the presented facial expression is perceived (Dimberg, 1988).

Ekman et al. divided facial expressions into units of fundamental actions called action units (AUs), and defined the relationship between the combination of AUs and emotional facial expression as the facial action coding system (FACS) (Fig. 4.24) (Ekman and Friesen, 1978). This has made it possible to estimate the emotional state from facial expressions. Many of the techniques for estimating emotions from facial images combine image processing technology with the FACS, which is the mainstream approach at present.

Russell has mapped 12 facial expressions on two axes 'pleasure–displeasure' and 'arousal' (two-axis spatial configuration of facial expressions) as shown in Fig. 4.25 (Russell, 1980). This suggests the possibility of estimating the subject's facial expression from his/her arousal level and level of pleasure/displeasure.

AU descriptions and related muscles

1 Inner Brow Raise	1. Frontalis muscle
2 Outer Brow Raise	2. Frontalis muscle
4 Brow Lowerer	4 Corrugator muscle
5 Upper Lid Raise	5 Orbicularis oculi muscle
6 Cheek Raise	6. Orbicularis oculi muscle
7 Lids Tight	7. Orbicularis oculi muscle
9 Nose Wrinkle	9 Levator labii suerioris alaquae nasi muscle
10 Upper Lip Raiser	10 Levator labii superioris muscle
11 Nasolabial Furrow	11 Zygomatic minor muscle
12 Lip Corner Puller	12 Zygomatic major muscle
13 Sharp Lip Puller	13 Levator anguli oris muscle
15 Lip Corner Depressor	15 Depressor angeli oris muscle
16 Lower Lip Depress	16 Depressor labii inferioris muscle
17 Chin Raiser	17 Mentalis muscle
20 Lip Stretch	20 Risorius muscle
23 Lip Tightener	23 Orbicular oris muscle
24 Lip Presser	24 Orbicular oris muscle
25 Lips Part	25 Relaxation of Mentalis
26 Jaw Drop	26 Relaxation of Masseter
28 Lips Suck	28 Orbicular oris muscle

Six emotional expressions are explained by combinations of AUs

Surprize : AU1 + AU2 + AU5 + AU26

Happiness : AU1 + AU2 + AU5 + AU6 + AU10
+ AU12 + AU13 + AU20 + AU25

Fear : AU1 + AU2 + AU4 + AU5 + AU7
+ AU15 + AU20 + AU25

Anger : AU2 + AU4 + AU5 + AU10 + AU23
+ AU24 + AU28

Sadness : AU1 + AU4 + AU7 + AU15

Disgust : AU4 + AU9 + AU10 + AU17

Fig. 4.24: FACS (Facial Action Coding System) (Ekman and Friesen, 1978).

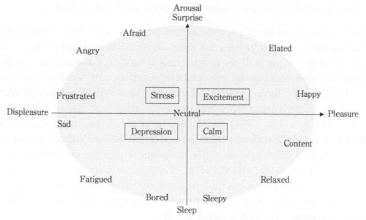

Fig. 4.25: Space for psychological determination of pleasure and arousal (Russell, 1980).

4.4.5.3 Techniques for Estimating Emotions Based on Facial Images

The following five processes are required for recognizing/estimating the subject's facial expression and emotional state from a facial image: (1) extracting the features of the face (matching the image and the respective parts of the face/2D-> 3D/tilt compensation); (2) extracting the deformation of the face (setting the facial feature points and measuring their changes); (3) describing the deformation based on facial expression elements (AU, etc.); (4) recognizing basic emotions (FACS, etc.); and (5) recognizing complex emotions (intentions) (Mase, 1997). This series of processes needs to satisfy three requirements: correctness (high accuracy); stability (robustness); and flexibility (adaptive learning).

Representative techniques for recognizing facial expressions from facial images are the Eigenface method (Turk and Pentland, 1991) and a technique using optical flows (Mase, 1991). The Eigenface method compresses high-dimensional data on the gradation of a facial image into a low-dimensional data by using principal component analysis, etc., to obtain a feature vector, and identifies the facial expression by comparing it with the vector of the registered category. It is used for personal authentication. The method using optic flow recognizes the facial expression by estimating the respective muscle activities from changes in gradations within the detection windows placed at the positions of mimetic muscles of the facial image. Otsuka et al. have studied a technique to recognize facial expressions by tracking feature points on a facial image, obtaining motion vectors through Delaunay meshes (wire frames), and estimating the contraction rates of the respective mimetic muscles based on AU (Otsuka and Ohya, 1999).

Many of the techniques for estimating facial expressions, based on facial images, use two-dimensional image information taken with a single camera. However, since changes in facial expressions occur on the order of several millimeters, measurement errors are likely to occur due to head movement. In order to establish high-accuracy technology for facial-expression recognition, it is necessary to establish estimation technology that uses three-dimensional information.

4.4.5.4 Relationship Between Facial Expression and Driver States

Compared to studies on facial expression and emotion, very few studies have been conducted on the relationship between facial expression and driver states, such as sleepiness or fatigue. This is likely to be because the FACS has no facial expression descriptions for sleepiness or fatigue, and characteristics of facial expressions in a sleepy or fatigue state have yet to be elucidated. However, humans can identify a person's degree of sleepiness from that person's facial expression. Wierwille et al. (Wierwille and Ellsworth, 1994) and Kitajima et al. (Kitajima et al., 1997) have proposed a technique for trained persons to observe a subject's facial image and rate the subject's degree of sleepiness. This technique has some problems concerning the work required for rating and the level of reproducibility since the evaluation is made by a human. However it has an advantage of being able to identify more slight sleepiness than other indices; so it is adopted by many researchers as a reference index for sleepiness detection technique (see 4.3.3.3).

4.4.5.5 Application of Facial Expressions to Automobile and Future Challenges

At present, research and development are being conducted on technology to estimate the driver state by processing the driver's facial image. For example, Hachisuka et al. have developed a technique to automatically rate the degree of sleepiness based on a facial image on a scale of six levels (Fig. 4.26) (Hachisuka et al., 2011). However, not a few challenges remain to be tackled before its practical application. In order to put into practice facial expression-sensing technology that has the potential for wide application, it is essential to resolve challenges, such as dealing with ambient light and glasses/sunglasses, developing large-scale learning data for distinguishing facial expressions, and advancing optimization of the image processing and distinguishing algorithms, one by one.

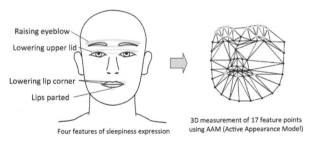

Fig. 4.26: System for estimating the degree of sleepiness from an image.

4.4.6 Biochemical Reactions

Psychological stress is known to affect the endocrine and the immune systems. Biochemical responses, observed in neuroendocrine and immune systems, are measured by collecting body fluids, such as blood, urine, or saliva. When measuring biochemical responses as one of the biological reactions in the field of ergonomics, the fluids often used are saliva and urine, which can be collected non-invasively, particularly saliva, which can be collected frequently.

A schematic of stress-related biological responses is shown in Fig. 4.27. The route can be largely divided into the hypothalamic pituitary adrenal system (HPA system) and the sympathetic adrenomedullary system (SAM system). A response of the HPA system increases the secretion of cortisol from the adrenal cortex into the blood, and results in responses, such as activation of energy production through metabolic stimulation, an increase in cardiac contractility and blood pressure, and an increase in gastric acid secretion. A response of the SAM system increases the secretion of catecholamine (such as adrenaline) from the adrenal medulla into the blood. The stress response of the SAM system is extremely sharp, and it triggers emergency reactions, such as increased energy supply, increased respiration, and increased cardiac output. Since the immune system changes by being affected by the HPA and the SAM systems, stress affects immune function.

In recent years, with an indication that hormones of the HPA system also activate the SAM system, it has been inferred that the two systems related to stress responses work coordinately by forming a positive feedback loop with each other. On the other hand, as an adjusting mechanism for preventing the systems from going out of control, both these two systems adjust through negative feedback, such as being suppressed by cortisol. Substances produced by immuno-competent cells are known to affect the central nervous system. The central nervous system, autonomic nervous system, endocrine system, and immune system are considered to be a composite system and react to stress and maintain homeostasis.

When measuring urine constituents, high-performance liquid chromatography (HPLC) is used, and endocrine indices—catecholamine and cortisol—are measured. Sluiter et al. evaluated long-distance drivers' fatigue and recovery by measuring adrenaline (epinephrine), noradrenaline (norepinephrine), and cortisol in the urine (Sluiter et al., 1998). Taguchi et al. who focused on the cross-correlation between the subjective fatigue and various physiological measures, extracted adrenaline in the urine as an index of fatigue that accumulated during long hours of driving, and

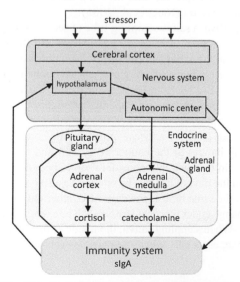

Fig. 4.27: Biological responses associated with stress.

evaluated the influence of vertical vibrations on a driver's fatigue while driving by using this index (Taguchi and Inagaki, 1999).

From saliva, cortisol, which is an HPA system endocrine index, and secretory immunoglobulin A (sIgA), which is an index of immune function, are measured. sIgA is generally recognized to decrease under chronic stress and increase under acute stress. However, changes in sIgA are considered to be influenced by the stress controllability and the person's character and behavioral patterns. An increase in sIgA is sometimes regarded as an index of health benefits since there is, for example, a study indicating that positive emotional states and music enhance immunity and increase sIgA (McCraty et al., 1996). The concentration of salivary cortisol is correlated with the concentration of free cortisol, which is one type of cortisol in blood that could serve as a stress index (Obmiński et al., 1997), and has been frequently used as an index for evaluating mental stress. Fouladi et al. compared salivary cortisol on a work day and leisure day, and reported that industrial noise has an effect on salivary cortisol elevation (Fouladi et al., 2012).

Catecholamine, which is a SAM system endocrine index, is difficult to measure from saliva. However, since methods to measure chromogranin A (CgA) and α-amylase, which are secreted in the saliva by the effect of the sympathetic nervous system, have been established, these substances are expected to serve as mental stress indices that promptly react to sympathetic activity.

Takatsuji et al. compared the salivary cortisol, sIgA, and CgA of nursing students before and after an examination, and found that sIgA and CgA increased due to acute stress from the examination, but cortisol did not increase (Takatsuji et al., 2008). Sakakibara et al. used a driving simulator to analyze salivary CgA and sIgA, subjective assessment, and driving behavior indices for driving tasks with different difficulty levels. They concluded that salivary CgA reflects the feeling of tension during driving and sIgA reflects psychological factors, such as the fun of driving and aggressiveness (Sakakibara and Taguchi, 2003). Nater et al. reported that salivary α-amylase increased when comparing between before and after the Trier Social Stress Test (TSST), which is one type of mental stress test (Nater et al., 2005).

When sampling saliva, a saliva sampling tube (a set of a sponge and tube, known as Sarstedt's Salivette) is used. After putting a sponge into the subject's mouth and letting it soak up saliva, the saliva is centrifuged. The obtained saliva is kept frozen until the time of measuring. For measuring salivary components, the Enzyme Linked Immunosolvent Assay (ELISA), that uses a measuring kit corresponding to each component is widely known and some companies provide measurement services. For salivary α-amylase, instruments that can measure the saliva sampled with a disposable chip in about 60 seconds (salivary amylase monitors) are commercially available.

When measuring indices of the endocrine system and the immune system, the following points should be taken into consideration (Obayashi, 2013). Many indices change depending on the time of the day, so when measuring an index once per day, it must be measured at the same time of the day. Also, attention must be paid to the fact that some indices change under physical stress (such as exercise stress) or are affected by a drug or disease. When sampling saliva, the inside of the mouth needs to be rinsed with water in advance, especially when the participant has just had food or drink or is a smoker.

References

Abe, T., T. Nonomura, Y. Komada, S. Asaoka, T. Sasai, A. Ueno and Y. Inoue. (2011). Detecting deteriorated vigilance using percentage of eyelid closure time during behavioral maintenance of wakefulness tests. International Journal of Psychophysiology, 82(3): 269–274.

Åkerstedt, T. and M. Gillberg. (1990). Subjective and objective sleepiness in the active individual. International Journal of Neuroscience, 52: 29–37.

Aoi, M., M. Kamijo and H. Yoshida. (2011). Relationship between facial expression and facial electromyogram (f-EMG) analysis in the expression of drowsiness. pp. 65–70. *In*: Biometrics and Kansei Engineering (ICBAKE), 2011 International Conference on, IEEE.

Arand, D., M. Bonnet, T. Hurwitz, M. Mitler, R. Rosa and R.B. Sangal. (2005). The clinical use of the MSLT and MWT. Sleep, 28(1): 123–144.

Asakawa, K. (2010). Flow experience, culture, and well-being: How do autotelic Japanese college students feel, behave and think in their daily lives? Journal of Happiness Studies, 11(2): 205–223.

Basner, M. and D.F. Dinges. (2011). Maximizing sensitivity of the psychomotor vigilance test (PVT) to sleep loss. Sleep, 34(5): 581–591.

Basner, M. and J. Rubinstein. (2011). Fitness for duty: A 3-minute version of the Psychomotor Vigilance Test predicts fatigue-related declines in luggage-screening performance. Journal of Occupational and Environmental Medicine/American College of Occupational and Environmental Medicine, 53(10): 1146–1154.

Basner, M., D. Mollicone and D.F. Dinges. (2011). Validity and sensitivity of a brief psychomotor vigilance test (PVT-B) to total and partial sleep deprivation. Acta Astronautica, 69(11-12): 949–959.

Basner, M. and D.F. Dinges. (2012). An adaptive-duration version of the PVT accurately tracks changes in psychomotor vigilance induced by sleep restriction. Sleep, 35(2): 193–202.

Bennett, L., J. Stradling and R. Davies. (1997). A behavioural test to assess daytime sleepiness in obstructive sleep apnoea. Journal of Sleep Research, 6(2): 142–145.

Blix, A.S., S.B. Stromme and H. Ursin. (1974). Additional heart rate—An indicator of psychological activation. Aerospace Medicine, 45(11): 1219–22.

Brown, S.L. and G.E. Schwartz. (1980). Relationships between facial electromyography and subjective experience during affective imagery. Biological Psychology, 11(1): 49–62.

Brozek, J. and A. Keys. (1944). Flicker-fusion frequency as test of fatigue. Journal of Industrial Hygiene and Toxicology, 26(5): 169–174.

Cacioppo, J.T., J.S. Martzke, R.E. Petty and L.G. Tassinary. (1988). Specific forms of facial EMG response index emotions during an interview: From Darwin to the continuous flow hypothesis of affect-laden information processing. Journal of Personality and Social Psychology, 54(4): 592.

Cajochen, C., M. Münch, S. Kobialka, K. Kräuchi, R. Steiner, P. Oelhafen, S. Orgül and A. Wirz-Justice. (2005). High sensitivity of human melatonin, alertness, thermoregulation, and heart rate to short wavelength light. Journal of Clinical Endocrinology and Metabolism, 90(3): 1311–1316.

Calhoun, V.-D., J.-J. Pekar, V.-B. McGinty, T. Adali, T.-D. Watson and G.-D. Pearlson. (2002). Different activation dynamics in multiple neural systems during simulated driving. Human Brain Mapping, 16(3): 158–167.

Chua, E.C.P., W.Q. Tan, S.C. Yeo, P. Lau, I. Lee, I.H. Mien, K. Puvanendran and J.J. Gooley. (2012). Heart rate variability can be used to estimate sleepiness-related decrements in psychomotor vigilance during total sleep deprivation. Sleep, 35(3): 325–334.

Connor, J., R. Norton, S. Ameratunga, E. Robinson, I. Civil, R. Dunn, J. Bailey and R. Jackson. (2002). Driver sleepiness and risk of serious injury to car occupants: population based case control study. BMJ: British Medical Journal, 324(7346): 1125.

Crawford, A. (1961). Fatigue and driving. Ergonomics, 4(2): 143–154.

Csikszentmihalyi, M. (1975). Beyond Boredom and Axiety: Experiencing Flow in Work and Play. Jossey-Bass, San Francisco, USA.

Csikszentmihalyi, M. (1990). Flow: The Psychology of Optimal Experience, Harper and Row, New York, USA.

Czeisler, C.A., J.S. Allan, S.H. Strogatz, J.M. Ronda, R. Sánchez, C.D. Ríos, W.O. Freitag, G.S. Richardson and R.E. Kronauer. (1986). Bright light resets the human circadian pacemaker independent of the timing of the sleepwake cycle. Science, 233(4764): 667–671.

Darwin, C., P. Ekamn and P. Prodger. (1872). The Expression of the Emotions in Man and Animals, John Murray, London, UK.

Dement, W.C. and M.A. Carskadon. (1982). Current perspectives on daytime sleepiness: The issues. Sleep, 5(Suppl 2): S56–66.

Desmond, P.A. and G. Matthews. (1997). Implications of task-induced fatigue effects for in-vehicle countermeasures to driver fatigue. Accident Analysis and Prevention, 29(4): 515–523.

Diener, E. (2009). Subjective well-being. pp. 11–58. In: E. Diener (ed.). The Science of Well-being, Champaign. IL: Springer.

Dimberg, U. (1982). Facial reactions to facial expressions. Psychophysiology, 19(6): 643–647.

Dimberg, U. (1988). Facial electromyography and the experience of emotion. Journal of Psychophysiology, 2(4): 277–282.

Dimberg, U. (1990). Facial electromyography and emotional reactions. Psychophysiology, 27(5): 481–494.

Dimberg, U., M. Thunberg and K. Elmehed. (2000). Unconscious facial reactions to emotional facial expressions. Psychological Science, 11(1): 86–89.

Dinges, D.F. (1992). Adult napping and its effects on ability to function. pp. 118–134. In: C. Stampi (ed.). Why We Nap. Birkhäuster, Boston, USA.

Dinges, D.F., M.M. Mallis, G. Maislin and J.W. Powell. (1998). Evaluation of techniques for ocular measurement as an index of fatigue and as the basis for alertness management. U.S. Department of Transportation, National Highway Traffic Safety Administration, Report No. DOT HS 808 762.

Dinges, D.F., G. Maislim, R.M. Brewster, G.P. Krueger, R.J. Carroll et al. (2005). Safety: Older drivers, traffic law enforcement, management, school transportation; emergency evacuation, truck and bus, and motorcycles. Washington, Transportation Research: Pilot Test of Fatigue Management Technologies, 175–182.

Doran, S.M., H.P.A. Van Dongen and D.F. Dinges. (2001). Sustained attention performance during sleep deprivation: Evidence of state instability. Archives Italiennes de Biologie, 139(3): 253–267.

Dorrian, J., N. Lamond, A.L. Holmes, H.J. Burgess, G.D. Roach, A. Fletcher and D. Dawson. (2003). The ability to self-monitor performance during a week of simulated night shifts. Sleep, 26(7): 871–877.

Edinger, J.D., M.K. Means, A.D. Krystal, J.D. Edinger, M.K. Means and A.D. Krystal. (2013). Does physiological hyperarousal enhance error rates among insomnia sufferers? Sleep, 36(8): 1179–1186.

Ekman, P. and W.V. Friesen. (1971). Constants across cultures in the face and emotion. Journal of Personality and Social Psychology, 17(2): 124.

Ekman, P. and W.V. Friesen. (1978). Facial Action Coding System, Consulting Psychologists Press, Palo Alto, CA., USA.

Faigin, G. (2012). The Artist's Compelete Guide to Facial Expression, Watson-Guptill.

Field, T.M., R. Woodson, R. Greenberg and D. Cohen. (1982). Discrimination and imitation of facial expression by neonates. Science, 218(4568): 179–181.

Fouladi, D.B., P. Nassiri, E.M. Monazzam, S. Farahani, G. Hassanzadeh and M. Hoseini. (2012). Industrial noise exposure and salivary cortisol in blue collar industrial workers. Noise and Health, 14(59): 184–189.

Fredholm, B.B., K. Bättig, J. Holmén, A. Nehlig and E.E. Zvartau. (1999). Actions of caffeine in the brain with special reference to factors that contribute to its widespread use. Pharmacological Reviews, 51(1): 83–133.

Fredrickson, B.L. (2009). Positivity, New York: One World Publications.

Ftouni, S., S.A. Rahman, K.E. Crowley, C. Anderson, S.M. Rajaratnam and S.W. Lockley. (2013a). Temporal dynamics of ocular indicators of sleepiness across sleep restriction. Journal of Biological Rhythms,28(6): 412–424.

Ftouni, S., T.L. Sletten, M. Howard, C. Anderson, M.G. Lenné, S.W. Lockley and S.M. Rajaratnam. (2013b). Objective and subjective measures of sleepiness, and their associations with on-road driving events in shift workers. Journal of Sleep Research, 22(1): 58–69.

Fuller, R. (2005). Towards a general theory of driving behaviour. Accident Analysis and Prevention, 37: 461–472.

Gulian, E., G. Matthews, A.I. Glendon, D.R. Davies and L.M. Debney. (1989). Dimensions of driver stress. Ergonomics, 32(6): 585–602.

Hachisuka, S., K. Ishida, T. Enya and M. Kamijo. (2011). Facial expression measurement for detecting driver drowsiness. pp. 135–144. *In*: International Conference on Engineering Psychology and Cognitive Ergonomics, Springer, Berlin, Heidelberg, Germany.

Hanin, Y.L. (1978). A study of anxiety in sports. pp. 236–249. *In*: W. Straub (ed.). Sport Psychology: An Analysis of Athlete Behavior, Ithaca, NY: Movement Publications.

Hanin, Y.L. (2000). Individual Zones of Optimal Functioning (IZOF) Model: Emotion-performance relationships. pp. 65–89. *In*: Y.L. Hanin (ed.). *In*: Sport. Emotions in Sport, Champaign, IL: Human Kinetics.

Hansen, C.H. and R.D. Hansen. (1988). Finding the face in the crowd: An anger superiority effect. Journal of Personality and Social Psychology, 54(6): 917.

Hayashi, M., A. Masuda and T. Hori. (2003). The alerting effects of caffeine, bright light and face washing after a short daytime nap. Clinical Neurophysiology, 114(12): 2268–2278.

Henelius, A., M. Sallinen, M. Huotilainen, K. Müller, J. Virkkala and K. Puolamäki. (2014). Heart rate variability for evaluating vigilant attention in partial chronic sleep restriction. Sleep, 37(7): 1257–1267.

Hirata, Y., J. Nishiyama and S. Kinoshita. (2009). Detection and prediction of drowsiness by reflexive eye movements, eye movements and pupil fluctuation reflecting alertness and their neuronal mechanisms. Proceedings of 31st Annual Conference of the IEEE EMBS, 4015–4018.

Hoddes, E., V. Zarcone, H. Smythe, R. Phillips and W.C. Dement. (1973). Quantification of sleepiness: A new approach. Psychophysiology, 10(4): 431–436.

Holmqvist, K., M. Nystrom, R. Anderssson, R. Dewhurst, H. Jarodzka and J. Can de Weijer. (2011). Eye Tracking, Oxford University Press, 231–284.

Iizuka, H. and K. Kaida. (2011). The effect of slow wave activity deprivation during an afternoon short nap on performance. Sleep and Biological Rhythms, 9: 409.

Ikeda, M. and T. Takeuchi. (1975). Influence of foveal load on the functional visual field. Perception &.Psychophysics, 18(4): 255–260.

ISO 10075-1. (2017). Ergonomic principles related to mental workload—Part 1: General issues and concepts, terms and definitions.

Izard, C.E. (1994). Innate and Universal Facial Expressions: Evidence from Developmental and Cross-cultural Research. Psychological Bulletin, 115(2): 288–299.

Johns, M.W., A. Tucker, R. Chapman, K. Crowley and N. Michael. (2007). Monitoring eye and eyelid movements by infrared reflectance oculography to measure drowsiness in drivers. Somnologie, 11(4): 234–242.

Johns, M.W., R. Chapman, K. Crowley and A. Tucker. (2008). A new method for assessing the risks of drowsiness while driving. Somnologie, 12(1): 66–74.

Johns, M.W. (2010). A new perspective on sleepiness. Sleep Biol. Rhythms, 8(3): 170–179.

Kaida, K., M. Takahashi, T. Åkerstedt, A. Nakata, Y. Otsuka, T. Haratani and K. Fukasawa. (2006). Validation of the Karolinska sleepiness scale against performance and EEG variables. Clinical Neurophysiology, 117(7): 1574–1581.

Kaida, K., T. Akerstedt, G. Kecklund, J.P. Nilsson and J. Axelsson. (2007). The effects of asking for verbal ratings of sleepiness on sleepiness and its masking effects on performance. Clinical Neurophysiology, 118(6): 1324–1331.

Kaida, K., T. Akerstedt, G. Kecklund, J.P. Nilsson and J. Axelsson. (2007). Use of subjective and physiological indicators of sleepiness to predict performance during a vigilance task. Industrial Health, 45(4): 520–526.

Kaida, K., T. Åkerstedt, G. Kecklund, J.P. Nilsson and J. Axelsson. (2007). The effects of asking for verbal ratings of sleepiness on sleepiness and its masking effects on performance. Clinical Neurophysiology, 118(6): 1324–1331.

Kaida, K., Y. Takeda and K. Tsuzuki. (2012). The relationship between flow, sleepiness and cognitive performance: The effects of short afternoon nap and bright light exposure. Industrial Health, 50(3): 189–196.

Kaida, K., Y. Takeda and K. Tsuzuki. (2013). The effects of short afternoon nap and bright light on task switching performance and error-related negativity. Sleep and Biological Rhythms, 11(2): 125–134.

Kaida, N. and K. Kaida. (2015). Spillover effect of congestion charging on pro-environmental behavior. Environment, Development and Sustainability, 17(3): 409–421.

Kaida, N. and K. Kaida. (2016a). Facilitating pro-environmental behavior: The role of pessimism and anthropocentric environmental values. Social Indicator Research, Published Online: 17 March 2015.

Kaida, N. and K. Kaida. (2016b). Pro-environmental behavior correlates with present and future subjective well-being. Environment, Development and Sustainability, 18(1): 111–127.

Karasek, R. and T. Theorell. (1990). Healthy Work Stress Productivity and the Reconstruction of Working Life. Basic Books, 31–32, New York, USA.

Kitajima, H., N. Numata, K. Yamamoto and Y. Goi. (1997). Prediction of automobile driver sleepiness 1st report, Rating of sleepiness based on facial expression and examination of effective predictor indexes of sleepiness. Transactions of the Japan Society of Mechanical Engineers. Series C., 63(613): 3059–3066 (in Japanese).

Krieger, A.C., I. Ayappa, R.G. Norman, D.M. Rapoport and J. Walsleben. (2004). Comparison of the maintenance of wakefulness test (MWT) to a modified behavioral test (OSLER) in the evaluation of daytime sleepiness. Journal of Sleep Research, 13(4): 407–411.

Kuriyagawa, Y., M. Ohsuga and I. Kageyama. (2009). HR changes in driving scenes with danger and difficulties using driving simulator. pp. 396–403. *In*: International Conference on Engineering Psychology and Cognitive Ergonomics, Springer, Berlin, Heidelberg, Germany.

Lacey, B.C. and J.I. Lacey. (1978). Two-way of communication between the heart and the brain. American Psychologist, 33(2): 99–113.

Lal, S.K.L. and A. Craig. (2001). A critical review of the psychophysiology of driver fatigue. Biological Psychology, 55: 173–194.

Lazarus, R.S. and S. Folkman. (1984). Stress Appraisal, and Coping Springer Publishing, New York, USA.

Lenne, M.G., T.J. Triggs and J.R. Redman. (1997). Time of day variations in driving performance. Accident Analysis and Prevention, 29(4): 431–437.

Lim, J. and D.F. Dinges. (2008). Sleep deprivation and vigilant attention. Annals of New York Academy of Sciences, 1129(1): 305–322.

Lim, J. and D.F. Dinges. (2010). A meta-analysis of the impact of short-term sleep deprivation on cognitive variables. Psychological Bulletin, 136(3): 375–389.

Littner, M.R., C. Kushida, M. Wise, D.G. Davila, T. Morgenthaler, T. Lee-Chiong, M. Hirshkowitz, L.L. Daniel, D. Bailey, R.B. Berry, S. Kapen and M. Kramer. (2005). Standards of practice committee of the american academy of sleep medicine: Practice parameters for clinical use of the multiple sleep latency test and the maintenance of wakefulness test. Sleep, 28(1): 113–121.

Luczak, H. (1971). The use of simulators for testing individual mental working capacity. Ergonomics, 14(5): 651–660.

Mase, K. (1991). Recognition of facial expression from optical flow. IEICE Transactions, E74: 3474–3483.

Mase, K. (1997). Automatic recognition of facial expression. The Institute of Image Information and Television Engineers, 51(8): 1136–1139 (in Japanese).

Matsumoto, Y., K. Mishima, K. Satoh, T. Shimizu and Y. Hishikawa. (2002). Physical activity increases the dissociation between subjective sleepiness and objective performance levels during extended wakefulness in human. Neuroscience Letters, 326(2): 133–136.

Matthews, G., T.J. Sparkes and H.M. Bygrave. (1996). Attentional overload, stress, and simulated driving performance. Human Performance, 9(1): 77–101.

Matthews, G., L. Dorn, T.W. Hoyes, D.R. Davies, A.I. Glendon and R.G. Taylor. (1998). Driver stress and performance on a driving simulator. Human Factors, 40(1): 136–149.

Matthews, G. (2002). Towards a transactional ergonomics for driver stress and fatigue. Theorical Issues in Ergonomics Science, 3(2): 195–211.

May, J.F. and C.L. Baldwin. (2009). Driver fatigue: The importance of identifying causal factors of fatigue when considering detection and counter measure technologies. Transportation Research Part F., 12: 218–224.

McCanne, T.R. and J.A. Anderson. (1987). Emotional responding following experimental manipulation of facial electromyographic activity. Journal of Personality and Social Psychology, 52(4): 759.

McCraty, R., M. Atkinson, G. Rein and A.D. Watkins. (1996). Music enhances the effect of positive emotional states on salivary IgA. Stress Medicine, 12(3): 167–175.

Miura, T. (1992). Visual search in intersections—An underlying mechanism. IATSS Research, 16(1): 42–50.

Monk, T.H. (1987). Subjective ratings of sleepiness—The underlying circadian mechanisms. Sleep, 10(4): 343–353.

Morishima, K., H. Ura, T. Chihara, H. Daimoto and K. Yamanaka. (2015). Estimation of useful field of view by machine learning based on parameters related to eye movement. New ergonomics perspective: Selected Papers of the 10th Pan-Pacific Conference on Ergonomics, Tokyo, Japan, 25–28 August 2014, CRC Press.

Morris, J.S., A. Öhman and R.J. Dolan. (1998). Conscious and unconscious emotional learning in the human amygdala. Nature, 393(6684): 467.

Nater, U.M., N. Rohleder, J. Gaab, S. Berger, A. Jud, C. Kirschbaum and U. Ehlert. (2005). Human salivary alpha-amylase reactivity in a psychosocial stress paradigm. International Journal of Psychophysiology, 55(3): 333–342.

Obayashi, K. (2013). Salivary mental stress proteins. Clinica Chimica Acta, 425: 196–201.

Obinata, G., T. Usui and N. Shibata. (2008). On-line method for evaluating driver distraction of memory-decision workload based on dynamics of vestibulo-ocular reflex. Review of Automotive Engineering, 29(4): 627–632.

Obmiński, Z., M. Wojtkowiak, R. Stupnicki, L. Golec and A.C. Hackney. (1997). Effect of acceleration stress on salivary cortisol and plasma cortisol and testosterone levels in cadet pilots. Journal of Physiology and Pharmacology, 48(2): 193–200.

Oken, B.S., M.C. Salinsky and S.M. Elsas. (2006). Vigilance, alertness, or sustained attention: Physiological basis and measurement. Clinical Neurophysiology, 117(9): 1885–1901.

Omi, T. (2016). Detecting drowsiness with the driver state monitor's visual sensing. Denso Technical Review, 21: 93–102 (in Japanese, English abstract available).

Ong, J.L., C.L. Asplund, T.T. Chia and M.W. Chee. (2013). Now you hear me, now you don't: Eyelid closures as an indicator of auditory task disengagement. Sleep, 36(12): 1867–1874.

Otsuka, T. and J. Ohya. (1999). Extracting facial motion parameters by tracking feature points. pp. 433–444. In: Advanced Multimedia Content Processing. Springer, Berlin, Heidelberg, Germany.

Phillips, R.O. (2015). A review of definitions of fatigue and a step forwards to a whole definition. Transportation Research Part F., 29: 48–56.

Phipps-Nelson, J., J.R. Redman, D.J. Dijk and S.M. Rajaratnam. (2003). Daytime exposure to bright light, as compared to dim light, decreases sleepiness and improves psychomotor vigilance performance. Sleep, 26(6): 695–700.

Picton, T.-W., S. Bentin, P. Berg, E. Donchin, S.A. Hillyard, R. Jr. Johnson, G.A. Miller, W. Ritter, D.S. Ruchkin, M.D. Rugg and M.J. Taylor. (2000). Guidelines for using human event-related potentials to study cognition: Recording standards and publication criteria. Psychophysiology, 37(2): 127–152.

Priest, B., C. Brichard, G. Aubert, G. Liistro and D.O. Rodenstein. (2001). Microsleep during a simplified maintenance of wakefulness test. A validation study of the OSLER test. American Journal of Respiratory and Critical Care Medicine, 163(7): 1619–1625.

Recarte, M.A. and L.M. Nunes. (2000). Effects of verbal and spatial-imagery tasks on eye fixations while driving. Journal of Experimental Psychology Applied, 6(1): 31–43.

Reyner, L.A. and J.A. Horne. (1997). Suppression of sleepiness in drivers: Combination of caffeine with a short nap. Psychophysiology, 34(6): 721–725.

Reyner, L.A., S.J. Wells, V. Mortlock and J.A. Horne. (2012). Post-lunch sleepiness during prolonged, monotonous driving—Effects of meal size. Physiology and Behavior, 105(4): 1088–1091.

Ruiz, M.C., J.S. Raglin and Y.L. Hanin. (2015). The individual zones of optimal functioning (IZOF) model (1978–2014): Historical overview of its development and use. International Journal of Sport and Exercise Psychology, 15(1): 1–23.

Russell, J.A. (1980). A circumplex model of affect. Journal of Personality and Social Psychology, 39(6): 1161.

Russell, J.A. and J.M. Fernández-Dols (eds.) (1997). The Psychology of Facial Expression, Cambridge University Press, USA.

Sachs, W. (1984). Die Liebe Zum Automobil, Ein Ruckblick in die Geschichte unsererWunsche, Rowohlt Verlag GmbH (W. Sachs, Don Reneau trans). For Love of the Automobile, Looking Back into the History of Our Desire, University of Calfornia Press, 1992.

Sagaspe, P., J. Taillard, G. Chaumet, C. Guilleminault, O. Coste, N. Moore, B. Bioulac and P. Philip. (2007). Maintenance of wakefulness test as a predictor of driving performance in patients with untreated obstructive sleep apnoea. Sleep, 30(3): 327–330.

Sakakibara, K. and T. Taguchi. (2003). Evaluation of driver's stress state based in biochemical components in saliva. Human Interfaces, 5(1): 95–102 (in Japanese with English abstract).

Salanova, M., A.B. Bakker and S. Llorens. (2006). Flow at work: Evidence for an upward spiral of personal and organizational resources. Journal of Happiness Studies, 7: 1–22.

Saper, C.B., T.C. Chou and T.E. Scammell. (2001). The sleep switch: Hypothalamic control of sleep and wakefulness. Trends in Neurosciences, 24(12): 726–731.

Schlegel, R.E. (1993). Driver mental workload. pp. 359–382. In: B. Peacock and W. Karwowski (eds.). Automotive Ergonomics, Taylor & Francis, London, UK..

Selye, H. (1976). The Stress of Life Revised Edition. McGraw-Hill Book Company, New York, USA.

Shin, D., H. Sakai and Y. Uchiyama. (2011). Slow eye movement detection can prevent sleep-related accidents effectively in a simulated driving task. Journal of Sleep Research, 20(3): 416–424.

Shinoda, H., M.M. Hayhoe and A. Shrivastava. (2001). What controls attention in natural environments. Vision Research, 41(25-26): 3535–3545.

Shirakawa, T., K. Tanida, J. Nishiyama and Y. Hirata. (2010). Detect the imperceptible drowsiness. SAE International Journal of Passenger Cars—Electronic and Electrical Systems, 3(1): 98–108.

Sluiter, J.K., A.J. van der Beek and M.H. Frings-Dresen. (1998). Work stress and recovery measured by urinary catecholamines and cortisol excretion in long distance coach drivers. Journal of Occupational and Environmental Medicine, 55(6): 407–413.

Smit, A.C., Van J.A.M. Gisbergen and A.R. Cools. (1987). A parametric analysis of human saccades in different experimental paradigms. Vision Research, 27(10): 1745–1762.

Taguchi, T. (1998). Evaluation of fatigue during car driving. Toyota CRDL Review, 33(4): 25–31 (in Japanese with English abstract).

Taguchi, T. and H. Inagaki. (1999). Influence of vertical vibrations on driver's fatigue during long-distance running-assessment in terms of stress hormones. Transaction of Society of Automotive Engineers of Japan, 30: 93–98 (in Japanese with English abstract).

Taillard, J., A. Capelli, P. Sagaspe, A. Anund, T. Akerstedt and P. Philip. (2012). In-car nocturnal blue light exposure improves motorway driving: a randomized controlled trial. PLoS One, 7(10): e46750.

Takahashi, M. and H. Arito. (2000). Maintenance of alertness and performance by a brief nap after lunch under prior sleep deficit. Sleep, 23(6): 813–819.

Takatsuji, K., Y. Sugimoto, S. Ishizaki, Y. Ozaki, E. Matsuyama and Y. Yamaguchi. (2008). The effect of examination stress on salivary cortisol, immunoglobulin A, and chromogranin A in nursing students. Biomedical Research, 29(4): 221–224.

Tsunashima, H. and K. Yanagisawa. (2009). Measurement of brain function of car driving using functional near-infrared spectroscopy (fNIRS). Computational Intelligence and Neuroscience, Article ID 164958.

Turk, M.A. and A.P. Pentland. (1991). Face recognition using eigenfaces. pp. 586–591. In: Computer Vision and Pattern Recognition, 1991, Proceedings CVPR'91. IEEE Computer Society Conference on, IEEE.

Ueno, A., S. Kokubun and Y. Uchikawa. (2007). A prototype real-time system for assessing vigilance levels and for alerting the subject with sound stimulation. International Journal of Assistive Robotics and Mechatronics, 8(1): 19–27.

Van Dongen, H.P., G. Maislin, J.M. Mullington and D.F. Dinges. (2003). The cumulative cost of additional wakefulness: Dose-response effects on neurobehavioral functions and sleep physiology from chronic sleep restriction and total sleep deprivation. Sleep, 26(2): 117–126.

Veltman, J.A. and A.W.K. Gaillard. (1996). Physiological indices of workload in a simulated flight task. Biological Psychology, 42(3): 323–342.

Wierwille, W.W. and L.A. Ellsworth. (1994). Evaluation of driver drowsiness by trained raters. Accident Analysis and Prevention, 26(5): 571–581.

Williamson, A., D.A. Lombardi, S. Folkard, J. Stutts, T.K. Courtney and J.L. Connor. (2011). The link between fatigue and safety. Accident Analysis and Prevention, 43(2): 498–515.

Yajima, K., K. Ikeda, M. Oshima and T. Sugi. (1976). Fatigue in automobile drivers due to long time driving. SAE Technical Paper 760050.

Yamamoto, K. (2001). Development of alertness-level-dependent headway distance warning system. JSAE Review, 22(3): 325–330.

Yamanaka, K., H. Nakayasu, T. Miyoshi and K. Maeda. (2006). A study of evaluating useful field of view at visual recoginition task. Transaction of the Japanese Society of Mechnical Engineers, Series (C), 72(719): 244–252 (in Japanese).

Yerkes, R.M. and J.D. Dodson. (1908). The relation of strength of stimulus to rapidity of habit-formation. Journal of Comparative Neurology and Psychology, 18(5): 459–482.

Zohar, D., O. Tzischinsky, R. Epstein and P. Lavie. (2005). The effects of sleep loss on medical residents' emotional reactions to work events: A cognitive-energy model. Sleep, 28(1): 47–54.

5

Driver and System Interaction

5.1 Mental Workload while Using In-vehicle System

5.1.1 Workload Measurement Using Questionnaires

Mental workload as a human state is discussed in Chapter 4. Assessment of driver mental workload has been actively studied since in-vehicle system was introduced because drivers have to perform in-vehicle tasks in addition to driving the vehicle. In order to evaluate the driver's mental workload, various methods using questionnaires have been developed.

5.1.1.1 Cooper-Harper Rating Scale

The Cooper-Harper Rating Scale evaluates demands on pilots involved in handling the quality of aircraft (Cooper and Harper, 1969). An integer rating value from 1 to 10 is obtained by choosing 'yes' or 'no' to each question in a decision tree that asks about the demands concerning the handling of an aircraft and its performance during a flight. Under this scale, a higher value means a heavier workload. Since this scale can only be used for the evaluation of aircraft's handling quality, it was amended and reformed into the Modified Cooper-Harper Rating Scale, which can also evaluate other types of work using the same decision tree structure (Skipper et al., 1986) (Fig. 5.1).

The Cooper-Harper Rating Scale is recommended for use when the main cause of workload is attributed to handling difficulties in the subject system. It is difficult to use this scale for evaluation of an element or a part of the subject system. For this reason, the Cooper-Harper Rating Scale should rather be used in a test environment where the general mental workload during the handling of the subject system is questioned.

5.1.1.2 NASA-TLX

The NASA Task Load Index (NASA-TLX) was developed by NASA's Ames Research Center as a system for subjective workload metrics. It rates workload based on six subscales (Hart and Staveland, 1988), namely, mental demand, physical demand, time pressure, performance, effort, and frustration level (Table 5.1). This is based on the frame work of task demands (mental, physical and time), mental resources required (effort) for providing performance to meet the demands and the subjective evaluation

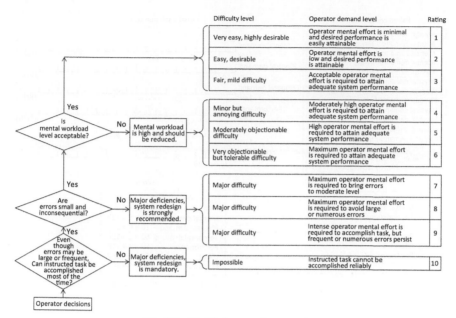

Fig. 5.1: Modified cooper-harper scale.

Table 5.1: Subscales of NASA-TLX and their Descriptions.

TITLE	ENDPOINTS	DESCRIPTIONS
MENTAL DEMAND	Low/High	How much mental and perceptual activity was required (e.g. thinking, deciding, calculating, remembering, looking, searching, etc) ? Was the task easy or demanding, simple or complex, exating or forgiving?
PHYSICAL DEMAND	Low/High	How much physical activity was required (e.g. pushing, pulling, turnimg, controlling, activating, etc.)? Was the task easy or demanding, slow or brisk, slack or strenuous, restful or laborious?
TEMPORAL DEMAND	Low/High	How much time pressure did you feel due to the rate or pace at which the task or task elements occurred? Was the pace slow and leisurely or rapid and frantic?
PERFORMANCE	Good/Poor	How successful do you think you were in accomplishing the goals of the task set by the experimenter? How satisfied were you with your performance in accomplishing these goals?
EFFORT	Low/High	How hard did you have to work(mentally and physically) to accomplish your level of performance?
FRUSTRATION LEVEL	Low/High	How insecure, discouraged, irritated, stressed, and annoyed versus secure, gratifiled, content, relaxed, and complacent did you feel during the task?

of own task performance (frustration level). Mean weighted workload scores (WWL scores) are calculated based on the relative importance of each of the six subscales decided by pair-wise comparison and the raw scores for each subscale obtained by the Visual Analog Scale.

Figure 5.2 shows an example of WWL score with the calculation based on the NASA-TLX subscales. Participants conduct pair-wise comparison of the six subscales using all of the 15 combinations. The importance of a subscale is obtained as the number of times the participants choose it as an important subscale that makes a

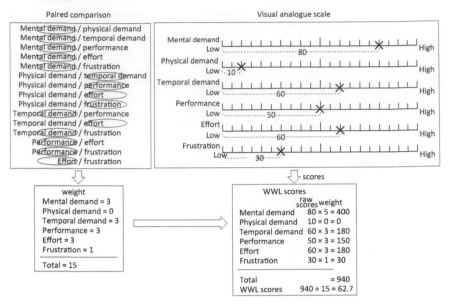

Fig. 5.2: Calculation of WWL scores based on pair-wise comparison and ratings using VAS.

greater contribution to mental workload than the other subscale regarding the subject task. Participants are also asked to evaluate the subject task by marking a position along a visual analogue scale with reference to the descriptions on the edges of the scale. Values ranging from 0 to 100 are extracted from the marked positions for use as raw scores for each scale. WWL scores for each scale are obtained by multiplying the importance value by the raw score and dividing the result by 15 (the maximum importance value).

Annie Pauzie of INRETS (currently IFSTTAR) modified the subscales and descriptions of the NASA-TLX and developed the Driving Activity Load Index (DALI) for workload evaluation of automobile driving tasks (Pauzié, 2008a). The DALI is used for the evaluation of workload while driving and talking on the cell phone and while using a car navigation system (Pauzié, 2008b). Subscales and descriptions of the DALI are shown in Table 5.2.

5.1.1.3 SWAT

The Subjective Workload Assessment Technique (SWAT) is an evaluation method developed by the USAF Armstrong Aerospace Medical Research Laboratory. It evaluates workload based on three dimensions (Reid and Nygren, 1988). The three discrete dimensions, namely, time load, mental effort load and psychological stress load, are at three discrete levels, i.e., specifically low, medium and high. With these three dimensions and three levels, 27 combinations are obtained. The rank order data obtained from the participants are converted to scales ranging from 1 to 100, which are the mental workload scales based on the SWAT. Table 5.3 shows descriptions for each dimension at each level of the SWAT.

As an example is shown in Fig. 5.3, the mental workload assessment process of the SWAT begins with preparing 27 cards (Cards A to Z and Card ZZ), on each

Table 5.2: Subscales of DALI and their Descriptions.

Title	Endpoints	Description
effort of attention	low/high	to evaluate the attention required by the activity - to think about, to decide, to choose, to look for and so on
visual demand	low/high	to evaluate the visual demand necessary for the activity
auditory demand	low/high	to evaluate the auditory demand necessary for the activity
temporal demand	low/high	to evaluate the specific constraint owing to timing demand when running the activity
interference	low/high	to evaluate the possible disturbance when running the driving activity simultaneously with any other supplementary task sych as phoning, using systems or radio and so on
situation stress	low/high	to evaliate the level of constraints/stress while conducting the activity such as fatigue, insecure feeling, irritation, discouragement and so on

Table 5.3: SWAT Dimensions (Reid et al., 1989).

Time Load:
1. Often have spare time. Interruptions or overlap among activities occur infrequently or not at all.

2. Occasionally have spare time. Interruptions or overlap among activities occur frequently

3. Almost never have spare time. Interruptions or overlap among activities are very frequent, or occur all the time.

Menrtal Effort Load:
1. Veyr little conscious mental effort or concentration required. Activity is almost automatic, requiring little or no attention.

2. Moderate conscious mental effort or concentration required. Complexity of activity is moderately high due to uncertainty, unpredictability, or unfamiliarity. Considerable attention required

3. Extensive mental effort and concentration are necessary. Very complex activity requiring total attention.

Stress Load:
1. Little confusion, risk, frustration, or anxiety exists and can be easily accommodated.

2. Moderate stress due to confusion, frustration, or anxiety noticeably adds to workload. Significant compensation is required to maintain adequate performance.

3. High to very intense stress due to confusion, frustration, or anxiety. High to extreme determination and self-control required.

of which one of the 27 combinations is written (letters are assigned to cards at random). Test participants are asked to read the description on each card and sort the cards in increasing order of mental workload. This process is called 'SWAT Card Sort'. By performing conjoint analysis of the results of sorting, the scales are assigned to each card. After that, participants are asked to evaluate the subject task

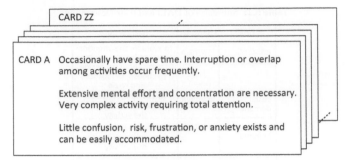

Fig. 5.3: Example of SWAT card (Reid et al., 1989).

Table 5.4: Example of Scales for Time, Effort and Stress of a Prototype Model.

Rank Order	Card Label	Descriptor Combination			Scale Values
		Time	Effort	Stress	
1	N	1	1	1	0.0
2	B	1	1	2	24.4
3	W	1	1	3	51.4
.	.	.	.	.	.
.	.	.	.	.	.
26	T	3	3	2	73.0
27	I	3	3	3	100.0

in terms of the level of time load, mental effort and psychological stress as shown in Table 5.4. Scale values, ranging from 0 to 100, can be obtained based on combinations of the results. Some problems regarding the SWAT have been pointed out, including the time required for the SWAT Card Sort and complexity of the scale calculation process, which lead to the proposal of simplified SWAT.

5.1.1.4 Workload Profile Method (WP)

The Workload Profile (WP) method is a subjective workload assessment method proposed by Tsang and Velazquez (Tsang and Velazquez, 1996) based on multi-dimensional scales that reflect Wickens' multiple resource model (Wickens and Hollands, 1999). After carrying out all the subject tasks, participants are asked to state the proportion of attentional resources spent on each task (effort dimension in Fig. 4.1.3). This process uses a workload profile rating sheet as shown in Table 5.5. In this rating sheet, conducted tasks are indicated in rows in a random order, and eight workload dimensions (scales) are indicated in columns across the entire sheet. Descriptions for each workload dimension are shown in Table 5.6. Test participants are allowed to refer to the definitions of the workload dimensions when making the ratings. The participants are asked to express the proportion of attentional resources spent on a specific dimension during the task as a number between 0 and 1 and write it in the relevant cell of the evaluation sheet. Here, 0 means that there was no demand for the subject workload dimension and 1 means that the task required the maximum attentional resource. For example, if the subject task does not involve any auditory task or sound response, the remaining six dimensions are submitted for analysis. Evaluation values for individual dimensions are summed up for each task.

Table 5.5: Workload Profile Rating Sheet (Tsang and Velazquez, 1996).

	Workload Dimension							
	Stage of Processing		Code of Processing		Input		Output	
Tasks	Perceptual/ Central	Response	Spatial	Verbal	Visual	Auditory	Manual	Speech
TR1								
TR1-SB4								
TR1-SB2								
SB2								
TR2-SB2								
TR2								
TR2-SB4								
SB4								

Table 5.6: Workload Dimensions in the Workload Profile and their Descriptions (Tsang and Velazquez, 1996).

1. *Stages of processing*
   (1) *Perceptual/central processing.* These are attentional resources required for activities like perceiving (detecting, recognizing, and identifying objects), remembering, problem-solving, and decision making.
   (2) *Response processing.* These are attentional resources required for response selection and execution. For example, there are three foot pedals in a standard shift automobile; to stop the automobile, we have to select the appropriate pedal and step on it.
2. *Processing codes*
   (1) *Spatial processing.* Some tasks are spatial in nature. Driving, for example, requires paying attention to the position of the car, the distance between the current position of the car and the next stop sign, the geographical direction that the car is heading, etc.
   (2) *Verbal processing.* Other tasks are verbal in nature For example, reading involves primarily processing of verbal, linguistic materials.
3. *Input modality*
   (1) *Visual processing.* Some tasks are performed based aon the visual information received. For example, playing basketball requires visual monitoring of the physical location and velocity of the ball. Watching TV is another example of a task that requires visual resources.
   (2) *Auditory processing.* Other tasks are performed based on auditory information. For example, listening to the person on the other end of the telephone is a task that requires auditory attention. Listening to music is another example.

4. Output modalities
   (1) Manual responses. Some tasks require considerable attention for producing the manual response as in typing or playing a piano.
   (2) Speech responses. Other tasks require speech responses instead. For example, engaging in a cnversation requires attention for producing the speech responses.

The total value is the workload evaluation value based on the WP method (Tsang and Velazquez, 1996) states that the entry of evaluation values in the WP evaluation sheet took 15 to 30 minutes, including the time for explaining the workload dimensions and for evaluating multiple dimensions.

5.1.1.5 Rating Scale Mental Effort (RSME)

The Rating Scale Mental Effort (RSME) is a subjective workload evaluation method developed by Zijlstra (Zijlstra, 1993). It evaluates the workload based on the single-

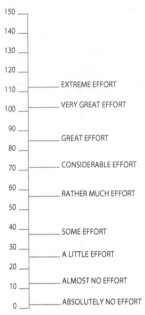

Fig. 5.4: Evaluation scale of the RSME (De Waard, 1996).

dimension scale that represents the demanded mental effort for conducting a task (see 4.1.3). As shown in Fig. 5.4, a scale of 150 mm marked every 10 mm with nine labels in between is used to express the level of mental effort in language. These labels range from 'Absolutely No Effort' (near the zero point on the scale) to 'Extreme Effort' (near the 112 point on the scale) and these are not always positioned at regular intervals. While this example scale uses English labels (De Waard, 1996), the labels of the original scale developed by Zijlstra are written in Dutch. After completing the subject task, test participants are asked to indicate the level of the required mental effort for the task by marking on the evaluation scale based on the scale and the labels that explain the level of mental effort. The marked positions are extracted as a number between 0 and 150, and are used as workload evaluation scores.

Since RSME is a single-dimension scale, its workload evaluation process takes less than one minute, which is an advantageous feature when compared with other multi-dimensional scales, like NASATLX and SWAT that take a longer time. Meanwhile, there is a report that its sensitivity becomes low in some countries and regions because the expression of mental effort level used in the labels are interpreted differently, depending on the culture (Widyanti et al., 2013). The labels should be translated into appropriate expressions for each country or region and the position and number of the labels should also be adjusted (Widyanti et al., 2013) gives the procedures for such adjustments.

5.1.2 *Mental Workload Assessment Using the Subsidiary Task Method*

The subsidiary task method asks participants to conduct the main task, that is, driving, and another task that is not directly related to driving (a subsidiary task) simultaneously. Mental workload during driving is estimated by measuring and analyzing the deterioration in the driver's performance in the subsidiary task and those of driving behavior caused by fluctuations in the driving demand.

The first application of the subsidiary task method in the field of automotive human factors was made in the 1960s. Brown et al. asked the participants to drive while dictating and calculating the numbers provided through auditory means (Brown and Poulton, 1961). It was found that the driver's correct responses to the auditory task fluctuated according to the driving environment. By this means, Brown et al. showed that spare mental capacity could be measured using a subsidiary task (effort dimension in Fig. 4.1.3).

5.1.2.1 Two Types of Subsidiary Tasks

Brown et al. conducted an experiment which asked participants to determine whether they could drive around obstacles while speaking on a car telephone (Brown et al., 1969). The results of this experiment showed that participants were slower in determining whether they could drive around obstacles when speaking on a phone and they also tended to make inaccurate judgments. These types of studies focused on how distractions other than driving affect driving behavior.

On the other hand, the Peripheral Detection Task (PDT) focuses on changes in performance in a subsidiary task (Harms and Patten, 2003; Jahn et al., 2005). In this task, an LED light is located in the driver's peripheral vision and it turns on at random times. The driver must respond as soon as possible on noticing that the light is on. It becomes harder for drivers to detect visual stimulation when driving is difficult due to complicated driving situations and when they are speaking on a cell phone during driving.

These two types of subsidiary task methods have different purposes. The first method aims to determine the impact on driving behavior when the driver is given additional non-driving related tasks. For example, in an experiment to determine the impact of cellular phone use while driving, participants are asked to have a conversation on a phone or a memory task is given instead of speaking on the phone. The second method aims to estimate the driver's cognitive status based on the performances of the subsidiary task. In this case, the subsidiary task should be something that does not place additional burden on the driver and neither affects the driving behavior. Specifically, very simple tasks, like visual detection tasks (Ogden et al., 1979), including PDT, should be used (see also 5.3.2.2).

5.1.2.2 Psychological Concepts Related to the Subsidiary Task Method

To understand the results obtained through the subsidiary task method, the driver's behavior needs to be analyzed based on the models of information processing and attentional resources (for example, Fig. 5.5). the driver selects and obtains the necessary information from the driving environment (the function of selective attention), allocate attentional resources to process the information (function of divided attention), make a judgment, and drive a car by moving his or her body. Analysis should be done to determine what kind of impact there is on which part of this information-processing process when the driver faces a complicated traffic situation or when the mental workload is increased by conducting non-driving related tasks, such as using in-vehicle devices and having a conversation.

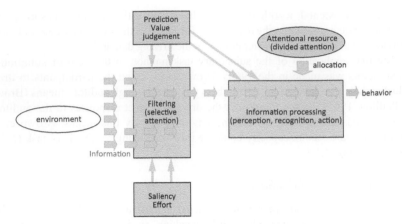

Fig. 5.5: Human information processing.

For example, when the load task is to speak on a cell phone, drivers need to give a certain level of attention to the conversation. Driving quality deteriorates when the conversation becomes complicated and the amount of allocable attentional resources for driving decreases. Moreover, it is expected that adverse effects from having a conversation have a great impact on high-level cognitive processes that need more attention and consciousness, such as risk prediction, route selection and recognition of the situation, whereas they have less impact on driving operations which the drivers are accustomed to and which are done automatically and unconsciously.

Here, attention is deemed as a kind of amount of energy needed for behavior and recognition, which is called 'attentional resources'. Since there is a certain limit on available attentional resources, the reserve capacity becomes low when the driver needs to pay greater attention to driving or when paying attention to tasks other than driving. The amount of spare mental capacity is important (see also 5.3.2.2).

Figure 5.6 shows the relationships between the demand (the difficulty of tasks), the amount of attentional resources and driving performance (modified from Wickens et al. (Wickens and Hollands, 1999)). Driving performance does not significantly deteriorate even when the spare mental capacity is not high. This is because driving is the main task to which the drivers allocate attentional resources preferentially. In other words, while it is difficult to determine the change in the level of spare mental capacity solely based on driving performance when the demand is not extremely high, subsidiary tasks, which have lower priority compared to driving, are more sensitive to the level of spare mental capacity and thus can be used as an index of spare mental capacity. However, the driving performance significantly deteriorates, even though it is the highest priority task, when the demand for attentional resources exceeds the limit (see also 1.7.4).

Meanwhile, the multiple resource theory (Wickens and Hollands, 1999) points out that drivers use different types of attentional resources depending on the properties of the task. For example, since a visual task and auditory task require different kinds of attentional resources, they hardly interfere with each other. However, when multiple tasks are given, they require the executive function for conducting these tasks simultaneously. Attentional resources will be required in such a case as well.

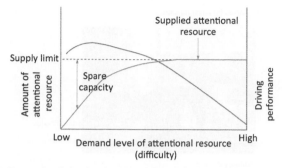

Fig. 5.6: Demand and supply of attentional resources and performance in driving.

5.1.2.3 Example of Application of Subsidiary Task Method

To assess the mental workload during driving and operating of car devices, Shinohara et al. developed a stimulus detection task that asks participants to detect visual, auditory and haptic stimuli given at random (Shinohara et al., 2012). A study by Fujii et al. asked participants to conduct four types of operations on a car navigation system (control conditions, move a hand, scroll on the map, and enter a phone number) while driving on a driving simulator or driving an actual car, and they also carried out a stimulus detection task simultaneously (Fig. 5.7) (Fuiji et al., 2014). The reaction time and the error rate were analyzed as the performance measures in the subsidiary task.

The results showed that reaction time (Fig. 5.8a) becomes longer when the operation of the car navigation system was more complicated. The subsidiary task method assumes that the reaction time reflects the level of spare mental capacity at the time when the stimulus was given. The results suggest that the spare mental capacity decreased as the operation on the car navigation system gets more and more complicated. In addition, the results were similar in all the visual, auditory and haptic tasks, suggesting that the complicated operation of the car navigation system had an impact, not only on the driver's vision, which is a particularly important sensory modality for driving, but also on all the attentional resources, and eventually deteriorated the participant's ability to select and process information from the outer environment.

As for the error rate (Fig. 5.8b), the visual stimulus in particular was greatly affected by the difficulty level of the operation of the car navigation system. The chance for test participants to miss the visual stimuli was higher when entering a phone number as they had to look away from the front for a long time. Therefore, the error rate for visual stimulus reflects not only the mental load, but also the visual load. It was also seen that participants could not detect stimuli even though they were looking ahead because their effective visual field narrowed with decrease in the spare resources due to controlling the car navigation system. Meanwhile, the error rates were also higher for auditory stimulus and haptic stimulus when the operation of the car navigation system was more complicated, even though what they were looking at was supposed to have nothing to do with the detection of these stimuli. Therefore, it is understood that difficulty in the operation of the car navigation system affected the participants' overall attentional resources. Multi-model detection task has been discussed in ISO TC22/SC39/WG8 working group (Bruyas and Dumont, 2013) and

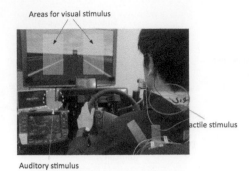

Fig. 5.7: Multi-modal detection task.

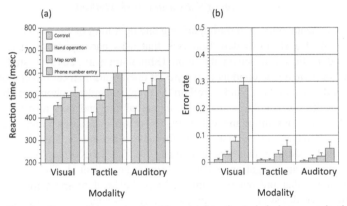

Fig. 5.8: Reaction times and error rate of multi-model detection task during car navigation tasks.

was standardized as DRT (Detection Response Task) in ISO 17488 in 2016 (ISO 17488, 2016).

The big issue concerning mental workload measurement using the subsidiary task method is that the combination of driving and other subsidiary tasks that have nothing to do with driving is an unnatural situation, which may affect one's driving behavior. For this reason, drivers may not be able to show normal driving behavior when the subsidiary task method is applied. While this method can be used in an artificial test environment for an assessment to study the use of new car devices, it is difficult to measure mental workload on general drivers in real driving settings. However, there are ways to apply mental workload assessment methods based on the basic idea of the subsidiary task method, such as one that monitors changes in a specific action in various driving situations, deeming an action taken in the real driving situation (for example, checking the mirrors) as an embedded subsidiary task.

5.1.3 Workload Measurement Based on Driving Performance

5.1.3.1 Overview

Major workload measurement methods include subjective assessment using questionnaires (NASA-TLX, Cooper Harper, RSME, etc., as described in 5.1.1), physiological measurement (based on the heart rate, electro-dermal activity and level of cortisol as described in 4.4.4 and 4.4.6), and measurement methods based on task performance (performance dimension in Fig. 4.3). This section will explain a workload measurement method based on driving performance, which is categorized in measurement methods based on task performance as with the subsidiary task method (see 5.1.2).

The greatest advantage of the workload measurement method based on driving performance is that it can assess *during* a drive the workload of the driving task, that is, the main task, as an expression of the amount of resource input for the driving task. Another advantage is that it can keep drivers in their natural state during measurement because it is a non-invasive and non-contact (without attaching a sensor) method.

The workload measurement based on driving performance is carried out by quantifying the smoothness of steering and brake-reaction time. In particular, the methods that quantify the smoothness of car steering can effectively determine changes in workload as they can measure it continuously during a drive. Examples of such methods that have been proposed include the one using steering input frequency and the one using the number of steering reversal (McLean and Hoffmann, 1975). There is also a similar method that uses the lateral position of vehicle within the driving lane, but the sensitivity of workload measurement is generally higher when using drivers' steering input than when using vehicle behavior.

Below, workload measurement methods based on driving performance are explained; namely, the steering entropy method and the real-time steering entropy method.

5.1.3.2 Steering Entropy (SE) Method

The SE method developed by Boer et al. in the 1990s quantifies the smoothness of steering as information entropy. They compared a control run and a test run using the same road and the same speed range (Nakayama et al., 1999). The results showed that steering entropy (Hp) becomes higher as the control of in-car information devices gets more complicated (Fig. 5.9).

Hp is calculated in the following four steps:

Step 1: Prepare steering angle data obtained at 150 ms intervals. If the data contains noise or if the resolution of the steering angle sensor is low, obtain data at 50 ms intervals and then use the average of every three points as the data obtained at 150 ms intervals.

Step 2: Calculate the prediction errors (Fig. 5.10). First, calculate the predicted steering angle (θ_p) by using the values of steering angles at three previous points ($n-3$, $n-2$, $n-1$) from the present point (n) and by applying the quadratic Taylor expansion centered on the time of $n-1$ point.

$$\theta_p = \theta(n-1) + (\theta(n-1) - \theta(n-2))$$
$$+ \frac{1}{2}((\theta(n-1) - \theta(n-2)) - (\theta(n-2) - \theta(n-3)))$$

PE is the difference between the predicted steering angle and the actual steering angle and is calculated by the following formula:

$$PE(n) = \theta(n) - \theta p(n)$$

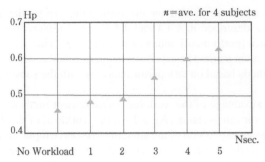

Fig. 5.9: Relationship between workload and entropy (Hp).

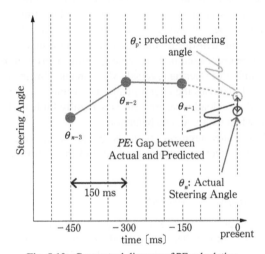

Fig. 5.10: Conceptual diagram of PE calculation.

Since a steering angle is a signed signal (clockwise or counterclockwise), PE is also a signed value.

Step 3: Create a PE diagram. First, calculate the 90th percentile value of the PE (hereinafter referred to as α). If steering is smooth, the kurtosis of PE distribution is large and the value of α is small. Next, create a PE distribution with the bin width set at a multiple of α. The percentage of the PE distribution in each bin is represented as $q_1, q_2, \ldots, q_9$ (Fig. 5.11).

Step 4: Calculate Hp. The Hp value is calculated by the following formula using the percentage of the PE distribution in each bin ($q_1, q_2, \ldots, q_9$).

$$Hp = -\sum_{i=1}^{9} q_i \log_9 q_i \ (i = 1 \sim 9)$$

To compare Hp between a control run and a test run, the Hp value of the control run is calculated first by following steps 1 to 4 and then the Hp value of the test run is calculated following the same steps. When calculating the Hp value of the test run, α is the value calculated for the control run.

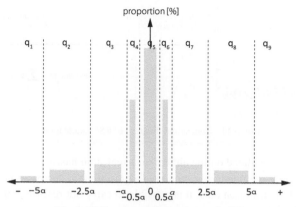

PE: prediction error of steering angle (%)

Fig. 5.11: Conceptual diagram of PE distribution.

For example, if workload of the test run is higher than that of the control run, the steering lacks smoothness. The kurtosis of the PE distribution is small and Hp increases in this case.

SE was used to measure changes in workload caused by tasks other than driving, including having conversation, calculation, operating a navigation system and using a cellular phone (Nakayama et al., 1999). Since the SE method evaluates the workload by comparing the control run and test run, which are conducted based on the same conditions, including the road and speed, it is suitable for workload measurement using a driving simulator or a test track, which allows testers to easily control the driving conditions. On the other hand, it is difficult to measure workload during real work on a real-time basis using the SE method. To improve this point, the real-time steering entropy method as described below was developed.

5.1.3.3 Real-time Steering Entropy (RSE) Method

The RSE method is an improved version of the SE method and can conduct measurement almost on a real-time basis (Kondoh et al., 2015). The characteristics of the RSE method compared with the SE method are as follows:

- It does not require any prescribed driving course.

Since the RSE method detects straight roads automatically when evaluating workload, there is no need to prescribe a driving course. Although this means that it cannot assess workload when the speed is low or when the car is on a sharp curve, there is no practical issue because the percentage of straight roads is high in a normal traffic environment.

- No distinction needs to be made between a control run and a test run.

The RSE method only needs a single run to evaluate changes in an individual's workload using relative entropy (RHp) obtained by comparing the PE distribution over a relatively long term to serve as a baseline for steering characteristics, and the recent PE distribution over a short term reflects the changes in steering characteristics (Fig. 5.12). It should be noted that it cannot assess workload in the initial term, right after the beginning of driving.

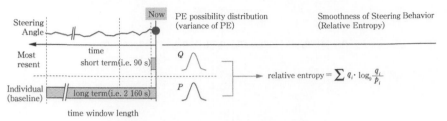

Fig. 5.12: Conceptual diagram of RSE calculation.

- Workload is calculated online almost on a real-time basis.

Since the RSE method uses a recursive algorithm that decreases computing load for updating PE distribution data, it is possible to carry out real-time processing with the current vehicle CPU.

RHp is calculated in the following five steps:

Step 1: Measure the steering angle data at 50 ms intervals. Calculate the mean value of the previous three points to obtain steering angle values at 150 ms intervals.

Step 2: Calculate PE. It is calculated by the same means as the SE method. The predicted steering angle is calculated using the steering angles at the previous three points and then the cap between the obtained value and the actual steering angle is calculated.

Step 3: Two sets of PE distributions are updated (*L*: PE distribution over a long term; *S*: PE distribution over a short term) only when the car is running in a measurable environment (linear road). While the SE method records all data obtained through a run and then calculates the PE distribution offline, the RSE updates the PE distribution online, using a recursive computing method as shown in Fig. 5.13, while minimizing memory usage and computing load.

Step 4: Calculate Hp. The Hp values of each of the PE distributions over a long term (*L*) and a short term (*S*) are calculated.

$$Hp_l = -\sum_{i=1}^{9} p_i \log_9 p_i$$

$$Hp_s = -\sum_{i=1}^{9} q_i \log_9 q_i$$

Step 5: Calculate RHp. It is calculated based on the PE distributions updated through Step 3 and the two Hp values calculated in Step 4 (Fig. 5.14).

$$RHp = \text{sign}(Hp_s - Hp_l)\sum_{i=1}^{9} q_i \log_9 \frac{q_i}{p_i}$$

The RSE method was applied to analyze the relationship between the increase in RHp, which was calculated with driving behavioral data collected on a public road, and driver behavior was observed by videos. As a result, 40 per cent of the drivers, whose RHp exceeded a prescribed threshold, were using a cell phone or having a

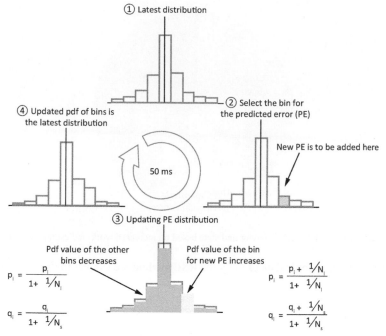

Fig. 5.13: Recursive calculation to obtain PE distribution.

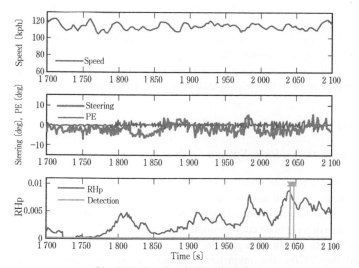

Fig. 5.14: Results of RHp calculation.

conversation while driving, 22 per cent were operating a navigation system or eating and drinking while driving, and 30 per cent had low arousal level (Fig. 5.15). This means that the majority of non-driving related tasks seen in real-driving settings were detected using the RSE method (Dingus et al., 2006).

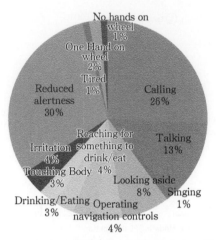

Fig. 5.15: Detected driver's behavior and state based on increased workload estimated by RSE.

Since steering angle data used by the RSE method are subject to the influence of the driver's posture during driving, the position of hands on the steering wheel, road conditions and other factors, measures to remove noise should be taken in order to measure workload in a practical manner.

5.1.3.4 Summary

Workload measurement methods quantify the relative relationship between a task demand and driver's attentional resources allocated to a task. In measuring driver's workload, it is important that the amount of resource input to the driving task, the main task, is assessed independently. In this regard, the workload measurement methods based on driving performance discussed in this section are very useful as they can assess resource input to the driving task in an objective manner.

5.2 HMI of In-car Information Systems

5.2.1 Interaction with a System

The number of functions installed in a car has been rising. In addition to providing information on the conditions of the car, in-car information systems can also obtain external information from the Internet. It is important to ensure that such information is conveyed to drivers in an easy-to-understand manner and that drivers' intentions are communicated to cars in a simple way.

Until the 1970s, the mainstream devices were mechanical, single-function devices; now there are cellular phones, car navigation systems and VICS (Vehicle Information and Communication System), which provide constant updates, rather than fixed information. Such information should be communicated to drivers in a comprehensible way and it is also necessary to display such information on in-car multi displays.

With respect to the location of manual controls, meters and indicators in motor vehicles (especially those of passenger cars), ISO 4040 provides rules for manual controls and the combination of functions in multifunction control devices as well as visibility requirements for meters and warning indicators. As for the installation and display of graphic display devices, the Japan Automobile Manufacturers Association released the guidelines in 2004 (see 5.3.3).

5.2.1.1 Design of Interaction

In-car interaction design means the designing of interactive behavior between in-car systems and drivers. Cars are equipped with meters, navigation systems, air conditioner control panels, HUD, etc., but not necessarily all of them. Interactive behavior among the devices needs to be designed differently according to the combination of such in-car devices. Interaction between a driver and a system tends to be complicated when the driver has to operate complicated procedures when using the system. As the number of interactions is increasing with highly functional devices, there is a need to design interactions from the driver's point of view.

Analysis, evaluation and designing methods for all kinds of interaction systems, including those for cars, are provided in (ISO 9241-110, 2006). Since it does not refer to the marketing and merchantability aspects of interaction systems, user interface should be designed based on the results of marketing research.

5.2.1.2 Tactile Feedback

Due to the need to ensure that car devices require drivers to look at them for as short a time as possible, single switches have been used. Of late, however, the number of car increases with a display device that can display many menu items due to the increase in the number of installed functions. A touch-screen device has been introduced for menu selection, however, it is not excellent in terms of usability since it is not capable of giving any feedback, leaving the driver unsure of whether the menu item has been successfully selected. In recent years, an increasing number of developers use devices with a screen that features a feedback mechanism, allowing drivers to feel whether they have successfully selected the menu item. Some of the devices for providing an operational feeling use Piezo actuators and capacitance.

Others provide an operational feeling by letting drivers operate the menu displayed at a remote distance with a controlling device located near their hands. There is a wide range of devices that are intended to achieve this purpose, from those with multiple switches to haptic devices that give tactile feedback to drivers when a cursor moves on the screen (Fig. 5.16). All of these devices contribute to enhancing safety as they allow drivers to keep their eyes looking forward.

The haptic sense of human skin is most sensitive to frequencies between 100 Hz and 300 Hz. An operational feeling should be generated based on this characteristic by such means as introducing switches that give a sense of clicking and applying tactile feedback technologies.

5.2.1.3 Audio Interface

The audio interface is a user-friendly device for operating the system while driving. When using audio input, devices should be designed bearing in mind the languages and dialects to be supported for individual regions as well as the noise from driving and radio since the devices are used while the car is in motion. Counter measures for

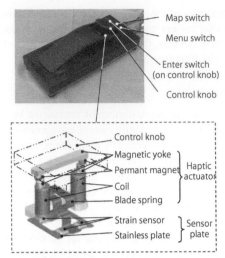

Fig. 5.16: Three-axis haptic jyostick (Hatanaka et al., 2012).

noise should especially be considered based on a test using target vehicles since the location of microphones affects the SN ratio greatly.

Drivers, who are not accustomed to audio input, have not memorized prepared audio commands yet and are not sure what to say. For this issue, it is needed to make it easier for drivers to use the audio recognition system by providing audio guidance or list of audio commands on a display.

5.2.1.4 Integrated Controller

An increasing number of cars are equipped with an integrated controller that unifies the control of the navigation function, radio function, audio function and display of vehicle condition, in order to enhance visibility and usability. In most cases, the display at the center stack of a car or the display in the meter cluster is concurrently used as a display for the integrated controller. Information should be displayed according to the priority of information when different functions request to display information simultaneously.

5.2.1.5 Internet Connection of In-car Devices

With respect to cars equipped with information communication systems that are connected to the Internet (connected cars), one must be careful for the difference in the characteristics of in-car ECU (Electronic Control Unit) and those of PCs and mobile information terminals. These two are significantly different in terms of the required characteristics. For example, in-car ECUs are subject to more strict non-functional requirements, including real-time processing capacity (time restriction), safety, reliability, quality and cost. Their development process and production cycle are also longer. Therefore, some unique issues will rise when building an in-car information communication network system to achieve a constant Internet connection like that of other information devices (Iwai, 2013).

Many countries are introducing compulsory laws, regulations, policies, licenses and guidelines to promote or restrict connected cars. Developers are recommended to refer to the *Distracted Driving Laws* published by NHTSA (US) in September 2014 when designing devices in order to learn about what to pay attention to regarding the use of cellular phones and the sending and receiving of e-mail (http://www.ghsa.org/html/stateinfo/laws/cellphone_laws.html.).

5.2.2 Route Navigation and Map Display

The car navigation function that guides drivers to their destinations safely by providing route navigation and displaying maps is recognized as one of the most major functions of in-car information systems. This function implements unique display and guidance technologies that obtain guidance information in an effective manner without threatening driving safety.

5.2.2.1 Volume of Graphic Information

While displayed maps and routes provide useful information that supports driving, the display of complicated information may cause drivers to gaze at the screen while driving. The guidelines developed by the Japan Automobile Manufacturers Association (JAMA Guideline version 3.0) provide that the amount of information of the entire information should be optimized by changing the content of information in accordance with the running conditions of the car. The optimal information volume is determined by the total glance time required to understand the information on the screen, and which was assessed in a series experiments conducted when developing the guidelines (see 5.3.3). Examples of changing displayed information, according to the running condition, are described below. In the case of map display, minor roads (residential roads less than 4 m wide) are not indicated (Fig. 5.17). In the case of route display, information that is useful for driving, such as traffic information, is allowed for indication while driving, but videos unrelated to driving and details of facilities are not allowed.

5.2.2.2 Mental Map

Mental maps that drivers have affect the selection of the route to their destinations. Mental maps are also called 'cognitive maps'. These maps are human internal representation developed and referred to when recognizing one's location and the

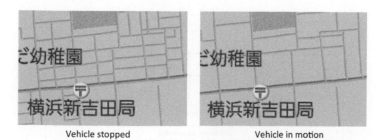

Vehicle stopped　　　　　　　　Vehicle in motion

Fig. 5.17: Road display during a stop and while driving.

direction to move to arrive at a particular location. Humans recognize the locations of themselves and other objects in space based on the structure of the said internal expression. To categorize and organize the variety of information concerning the driver's route selection, the fundamental elements of cognitive maps developed by Lynch are useful (Lynch, 1960). Lynch gives five fundamental elements of human cognitive maps, namely, landmarks, paths, nodes, districts and edges (Fig. 5.18). With respect to the mental maps used for route selection, 'path' is a course where people can move, such as roads and railway; 'nodes' mean intersections of the paths and road entrances and exits, and 'landmarks' mean features that are used as referring objects. 'Districts' are two-dimensional spaces that are represented by a single concept or a single name, such as parks and downtown districts. 'Edges' are boundaries that people cannot cross, such as walls and rivers.

Mental maps are transfigured into procedural knowledge, network knowledge and map-like knowledge based on landmark knowledge through behavior experience and learning (Freundschuh, 1989). A conceptual diagram of such transformation is shown in Fig. 5.19.

Landmark knowledge refers to a collection of knowledge that associates features with major objects (landmarks). This knowledge allows humans to reach their targets even when they have no knowledge regarding the paths in some cases.

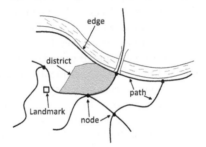

Fig. 5.18: Fundamental elements of mental maps.

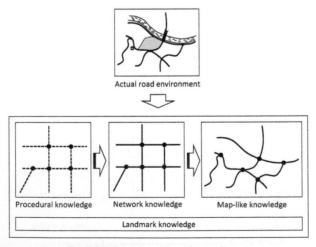

Fig. 5.19: Transformation of mental maps.

Procedural knowledge is formed as a memory of a landscape and of human behavior in it. This knowledge refers to the scenes that include landmarks or the scenes in which characteristic behaviors were involved, such as turning right or left. Although humans cannot express such knowledge as a map, they remember their behavior from the past once they start for their destinations and as the scenes develop in front of their eyes, which eventually lead them to their destinations.

Network knowledge refers to a collection of knowledge related to networks, such as the connections of and relationship between nodes and paths. It is topological knowledge about connective relationships. Here, distance and directions are distorted or are lacking in comparison to the actual environment.

Map-like knowledge is network knowledge that is accurate in terms of distance and directions. In our daily lives, space covered by map-like knowledge is very limited. Even in a space for which a person has map-like knowledge, most of the behaviors are defined by network knowledge and other knowledge in one of the previous phases.

Route guidance by car navigation systems is categorized into the following three types based on the types of knowledge: (1) turn-by-turn navigation based on procedural knowledge; (2) navigation using simple figures based on network knowledge; and (3) navigation on digital maps based on map-like knowledge.

5.2.2.3 Expression of Maps

Maps are expressed on a screen in any of the three forms: north-up, heading-up or bird's-eye view. 'North-up' refers to a two-dimensional expression where a map is always indicated with north up (Fig. 5.20a). The orientation of the map is fixed regardless of the direction the car is heading to, which makes it easier for drivers to understand which direction on the compass they are moving to.

'Heading-up' refers to a two-dimensional expression where the vertical axis of the map is always aligned with the direction drivers are heading to (Fig. 5.20b). The location mark is positioned at the bottom center of the display, thus allowing drivers to see a wider map of the area in the travelling direction. The directions of forked roads are visibly comprehensible.

'Bird's-eye view' refers to a steric expression, which gives a view seen from an elevated, oblique position (Fig. 5.21). As with heading-up maps, bird's eye view maps always display the driver's travelling direction. Similarly to heading-up maps, bird's eye-view maps allow drivers to comprehend the directions of forked roads intuitively. They can display facilities at a further distance than heading-up maps do.

(a)   (b)

Fig. 5.20: North-up map (a) and heading-up map (b).

Fig. 5.21: Bird's eye-view map.

5.2.2.4 Displaying Roads

Digital map data is provided in the form of network data (road network), which deems roads as links and intersections as nodes. Each node in a network contains coordinate data and each link contains the type, width, etc., of roads. Car navigation systems render and draw road networks on a real-time basis. They optimize the information volume by controlling the amount of the displayed roads under a certain level; they sometimes show major roads only and sometimes show more detailed and minor roads, depending on the scale of the map. Car navigation systems enhance visibility by drawing roads in different colors and widths according to the type, width and number of lanes of the roads. For large-scale representation (scale of 1/5000 or less), they use street maps. Contrary to road networks, street maps are composed of polygon data that can express the shape and width of roads and median strips, making it easier for drivers to ascertain the landforms (Fig. 5.22).

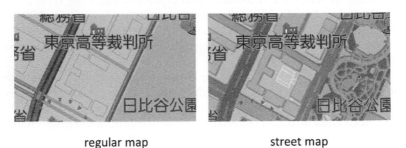

regular map   street map

Fig. 5.22: Comparison between a regular map and a street map.

5.2.2.5 Displaying Background

Objects other than roads, which are called the 'background', are also displayed along with roads. The background includes water areas representing the sea and rivers, green areas representing forests and parks, and house-shape symbols representing buildings. The visibility of maps is enhanced by drawing the background based on the polygon data that expresses the shape of such features or based on the texture data.

5.2.2.6 Presenting Text

Maps display overlapping text objects for the names of facilities, administrative districts, roads and intersections. Car navigation systems optimize the volume of character information by limiting the amount of displayed text according to the scale of the map and by adjusting text objects for appropriate representation when the display areas of multiple objects overlap.

Information that drivers need to obtain from text objects depends on the scale of the map. For example, a detailed map like a street map needs to display names of intersections, streets and facilities so as to provide landmarks during route navigation. On the other hand, a wide-area map, such as a map of an entire prefecture, is required to provide text objects that allow the driver to get an overview of the route and destination, such as the names of cities and expressways.

Text objects may overlap with each other even when the display of text is controlled by the above-mentioned measures. Especially, the orientation of the maps on car navigation systems is subject to change (such as when a heading-up map is used) and their text objects may overlap in one orientation even when there is no problem in another orientation. In order to avoid this, car navigation systems calculate the area that each text object occupies, then display the high-priority text in the top layer and either omit low-priority text or move it to the bottom layer (Fig. 5.23). If the text objects have the same priority, the systems decide the hierarchy among them based on the arrangement of these objects to select objects to display (Fig. 5.24: the text in the left figure is omitted in the right figure due to change in the orientation of the map).

Fig. 5.23: Control of overlapping text objects when the types of text are different.

Fig. 5.24: Control of overlapping text objects when the types of text are same.

5.2.2.7 Presenting Landmarks

Icons are overlaid to show the location of landmarks. Since text objects are sometimes omitted, icons should be displayed in a way that enables drivers to understand the type of landmarks by looking at the icons themselves. The designs of icons used for the car navigation systems are generally categorized into map symbols, pictograms, brand icons and selectable icons.

Map symbols refer to the icons that indicate the facilities and objects on the map in the broad sense. In the narrow sense, they refer to the standardized map symbols provided by public organizations of individual countries. In Japan, the Geospatial Information Authority of Japan stipulates the icons as shown in Fig. 5.25. Some map symbols suggest a specific role or function (for example, elementary/secondary schools and temples) but others do not (for example, prefectural government offices and city halls). The map symbols are useful for the identification of the types of facilities as they are standardized even if their shapes themselves have no meaning.

Pictograms are the techniques to express a specific information or concept as a picture without using text. They are mainly used as icons to indicate specified facilities in public spaces in a simple manner. Although they are not publicly standardized, they are useful for the identification of the types of facilities as they suggest the functions by a visible means.

As for brand icons, the brand marks of specific companies and organizations are used as the icons on the map. The well-known marks of facilities that have many chain stores (such as convenience stores and gas stations) fall under this category. While brand icons contain more information than map symbols and pictograms, their information is understood by drivers at almost the same speed.

Many car navigation systems have selectable icons, which drivers can add to any chosen point in order to mark the place in addition to the prescribed landmarks.

Individual icons explained above are usually displayed as two-dimensional symbols. However, especially famous facilities and facilities that have characteristic shapes are rendered on the maps based on 3D polygon data that includes the information of the detailed shape and size. Such landmarks are called '3D landmarks' (Fig. 5.26).

5.2.2.8 Displaying Remaining Distance/Estimated Required Time

Car navigation systems display the remaining distance and estimate the required time to reach the destination (or estimated arrival time) in various forms. The display patterns are generally categorized into the following two groups: If the distance is longer than a certain threshold and needs to be displayed in a continuous manner while running on the route (such as the distance to the destination), the distance is indicated in the text. To avoid redundant expression, the distance is indicated in a rounded number (for example, rounded to the closest 100 m). For those comparatively long routes, the estimated remaining time or estimated arrival time is displayed.

Prefecture   City hall   Secondary school   Temple

Fig. 5.25: Map symbols.

Fig. 5.26: A 3D landmark.

The display of distance for showing the location of forks (such as the distance from the car and the closest intersection) is given for a short time when the car comes close to that point. In such a case, the distance is displayed in a more detailed number than when it is displayed continuously. In addition to displaying the distance by numbers, many car navigation systems also display a gauge (bar graph) that indicates the remaining distance to the target point so that drivers can intuitively understand how the remaining distance is decreasing.

5.2.2.9 Displaying Routes

Below are examples of various types of route display by car navigation systems.

5.2.2.9.1 Turn by Turn Display

This type shows the distance to intersections and forks in the immediate area on the route and indicates the direction in which the driver should proceed (Fig. 5.27). The names of intersections and of the roads beyond the divergence are also displayed in some cases as auxiliary information. But since these give a limited amount of information by displaying the route in the simplest way, information can be obtained by looking at the display for a short time. As it does not require a high-resolution or large display, it is suitable for the instrument cluster and head-up displays.

5.2.2.9.2 Route Display

This type shows the route on the map. It ensures visibility by displaying in different colors, emphasizing road widths, and using steric expressions. In addition, it also distinguishes between general roads and expressways by using different colors and expressions so that drivers can ascertain the features of individual parts of the route they are taking (Fig. 5.28).

Fig. 5.27: Turn by turn display.

Fig. 5.28: Displaying route.

5.2.2.9.3 Traffic Lane Display

This type shows a picture that imitates the road markings before reaching an intersection or a fork ahead (Fig. 5.29). This helps drivers to select the lane in a smooth manner when the lane structure is complicated. Moreover, usability is also enhanced by emphasizing the recommended lane with reference to the direction of the route at a fork.

5.2.2.9.4 Crossing Macrograph

A map display shows many road connections, and text and icons that are not related to the drivers' target route, which may distract them in identifying which intersection they need to turn at. To address this issue, car navigation systems are programmed to show a magnified map called a 'crossing macrograph' automatically when the drivers are approaching a fork on the course of their route. This type of route display uses expressions focused on helping the drivers to make judgments at intersections by enlarging the intersections and hiding the roads that are not directly connected to the route, or by omitting unnecessary information including the shape of the buildings and text, or by displaying only the icons that provide landmarks near the intersection (Fig. 5.30).

When the structure of the lanes is complicated, like forks at the entrance of expressways or at junctions, making it impossible to provide sufficient information by showing a normal crossing macrograph, some car navigation systems provide an illustration that imitates the lane structure or landscape at the point (Fig. 5.31).

5.2.2.9.5 Highway Map

Some car navigations systems show facilities along the road in a row while the car is on an expressway or a toll road (Fig. 5.32). This type of map is called a highway map. It shows the names of the facilities, distance to the facilities, estimated arrival time, divergent direction, and information on ancillary facilities (gas station, restaurant, etc).

5.2.2.9.6 Manoeuver List

This type displays intersections and forks ahead on the route as a list if the lane structure is not simple (Fig. 5.33). It shows arrows indicating divergent directions, like the ones used by the turn-by-turn display, and distance to the points in a row. Some systems show the names of intersections and road markings, too.

Fig. 5.29: Traffic lane display.

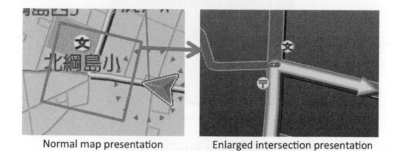

Normal map presentation   Enlarged intersection presentation

Fig. 5.30: Crossing macrograph.

Fig. 5.31: Illustration.

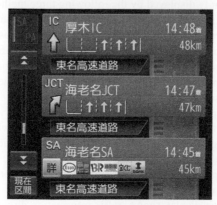

Fig. 5.32: Highway map.

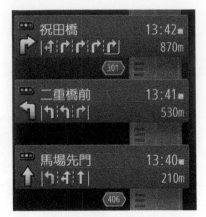

Fig. 5.33: Manoeuver list.

5.2.2.9.7 Guide Information to Support Safe Driving

The types of displays shown above are intended to guide drivers along the instructed routes. Apart from such types, some car navigation systems provide information intended to support safe driving. A typical example of this category is the display that provides information normally given by road signs, such as junctions, sharp curves, stops, and speed limits (Fig. 5.34). In recent years, more and more cars are using the display device of in-car information systems as a part of an advanced driving assistant system (ADAS). Various types of information to support safe driving in addition to the route guidance by using information obtained with car sensors, such as running velocity and acceleration speed, and various data obtained from video images by on-board cameras and from communication between a road and a car or between a car and a car are provided.

5.2.2.10 Display of Traffic Information

In-car devices can receive a wide range of traffic information through VICS (Vehicle Information and Communication System) and various telematic services. Such information is commonly displayed by the following three methods: text information display, simplified graphic information display and map information display.

A text display shows traffic information on the screen using text (Fig. 5.35). This simplest method is suitable for low-capacity displays. Since the number of displayable letters per screen is limited to optimize the information volume, the amount of information drivers can obtain is fewer than from other methods. Such text information can be also communicated vocally to the drivers.

A simplified graphic information display indicates received information as a diagram (Fig. 5.36). Examples include diagrams that show where a traffic jam or traffic control is taking place using a deformed image of roads in the immediate area, photographs that show the traffic situation of a specific point, and illustrations that show the details of regulations. The graphic information can convey a higher volume of information visually than a text information can.

A map information display shows traffic information directly on the map of car navigation systems (Fig. 5.37). Information regarding traffic jams and traffic controls

Fig. 5.34: Display of information for supporting safe driving.

Fig. 5.35: Text information display.

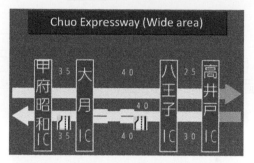

Fig. 5.36: Simplified graphic information display.

Fig. 5.37: Graphic information display (arrows indicate traffic jam).

associated with individual roads is drawn as a line along the road. Different colors are applied to such lines depending on the significance and type of traffic jam or traffic control. In addition, other information that is associated with a specific region or area, rather than with individual roads, such as information on regulation and weather warnings, is displayed as an icon on that spot or overlaid to show the subject zone. This allows drivers to readily understand the relationship between the current position and the point for which traffic information is provided.

5.2.3 Design of Menus

The menu allows drivers to initiate a command by selecting one of the menu items provided by the system in the form of visual symbols. If the number of commands is greater than the number of menu items that are displayed simultaneously, the menu is designed in a hierarchical manner. A user selects a menu item at a higher layer before moving to a menu at a lower level. Commands are initiated when a user eventually selects a menu item that is connected to a command. An example of a hierarchical menu is shown in Fig. 5.38. In this example, the first level contains three items, the second level six items for each item in the first level, and the third level six items for each item in the second level. Therefore, this menu can implement a total of 108 commands ($3 \times 6 \times 6 = 108$).

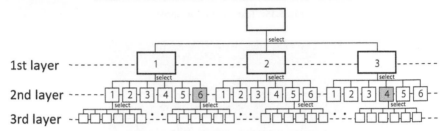

Fig. 5.38: An example of hierarchical menu.

5.2.3.1 Menu-based Interaction

Drivers achieve their objectives by initiating commands. Sometimes, drivers have knowledge on which menu item and in what order they should select in order to achieve their goals, and they select menu items by referring to this knowledge. In other cases, drivers do not have any such knowledge and they select menu items that seem to contribute to the achievement of their goals accordingly. In the former case, drivers have acquired knowledge concerning the menu from their growing experience of using it since they were in the situation of the latter case. When designing a menu, it is necessary to ensure that the menu will be easy to use for drivers in the latter case.

5.2.3.1.1 Fundamental Principles

A menu exists between the drivers' objectives and the commands to be used to achieve these objectives. It enables drivers to achieve their objectives by executing commands corresponding to individual menu items. In principle, the number of such objectives is equal to the number of commands. Drivers select one of these objectives as they intend to reach the command by selecting menu items sequentially. Drivers' mental representation of their objectives does not always conform to the visual symbolic representation of the menu item. For example, if the objective is to go home, drivers would use an expression like 'I want to go home' or "I'm going home'. On the other hand, menu items of car navigation systems indicate the same objective as 'destination' or 'home'.

As the fundamental principle, menu items presented to drivers should be affixed with appropriate labels, arranged properly, and organized in a hierarchical manner so as to ensure that drivers can achieve their various objectives by implementing the commands of the systems.

5.2.3.1.2 Presentation and Selection of Menu Items

Commands to access the functions provided by the system are organized in a hierarchy, and a hierarchical menu is designed in a way that corresponds to these commands. Labels on individual menu items in the hierarchically structured menu are provided in the form of text, pictograms or symbols. The menu items are arranged on the screen and presented to drivers as a drop-down menu, pop-up menu, cascade menu, pie menu or tile menu. Methods for selecting menu items include the following:

- Pointing use of an input device, such as a joystick
- Touching the target on a touch screen
- Pushing a button beside an indicated option
- Giving an instruction vocally

5.2.3.1.3 Strengths and Weaknesses of Menu-based Interaction

Speaking from the viewpoint of functionality, menu-based interaction strongly supports searching behaviors for browsing the functions provided by systems. This can effectively help drivers who are seeking a specific function or data items and drivers who want to find out what functions are available. Drivers who are experiencing the system for the first time can also easily find menu items that allow them to achieve their objectives as long as the menu items are organized hierarchically and labeled in a proper manner.

On the other hand, problems may arise when drivers try to access a part of the menu, which they cannot directly access. For example, look at Fig. 5.38. If the driver has selected the sixth menu item in the second level that is related to the first menu item in the first level, the driver cannot directly move to the fourth menu item in the second level that is related to the third menu item in the first level. Instead, the driver has to go back to the menu of the first level. As this example shows, menu-based interactions which require drivers to go through a hierarchical menu, do not provide especially effective means when they already know the position of the target command in the menu. This may cause drivers to look at the menu for a long time, which leads to the risk of distraction.

5.2.3.2 Design Guidelines

Style-specific design knowledge is often shared among designers in the form of guidelines. Guidelines outline the results of various studies for use by designers. For example, the following is one of the guidelines on menu hierarchies:

"The structure of a hierarchical menu should be broad and shallow, rather than narrow and deep." This expression was given by Shneiderman (Shneiderman, 1992) based on a study originally done by Miller (Miller, 1981). Miller asked users to evaluate various types of hierarchical menus with each menu having 64 items, whose depth of level ranges from 1 to 6. The width ranges from 64 to 2 accordingly. Miller discovered that the selection time and error rate were minimized when the menu was composed of two levels with eight options.

Table 5.7 is an example of the guidelines for a menu-based interface developed in this context.

Table 5.7: Menu Design Guidelines (Shneiderman, 1992).

- Use task semantics to organize menus (single, linear sequence, tree structure, acyclic and cyclic networks)
- Prefer broad-shallow to narrow-deep
- Show position by graphics, numbers, or titles
- Use items as titles for sub-trees
- Group items meaningfully
- Sequence items meaningfully
- Use brief items, beginning with the keyword
- Use consistent grammar, layout, terminology
- Allow type ahead, jump ahead, or other shortcuts
- Enable jumps to previous and main menu
- Consider online help, novel selection mechanisms, and optimal response time, display rate, screen size

---

5.2.3.3 Evaluation Methods for Menu Designs

The evaluation of menu designs can be conducted by checking if drivers can successfully achieve their objectives using the menu system. When the driver has little experience in using the subject menu system, the necessary condition for the menu system to be proved defect-free is that its hierarchical structure does not conflict with the driver's mental model. That is, it should ensure that drivers can select necessary menu items for achieving their objectives as long as they have enough time to choose menu items based on their objectives. It is also necessary to ensure that the drivers' mental representation of their objectives has a similar meaning to the representation of the menu items they need to select. When it is the first time for drivers to use an information device, it is known that they choose a label that seems closest to their objective among other labels provided on the interface; subsequently they select an interface object in the vicinity of that label. This strategy is called the Label Following Strategy. The hierarchical menu can be evaluated and improved based on this principle (Kitajima et al., 2006).

When drivers have abundant experience in using the subject menu and when they are able to access their target menu item without any conscious effort, they refer to their memories regarding their menu-item-selection actions that are connected to their objectives, rather than reading individual labels and looking for the one that has the closest meaning. In such cases, menu designs are evaluated based on the manipulation speed. This type of evaluation can be conducted by building the Keystroke Level Model (Card et al., 1983), which is one of the GOMS models (Newman and Lamming, 1995) for implementing objective-oriented routine tasks.

5.3 Assessment of Driver Distraction

5.3.1 Definition of Distraction

For more than 25 years, drivers were able to obtain a variety of information that graphic display devices, like car navigation systems (Nakamura, 2008) and portable devices, like PNDs (Portable Navigation Devices), supply. In such a situation, the problem of distraction, caused by interference in the driving tasks that are primarily

Fig. 5.39: Distracted drivers.

important in terms of safety due to the tasks required to obtain information presented inside vehicles, was pointed out (Fig. 5.39) (NHTSA, 2010; Green, 2004; Charlton, et al., 2013).

As a result of reviewing the studies related to distraction, the understanding of the concept varies widely from researcher to researcher without finding a unified definition of the word. Discussed below are the characteristics of human attention, which are important when thinking about the problem of distraction. After that, definitions of distraction are sorted out by comparison to other similar concepts.

5.3.1.1 Characteristics of Attention and Related Definitions

When thinking about the cause of traffic accidents, it is needless to say that inattentive driving increases the risk of accidents (Klauer et al., 2006). The phenomenon of inattention occurs not independently, but in the course of continuous driving. To understand the definitions of distraction related to attention-related errors that occur during continuous driving, it is important to know the characteristics of attention.

As human nature cannot sustain attention on a specific task for a long time, it becomes difficult for drivers to maintain their attention as their arousal level becomes lower, which eventually errors may occur (see 4.3.1). There are various levels in human arousal and some of them are suitable for maintaining proper driving behaviors. Sanders categorized these arousal levels into two groups: (1) tonic arousal, which refers to chronic arousal that shows comparatively small changes from the influence of individual features, like personality, in addition to the influence of sleep; and (2) phasic arousal, which refers to a tentative arousal affected by external stimulations and human efforts (Sanders, 1997).

Even if drivers maintain an optimum arousal level, they may fail to drive properly when the demand of the task becomes so high that it exceeds their attention. One of the causes of errors that occur despite the proper arousal level is mental stress (see 4.1.7).

The activity level of attention changes according to interactive relationships between external factors (such as the characteristics of a task) and human being's internal factors (such as personality and conditions). Demand is one of the external factors. It is defined as a requirement posed by an external task. On the other hand, effort is an internal factor. De Waard categorized human efforts into two groups: (1) state-related efforts, which maintain a certain arousal level when the demand is small, and (2) task-related efforts, which work when the demand is great (De Waard, 1996).

Errors associated with the selectivity of attention are also one of the factors that increase the risk of accident. That is, humans have difficulty in paying sufficient attention to driving itself when their attention is diverted to a task other than driving because humans are not capable of recognizing or considering many things at one time. Distraction is related to such selectivity of attention (see 6.1.1.1).

5.3.1.2 Distraction

5.3.1.2.1 Suggested Definitions

It was pointed out in the first half of the 2000s that the word 'distraction' had not been defined scientifically (Green, 2004). Since then, many researchers gave their ideas about a definition. Victor et al. defined distraction as improper information selection that results in overlooking safety-related information (Victor, 2008). Hoel et al. used the word to refer to the interference between the driving task and other tasks (Hoel et al., 2010). Regan et al. saw distraction as a cause, and defined the word as "the diversion of attention away from activities critical for safe driving toward a competing activity, which may result in insufficient or no attention to activities critical for safe driving (Regan et al., 2011)."

5.3.1.2.2 Relation to Inattention

Researchers have different opinions on whether inattention, which refers to a state of thinking about a matter that is not related to driving, should be included in the definition of distraction. For example, Hoel et al. assert that interference in a driving task by internalized thoughts belongs to the scope of inattention and they did not deem such state as distraction (Hoel et al., 2010). Meanwhile, Wallen Warner et al. included preoccupation with internal thoughts and daydreaming in the definition of distraction (Warner et al., 2008). Moreover, there are also different opinions on whether tasks outside the vehicle should be included as triggers of distraction, in addition to tasks inside the vehicle (Ranney, 2008).

Summarizing the relationship between inattention and distraction, Pettitt et al. stated that distraction is a form of inattention, but there are also other forms of inattention (Pettitt et al., 2005). Distraction can be defined as one of the many phenomena of inattention.

5.3.1.2.3 Relation to Arousal Level and Workload

Regan et al. pointed out key elements in defining distraction as follows (Regan et al., 2011):

- There is a diversion of attention away from driving, or safe driving;
- The competing activity may compel or induce the driver to divert attention toward it; and
- There is an implicit, or explicit, assumption that safe driving is adversely affected.

According to these elements, drivers give attention to non-driving-related tasks when distracted. Therefore, distraction is distinguished from the state where the driver's attention does not work due to reduced arousal level. In addition, distraction is also to be distinguished from deteriorated state due to alcohol, drug and fatigue, which affect the driver's condition, and from declined performance due to medical prescription (Tijerina, 2006). While the concept of distraction is based on an assumption that there are tasks other than driving, workload estimates the influence on drivers as a whole, rather than the interference of multiple tasks.

5.3.1.3 Conclusion

The definitions of distraction stated so far can be summarized as "a state of drivers whose consciousness or attention is shifted to a task other than the primary driving task while the driver's arousal level is maintained at a proper level, or decrease in the driver's performance due to such state," which is affected by both external factors including demand and internal factors (Fig. 5.40). However, it should be noted that the driving task itself cannot be prescribed clearly and thus it is difficult to strictly define distraction. For example, a route search seems to be a part of the driving task, while reading a map may induce distraction. There remains some ambiguity in the definition of distraction, as to whether each specific task is deemed as a source of distraction or not depends on what are defined as necessary tasks for driving. Therefore, it is important for individual researchers to clearly define the driving tasks and other tasks when studying distraction.

5.3.2 Assumptions for Distraction Assessment

5.3.2.1 Information Processing and Distraction

The workload and demand, which are a requirement from an external task, have a u-shaped relationship. Workload is high when the demand is high or when it is low (De Waard, 1996). Moreover, demand and driving performance are expected to have an inverted u-shaped relationship if performance is high when demand is high or when

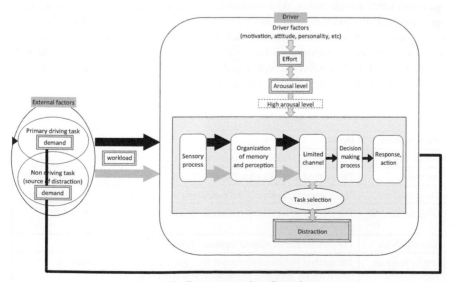

Fig. 5.40: Terms centered on distraction.

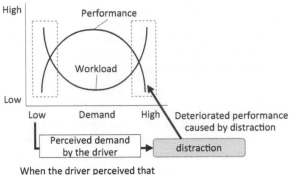

When the driver perceived that
driving demand is low, the driver starts
to perform a non-driving related task
(inducing the driver distraction)

Fig. 5.41: Relationship between distraction and performance.

it is low (De Waard, 1996). This relationship is historically known as Yerks-Dodson law (Yerks and Dodson, 1908). The distraction issue is deemed to be a result of an increase in demand that must be processed when drivers conduct the driving task or other tasks (Fig. 5.41).

5.3.2.2 Ideas and Types of Assessment Methods

5.3.2.2.1 Requirements for Assessment Methods

To estimate the influence of distraction, it is necessary to use an assessment method that can measure the state of increased workload due to high demand. Indices used for workload assessment include (1) subjective evaluation index, (2) performance index (see 5.1), and (3) physiological index (see 4.4). A distraction assessment method is also required to obtain these items. Wierwille and Eggemeier pointed out the requirements for workload measurement methods (Wierwille and Eggemeier, 1993). They can be applied to distraction measurement methods (Table 5.8). Some of the distraction assessment methods proposed so far fulfill those requirements, but others do not. So, methods should be selected with careful consideration of the characteristics of individual methods when assessing distraction.

Table 5.8: Requirements for Distraction Assessment Methods.

Requirement	Details
Sensitivity	The method is capable of measuring the actual distraction level.
Intrusion	The method itself does not change the performance among users and systems.
Diagnosticity	The method is capable of distinguishing the types and causes of distraction.
Global sensitivity	The method is capable of reflecting variations due to various factors that affect distraction.
Transferability	The method can be used to measure distraction in various fields.
Implementation requirements	The method fulfills requirements for equipment and instrumentations necessary for providing stimulation and collecting data.

5.3.2.2.2 Types of Assessment Methods

(1) *Primary task measurement and secondary (subsidiary) task measurement*: There are two distraction assessment methods: primary task measurement and secondary task measurement (O'Donnell and Eggemeier, 1986). The secondary task measurement aims to determine distraction caused by tasks other than driving based on the assumption that it can create a state of information overload by having drivers conduct two different tasks. Various assessment methods have been tried out under this category.

(2) *Assumptions and notes for the secondary task measurement*: The secondary task measurement is based on assumptions developed within a human information processing model. Major approaches that belong to this category include the filter model (bottle neck model) proposed by Broadbent (Broadbent, 1958) and the capacity model proposed by Kahneman (Kahneman, 1973). Broadbent's filter model assumes that the channels for information to pass in the course of human information process is not so large and that only information selected by filters can go through the channels. On the other hand, there is a counter argument that some of the unattended information can also pass the filters (Treisman and Gelade, 1980). The capacity model, developed by Kahneman, assumes that there is a limit on the capacity to process stimuli. It also assumes that individual mental activities use mutually shared capacity and if the capacity is allocated to a specific activity, the other activity cannot be executed sufficiently. In particular, difficult activities require a great part of capacity, making it impossible to carry out other activities due to shortage in resources for such activities. Among various capacity models, there is also a multiple resource model like the one proposed by Wichens that categorizes resources according to the modality of stimuli, types of presented issues (code), cognitive levels, and response to stimuli or tasks, which are assumed to be mutually independent (Wickens and Hollands, 1999).

Other information processing-related models include the perceptual cycle model by Neisser (Neisser, 1976), the spotlight model by Posner (Posner, 1980), and the feature integration theory by Treisman et al. (Treisman and Gelade, 1980).

As discussed above, human attention and information-processing are based on various assumptions, including limits on channels and capacity. These assumptions are also used by the secondary task measurement, which is a distraction assessment method based on the performance indices. For example, Senders gave the following four items as assumptions for secondary task measurement (Senders, 1970).

 (i) A user (driver) has a single channel.

(ii) The capacity of a channel is fixed (limited).

(iii) A single method is applied to measure the capacity.

(iv) The influence of distraction is additive.

When it is assumed that there is a limit on a single channel or on the capacity of information processing in accordance with the above-mentioned assumptions, the results of assessment will be different depending on which task the driver gives attention to among the driving task and other tasks. In this regard, the direction of attention should be considered when assessing distraction.

(3) *Secondary task measurement and dual task measurement*: Ogden et al. categorized assessment methods into two groups: (1) the secondary task ('subsidiary task' and

'auxiliary task') measurement, which asks drivers to mainly focus on one of the two tasks by giving instruction; and (2) the dual task ('concurrent task' and 'simultaneous task') measurement, which does not require drivers to prioritize any specific task (Ogden et al., 1979).

O'Donnell et al. categorized secondary task measurement into two groups: the subsidiary task measurement and the loading task measurement (O'Donnell and Eggemeier, 1986). The subsidiary task measurement instructs drivers to keep their performance in the primary task (defined by Knowles (Knowles, 1963) as a task subject to be assessed), and then estimate the decrease in their performance in another task (sub task) to estimate workload caused by the primary task. On the other hand, the loading task measurement uses an additional task that would deteriorate the performance in the primary task and drivers are asked to keep the performance in this additional task. The level of workload is estimated by comparing the driver's performance in the primary task. For example, Brown et al. used the subsidiary task measurement in their study (Brown and Poulton, 1961). They assumed that the driver's cognitive processing capacity is fixed and that their capacity while driving can be estimated by measuring the spare capacity. They measured workload by estimating the spare capacity based on the performance in the subsidiary task, which was conducted simultaneously while driving.

(4) *Primary task and subsidiary/additional task*: As discussed above, the secondary task measurement asks drivers to focus on one of the two tasks that are to be conducted simultaneously. It gives instruction to drivers regarding the allocation of their attentional resources to the main and secondary tasks (subsidiary/additional tasks). However, certain tasks tend to take priority due to their own characteristics in spite of the instructions. In this regard, Ogden et al. summarized the characteristics of the tasks that tend to be used as a primary task based on the following four criteria: (1) the dimension of active/passive tasks (active tasks tend to be used as a primary task); (2) centric tasks and peripheral tasks (centric tasks tend to be used as a primary task); (3) the dimension of experimenter-paced/subject-paced tasks (subject-paced tasks tend to be ignored due to experimenter-paced tasks) (Ogden et al., 1979).

Requirements for secondary tasks were given by Knowles as follows: (1) non-interfering (the task involves constant input and response demand that do not have a physical influence on or interfere with the performance in the primary task); (2) simple (the task has simple features); (3) self-paced (drivers/conductors are allowed to conduct the task at their pace); (4) continuity of measurement results (the results show the level of influence in a continuous manner); and (5) compatibility with the primary task (the task has appropriate features as a secondary task in terms of its relationship with the features of the primary task) (Knowles, 1963).

5.3.2.2.3 Conclusion

As shown above, the secondary task measurement should be conducted bearing in mind the instructions given to drivers, requirements for secondary tasks and the characteristics of tasks that tend to be used as a primary task. However, even when those conditions are controlled in a test, the influence of distraction may not always be reflected on the performance in one of the two tasks. That is, while the subsidiary task measurement and loading task measurement assess the influence of distraction based on the performance in the subsidiary task and the primary task, respectively, distraction may affect the performance in tasks other than those measured in the test. This however depends on the characteristics of the primary and secondary tasks that are conducted simultaneously and the drivers' interpretation of the given instructions. Therefore, it is desired to measure performance in other tasks, too, when applying the secondary task measurement. When conducting the dual task measurement, it is essential to measure performance in both the tasks since it does not require drivers to focus on one of the tasks.

As mentioned above, distraction assessment methods can be categorized according to the index they use: (1) subjective evaluation index, (2) performance index, and (3) physiological index. In addition to this, assessment methods are based on such features as ex-ante assessment (bench test)/direct assessment (driving behavior) and input-based (measurement of the initial phase of information processing)/response-based (measurement of driving behaviors, etc.) (Table 5.9). While adequacy and reliability are required in all assessment methods, the requirements vary depending on the position of individual assessment methods.

Moreover, distraction is generally categorized according to the influenced modality into visual distraction (caused by tasks that require the driver to look away from the road), manual distraction (caused when the driver takes a hand off the steering wheel), and cognitive distraction (caused by thinking about something other than the driving task). The assessment methods can be also categorized according to these types of distraction (NHTSA, 2010). However, when thinking about the task of using an audio system, the driver looks away from the front and takes a hand off the steering wheel to turn the dial, while thinking about the radio station to choose. As seen in this example, a task may involve all of the visual, hand manual and cognitive distractions. In most cases, it is difficult to clearly tell which type of distraction is being assessed.

In the next section, existing assessment methods are divided into two groups for the sake of convenience: (1) assessment methods for visual-manual distraction that involves tasks that require drivers to look at a device and manipulate it; and (2) assessment methods for cognitive distraction, including visual-manual distraction. Characteristics of individual assessment methods are discussed below:

5.3.3 Visual-Manual Distraction Assessment

5.3.3.1 Direct Assessment

The direct assessment method asks drivers to conduct a visual-manual task while driving on an actual road or on a test course or using a driving simulator to assess the distraction caused by the task by measuring the drivers' visual behavior, driving performance, subjective evaluation and physiological index. Among them, visual behavior and driving performance are the major indices for direct assessment.

5.3.3.1.1 Visual Behavior

Assessment method: Visual behavior is commonly measured by analyzing the driver's gaze direction and performance in device manipulation based on the enlarged images of the driver's face and display devices, which are recorded with multiple small video cameras fixed inside the vehicle in a position that does not obstruct driving (ISO 15007-1, 2002). Eye trackers are also used to measure the eye movement.

202 *Handbook of Automotive Human Factors*

Table 5.9: Characteristics of Major Assessment Methods.

Assessment Method and Index		Type	Major Required Equipments	Type of Measurable Distraction	Input-process (Measurement of the Initial Phase of Information Processing/Response-process (Measurement of Driving Behaviors, etc.)	Ease of Conduct	Adequacy (Relevance to Driving Safety)	Reproducibility
Direct assessment (while driving)	Visual behavior	Primary task measurement	• Small cameras or • Eye tracker • Test vehicle	• Visual/Visual-manual distraction	Input-based	△	○	○
	Driving performance	Primary task measurement	• Equipment for recording driving behaviors • Test vehicle	• Visual-manual distraction • Other than visual-manual distraction	Response-based	△	◎	○
Ex-ante assessment (bench test)	Occlusion method	Primary task measurement	• Equipment to obstruct vision (such as shutter goggles)	• Visual-manual distraction	Input process	◎	○	◎
	LCT method	Dual task measurement	• LCT software • PC	• Visual-manual distraction • Other than visual-manual distraction	Response process	○	△	△
	DRT method	Dual/triple task measurement	• Equipment for providing stimuli • Response button • Test vehicle/driving simulator	• Visual-manual distraction • Other than visual-manual distraction	Response process	△	Unknown	Unknown
Compatible index	Subjective evaluation index	Primary task measurement	• Questionnaires • Scale	• Visual-manual distraction • Other than visual-manual distraction	Overall	Depending on assessment methods	○	△
	Physiological index	Primary task measurement	• Biological instruments (e.g., heart rate)	• Visual-manual distraction • Other than visual-manual distraction	Overalls	△	△	△

Measures: Measures for visual behavior include the frequency of glancing at the display device and the glance duration during a continuous visual/manual task (Fairclough et al., 1993). Since most drivers look at the device repeatedly in the course of a continuous task, many studies obtain both the total glance time for looking at the display device (TGT: Total Glance Time) and the average time of a single glance (SGT: Single Glance Time) (Dingus et al., 1989).

Example of application: There are many studies that have assessed visual behavior as an index for visual-manual distraction. The criterion values for allowable visual-manual task during driving are established based on such studies. For example, the guidelines developed by the Japan Automobile Manufacturers Association (JAMA Guidelines) provide that "the total glance at the screen until the completion of a task must not exceed 8 seconds (JAMA in-vehicle display guideline version 3.0)." This value is based on the level of anxiety evaluated subjectively by drivers during driving and with reference to the lateral displacement of the car as an external criterion (Fig. 5.42) (Kakihara and Asoh, 2005). It is a value that stands on the safety during a drive. The guidelines developed by the Alliance of Automobile Manufacturers (AAM Guidelines) recommend that the SGT during a visual-manual task should not exceed 2 seconds and the total time of looking away from the road (comparable to TGT) (TEORT: Total Eye Off Road Time) should not exceed 20 seconds (Alliance of Automobile Manufacturers). These are based on the discussion that safety may deteriorate when drivers look away from the road for 2 seconds (French, 1990). The NHTSA proposes that SGT should be under 2 seconds and TEORT under 12 seconds when referring to these two guidelines (National Highway Traffic Safety Administration, 2013).

Note: While visual behavior is an adequate index in terms of safety (Rockwell, 1988; Zwahlen and DeBald, 1986), it is unclear on which index (SGT, TGT or glance frequency) the qualitative difference in visual-manual distractions is reflected (for example, differences in the visibility of the screen and in the response of the device) (Fairclough et al., 1993).

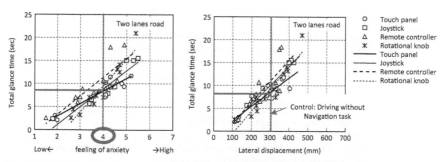

Fig. 5.42: Relationship between TGT and external criteria (e.g., a general road with two lanes) (Kakihara and Asoh, 2005).

5.3.3.1.2 Driving Performance

Assessment method: Visual-manual distractions can be assessed by measuring the quality of driving performance (Zaidel, 1992) on the actual road, based on the information obtained through the vehicle.

Measures: Measures can be generally categorized into longitudinal indices that are related to the accelerator and other pedals operations, and the lateral indices related to the steering wheel operations (Table 5.10) (see 6.2.1.2).

For the assessment of visual-manual distraction based on driving performance, the AAM guidelines propose to compare the driving performance during manual radio tuning, which is a conventional task, as a control condition to assess distraction for a target visual-manual task. In addition, distraction can be also assessed based on the comparison with the driving performance without any additional load (Kakihara and Asoh, 2005).

Example of application: Distraction assessment based on driving performance directly measures safety-related items, such as lateral displacement. This assessment method can be used not only for visual-manual distractions, but also for cognitive distractions. Due to such features, driving performance is used as an external criterion when examining the validity of other indices.

Note: While driving performance is a highly adequate index, it is important to understand on what element of the driving performance the qualitative differences in visual-manual distraction is reflected, as with the method based on visual behavior (Ranney et al., 2005).

It should also be noted that measurement of the driving performance takes a significant amount of time and effort as it is conducted on an actual road. When using the driving performance for a direct assessment method, the traffic condition at the test site needs to be taken into account when interpreting the assessment results. If an

Table 5.10: Example of Driving Performance Indices.

	Example of Index
Longitudinal	Average speed of the test vehicle Standard deviation of the speed of the test vehicle Maximum acceleration/deceleration of the test vehicle Average distance to the lead vehicle Standard deviation of the distance to the lead vehicle Average of the TTC (Time to Collision) between the test vehicle and the lead vehicle Standard deviation of the TTC between the test vehicle and the lead vehicle Average of the THW (Time Headway) between the test vehicle and the lead vehicle Standard deviation of the THW between the test vehicle and the lead vehicle Pedal reaction time in response to the deceleration of the lead vehicle Maximum pedal pressure in response to the deceleration of the lead vehicle Indexing based on the coherence technique (Brookhuis et al., 1994)
Lateral	Average lateral position of the test vehicle Standard deviation of the position of the test vehicle Range of the test vehicle's lateral positions (that is, lateral displacement) Maximum lateral acceleration of the test vehicle Average steering angle Standard deviation of the steering angle Range of the steering angle Maximum lateral acceleration of the steering angle TLC (Time to Line Crossing) Indexing based on the steering entropy method (Nakayama et al., 1999; Kondoh et al., 2015) (see 5.1.3)

event is used to obtain a driving performance measure, the situation when the event happens should be carefully examined. If visual-manual distraction is assessed based on the driver's reaction time to an event, such as the deceleration of the lead vehicle, the reaction time depends on whether the driver was looking at the roadway or while glancing at the display device. In this case, it is required to control the timing of the onset of deceleration. It should be also noted that repeated occurrence of a certain event may cause the driver to focus on the reaction to them. This can be a cause of bias in the result.

When measuring indices based on a continuous activity, such as car-following performance, the total required time for completing a task (TTT: Total Task Time) may affect the results. The effect on the performance measure may be small in a task with a short TTT. Therefore, it is important to examine the relationship between the length of TTT of visual-manual tasks and the driving performance measure.

5.3.3.2 Occlusion Method

Since the direct assessment methods take a significant amount of time and effort, a surrogate method that can assess visual-manual distraction in a simple experimental setup is needed. The occlusion method is one of such surrogate visual-manual distraction assessment methods.

Assessment method and measures: The occlusion method uses a crystal liquid shutter to simulate the drivers' state of looking at the road (shutter closed) and of looking at the in-vehicle device (shutter opened) while operating the in-vehicle device in a static state that does not involve driving. It measures visual-manual distraction based on the total time when the shutter was open during a task (TSOT: Total Shutter Open Time) (equivalent to the total glance time on the device) (Fig. 5.43).

Example of application: A surrogate assessment method needs to fulfill the following requirements: (1) it is highly adequate and is related to safety indices during a drive; and (2) it is capable of distinguishing the qualitative and quantitative difference among visual-manual tasks. The occlusion method fulfills both these two requirements (Ohtani et al., 2007).

For this reason, the JAMA Guidelines (JAMA in-vehicle display guideline version 3.0) and ISO (ISO 16673, 2007) use the occlusion method for visual-manual distraction assessment. However, there is a difference in how they apply this method. ISO fixes the shutter opening time at 1.5 seconds and the closing time at 1.5 seconds, the JAMA Guidelines fix at 1.5 seconds and 1.0 seconds, respectively. Results of a study on the actual road show that the mean +1SD of the duration of a glance outside

Shutter: Open   Shutter: Close

Fig. 5.43: Opening and closing a shutter used for the occlusion method.

during a visual-manual task was 1.0 seconds, and the mean +1SD value of the duration of glance at the device was 1.5 seconds (Asoh, 2002). The shutter opening/closing times used in the JAMA Guidelines are based on these values. Another reason that the JAMA Guidelines used this shutter opening/closing times was most relevant to the TGT during a run using an actual vehicle (Asoh, 2002).

The JAMA Guidelines do not allow to operate visual-manual tasks whose TSOT exceeds 7.5 seconds while driving. This TSOT value was equivalent to the TGT threshold (8 seconds) provided in the JAMA Guidelines (Akamatsu, 2008). In addition, the AAM Guidelines (Alliance of Automobile Manufacturers) and the NHTSA Guidelines (National Highway Traffic Safety Administration, 2013) restrict the use of tasks whose TSOT exceeds 15 seconds and 9 seconds, respectively.

Note: The occlusion method can be regarded as a highly adequate method with a great sensitivity for the demand of visual-manual tasks because of a close relationship with the TGT while driving an actual vehicle. However, further examination must be conducted to assess distraction from visual-manual tasks whose SGT exceeds 2 seconds.

5.3.4 Cognitive Distraction Assessment

In addition to the occlusion method that measures visual/manual distraction, there is a need to develop surrogate assessment methods for cognitive distractions that do not involve gaze shifting, such as distractions caused by vocal manipulation of devices. Below outlines surrogate methods for assessing cognitive distractions together with visual-manual distractions have been proposed. This section also discusses the physiological index, which is one of the indices used for the direct assessment method.

5.3.4.1 Lane Change Test (LCT Method)

The LCT method (Lane Change Test Technique) was developed by the ADAM Project in Europe. It is a surrogate method for assessing cognitive distractions along with visual-manual distractions (ISO 26022, 2010).

Assessment method: The LCT is a dual task assessment method that asks the driver to conduct a subject task while conducting simulated lane change test issues provided by a PC. Distraction caused by the subject task is estimated based on the performance measures of the simulated lane change. The LCT tasks require drivers to move to the lane instructed by the road sign as promptly and accurately as possible. The instruction in the sign appears in a pop-up window when the driver reaches the point of 40 m before the sign. The driver is asked to repeat the subject task as many times as possible while responding to the 18 LCT tasks during a trial (Fig. 5.44).

Assessment items: The results of the LCT method are obtained by calculating Mdev (mean deviation) and Adaptive Mdev, which will be discussed later, based on the deviation of the actual trajectory of the driver compared with the pre-fixed reference trajectory (Fig. 5.45).

Note: The assessment procedures standardized by the ISO use the dual task method that requires drivers to make their best efforts in both the tasks, rather than focusing on one of them. Therefore, it is important to obtain both other performance results in

Fig. 5.44: PC screen set for the LCT and a picture from the test.

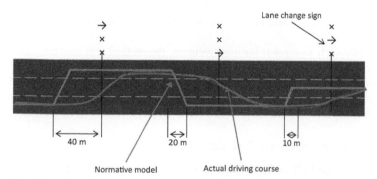

Fig. 5.45: The reference trajectory provided by the LCT and the driver trajectory.

the LCT tasks and those in the subject task, but the ISO states that distraction should be estimated based on the results in the LCT performance only, while the results in the subject task are positioned as reference data. However, it should be pointed out that the sensitivity of the LCT performances degrades when the drivers are highly learned to the dual task settings as they use a strategy to address the LCT tasks without looking at the display for the visual-manual task (Fig. 5.46) (Young et al., 2011). Surrogate assessment methods are required to have sensitivity to the difference of task demand between the tasks as well as to the indices related to the safety during driving an actual vehicle, but the LCT method is also questioned in this regard (Young et al., 2011; Ohtani et al., 2008). Moreover, while the LCT method is expected to be applied for the measurement of cognitive distractions, there is a report showing that there is learning effect that degrades the sensitivity among tasks (Ohtani et al., 2007; Petzoldt et al., 2014) and that the LTC method has low sensitivity to the difficulty of numerical memory tasks (Lei and Roetting, 2011).

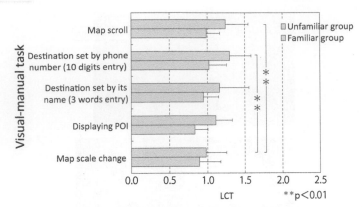

Fig. 5.46: Decreased sensitivity of the LCT due to the habituation of dual tasks (Ohtani et al., 2008).

For these reasons, measures to improve the LCT method are considered. For instance, the ISO provides a standardized method that calculates the adaptive reference trajectory based on the baseline runs conducted without any added task in order to enhance the sensitivity for the tasks by reducing the differences in the experimental conditions of the experiment sites. It then calculates the Mdev and Adaptive Mdev during individual tasks based on deviation from the adaptive reference trajectory (ISO 26022, 2010).

Despite such a proposal for improvement, the method still contains many issues regarding procedures and analysis as pointed out before, such as the procedures concerning the instructions and assessment indices for the dual tasks that affect the allocation of the driver's attentional resources, learning effects to the LCT task, and the timing of providing the subject task when given repeatedly during LCT.

5.3.4.2 Detection Response Task (DRT Method)

In addition to the LCT method, the detection response task (DRT) method is also an assessment method for visual-manual distractions and cognitive distractions. While the LCT method measures distractions based on general driving behaviors, the DRT method is focused on the detection response.

Assessment method: The DRT method uses a dual-task situation that requires drivers to respond to simple sensory stimuli while addressing the subject task or a triple-task situation that requires drivers to conduct similar dual tasks during driving on a driving simulator. The level of distraction is assessed by measuring detection response time to stimuli and error response rates.

When using the DRT method, the selection of the sensory modality for stimulation is the first step. The common sensory modalities used for the DRT method are visual, auditory, and tactile senses. When selecting the modality for receiving stimulation in a DRT, it is important to consider whether the tasks that are conducted simultaneously compete with each other if the same modality is used for both subject task and stimulation task, as Wichens (1984) assumed in the multiple resource model (Wichens and Hollands, 1999). The ISO working group worked on the DRT with three types of stimulation methods (Marie-Pierre and Laëtitia, 2013) (Fig. 5.47) and published the DRT standard (ISO 17488, 2016).

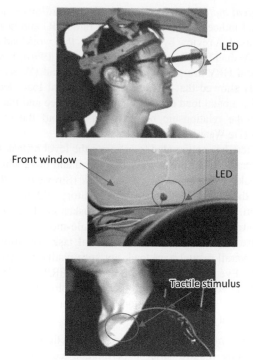

Fig. 5.47: Stimulation methods for the DRT considered by the ISO 17488 (ISO 17488, 2016).

Note: In the DRT method, the intervals between and frequency and strength of stimuli are important for the measurement of distractions. Results may vary, depending on the timing of stimulation in the course of a subject task. Drivers' attention tends to be concentrated on the performance of DRT when the frequency of the stimuli is high or when the stimuli are strong. It is important to consider the allocation of the drivers' attentional resources to the subject task. Moreover, as with the LCT method, it must be borne in mind the influences of the instructions given to the drivers regarding the priority among the tasks and of learning to multiple tasks condition, when conducting more than two tasks simultaneously. For example, it is necessary to interpret the results based not only on the performance in the DRT but also the performance in the subject tasks (for example, TTT and error rates during a task).

There is a report stating that the DRT method has an excellent sensitivity to detect the differences between the task demands that cause visual-manual distractions or cognitive distractions (Harms and Patten, 2003; Olssson and Burns, 2000). Further examination is needed regarding the requirements to enhance such sensitivity as well as the corresponding relationship to the indices related to safety during driving an actual vehicle, as discussed above.

5.3.4.3 Physiological Index

In addition to visual behavior and driving performance, the physiological index can be used for determining the influence of distractions.

Types of physiological indices and assessment items: Cardiovascular parameters can be the physiological indices. Among them, heart rate (HR) and heart rate variability (HRV) are widely used. A review by De Waard states that workload is more related to the decrease in HRV than to HR in principle (De Waard, 1996); however, in contrast to this, both HR and HRV are poorly related to workload (Wierwille et al., 1985). In addition, a study showed that the increase in physical load decreased HRV and increased HR, while mental load decreased only HRV (Lee and Park, 1990). Another study investigated the relationship between the HR and the task demand using frequency analysis (De Waard, 1996) (see 4.4.4.2).

Some researchers have attempted to estimate the level of task demand by EEGs (see 6.1.1.2). For example, the alpha wave decreased and theta wave increased when conducting dual tasks as compared to a single task (Sirevaag et al., 1988). Another study showed that the P300 of ERP evoked by auditory odd ball stimuli could be an index of distraction in dual task condition (Rakauskas et al., 2005). Another study attempted to estimate the workload based on the oculo-motor system (VOR) (Obinata et al., 2008). These physiological indices for dual task conditions are considered reflections of the overall mental workload, not specifically reflections of distraction of driving task. The Eye Fixation Related Potential (EFRP) that is ERP triggered by the timing of onset of eye fixation can be a direct physiological measure of distraction of driving task because EFRP is a reflection of the amount of visual attention to road environments (Takeda et al., 2012).

Note: While there are many attempts to assess distractions based on physiological indices, such methods require strict experimental control and data processing procedures in order to remove artifacts. It is often difficult to obtain useful information to improve in-vehicle systems due to difficulty in understanding the changes in a very complex biological system solely based on physiological indices (see 4.4.1.3).

5.3.5 Reference Tasks in Distraction Assessment

Some distraction assessments use reference tasks in order to measure distractions caused by conducting non-driving-related tasks. Such reference tasks are used for verification of the adequacy and reliability of various distraction assessment methods.

5.3.5.1 Item Recognition Task

Origin: Based on an assumption that human information processing involves several sub-processes and each sub-process needs a certain amount of time, Donders attempted to estimate the processing time required for each sub-process based on the difference between the reaction times for experimentally controlled sensory-motor tasks (subtractive method, Fig. 5.48) (Donders, 1969).

While the approach of Donders stands on the assumption that individual sub-processes are independent from each other, Sternberg tried to understand information processing for memory tasks without assuming such independency. In his studies, Sternberg used item recognition procedures (Sternberg, 1969).

Human memory involves two actions: Recalling what is stored in the memory and recognition that determines whether what is remembered belongs to a specific category.

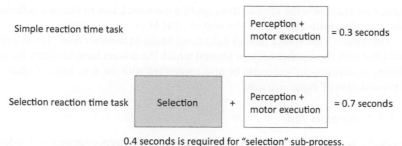

Fig. 5.48: Example of applying Donders' subtractive method.

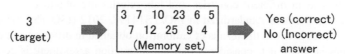

Fig. 5.49: Example procedures for Sternberg's item recognition task.

Assessment items: The item recognition task provides test stimuli (targets) and a memory set given by visual or auditory means to drivers and they are asked to tell whether the target exists within the items of the memory set as immediately and accurately as possible (Fig. 5.49).

The results of the task are calculated, based on the response time and error rates. The difficulty of the task is controlled by the number of items included in the memory set, intervals between the provision of the target and the provision of the memory set, and the order of the provision of the target and the memory set. The item recognition task is considered for not only use as a reference task but also for use as a workload measurement method (Driver workload metrics project, DOT HS 810 635, 2006).

5.3.5.2 N-back Task

Origin: In addition to Sternberg's item recognition task, the n-back task is considered for use as a reference task. The *n*-back task is related to human information processing and was used for the first time to determine the age differences in short-term memory (Kirchner, 1958).

Assessment items: The *n*-back task presents a sequence of stimuli in the forms of digits or letters by a visual or auditory means and drivers are asked to respond when a stimulus matches the digit or letter from *n* steps earlier (Fig. 5.50). Mehler et al. applied *n*-back task for driving situation (Mehler et al., 2009). In their studies, the stimuli were auditory presented and drivers were required to verbally recall the delayed digit to

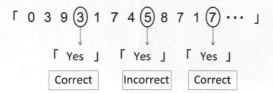

Fig. 5.50: Example of the *n*-back task (2-back).

imitate conversation in the vehicle. Ross applied the *n*-back task to examine influences of the cognitive loads on the LCT (Ross et al., 2014).

The results of the *n*-back task are calculated based on the error rates. The difficulty is adjusted with the number of steps beyond which the drivers have to match the digit or letter. In addition, difficulty can be also controlled with the time intervals between the presentation of digits or letters.

5.3.5.3 Calibration Task

Approach: In order to compare the results obtained in different countries and different laboratories, it needs to be ensured that these results are obtained under the same conditions. The calibration task is one of reference used for determining whether the results obtained in different countries and laboratories are obtained under similar conditions. This task is standardized in ISO/TS14198, 2012 (ISO/TS 14198, 2012). If the difference between the results of calibration obtained in different countries and laboratories is small, it is considered that the distraction assessment is comparable with different characteristics and attributions of participants, and even under different experiment conditions. There are two types of calibration tasks that are publicized as the ISO's technical specification (ISO TS14198): Critical Tracking Task (CTT) and Surrogate Reference Task (SuRT).

CTT: The CTT is a visual-manual task conducted on a PC that gives drivers a continuous tracking task. Specifically, it asks drivers to match the target bar with the reference line positioned at the center of the target area (horizontal line) by moving the target line with the up-and-down arrow keys on the PC. The difficulty of the CTT is controlled with the factor λ that defines the displacement of the target bar in relation to the input value of the drivers' key manipulation. The ISO recommends testers to use a constant value ($\lambda = 0.5$) when conducting a task.

SuRT: The SuRT (Surrogate Reference Task) is based on the idea that the feature extraction is automatic and parallel processing is assumed in the feature-integration theory proposed by Treisman and Gelade (Treisman and Gelade, 1980). The SuRT is also a visual-manual task conducted on a PC. Drivers are asked to search for a circle (target) of a certain size among multiple circles (distractors) and respond by moving the vertical bar to the position of that circle (Fig. 5.51). The difficulty of the SuRT is controlled with the similarity in diameter between the target and distractors (the task becomes difficult when they are of similar sizes) and the width of the vertical bar manipulated by the drivers (the task becomes difficult when the bar is narrow since it becomes tough for drivers to match it with the target). Regarding the procedures

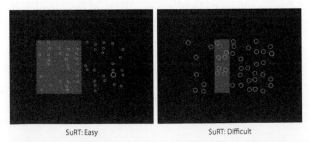

SuRT: Easy   SuRT: Difficult

Fig. 5.51: Example of the SuRT.

for the SuRT, ISO/TS14198 proposes that the SuRT be conducted at the drivers' pace when it is conducted in combination with other tasks.

Petzoldt et al. showed that there was learning effect in both the CTT and SuRT that affects the performance in the LCT (Petzoldt et al., 2011). In the revision process of the ISO, the *n*-back task is included in ISO/TS14198.

5.3.5.4 Conclusion

When using a reference task as secondary task, it has to fulfill the requirements for secondary task paradigm as mentioned above. Among the item recognition tasks, *n*-back task and calibration task (CTT and SuRT) fulfill the requirement in terms of the capability of continuous assessment and where it is important to understand other characteristics of these reference tasks when using them. For example, while the ISO proposes that the SuRT be conducted at the drivers' pace, drivers tend to be focused on the SuRT when its difficulty is of moderate level, which may increase the interference with the primary task. As a result, a quantitative difference may not be obtained even if a qualitative difference is obtained, when compared with the distraction caused by a task of higher difficulty level. Detailed examination is necessary on what kind of influence the results of various reference tasks have on the actual driving behaviors.

This section introduces several methods for measuring drivers' distraction. It is important to conduct multi-faceted assessment using an effective and efficient test battery composed of multiple methods. When doing so, assessment methods using a subjective evaluation index are useful although not discussed in this section. Major subjective evaluation indices used for workload measurement include NASA-TLX and SWAT (Subjective Workload Assessment Technique) (see 5.1.1). The anxiety scale that the JAMA Guidelines used to establish the TGT and TSOT criterion value is shown in Fig. 5.52. It is also necessary to determine the policy for the improvement of in-vehicle HMI designs based on drivers' introspective reports.

5.3.6 Use of Cellular Phone while Driving

Cellular phone use while driving has come under intense scrutiny as it is considered a typical driver-distraction situation. Therefore, there have been many research papers regarding this matter. NSC (National Safety Council) of US reviewed related literatures and statistical data that provide evidence of driver distraction (*NSC: Distracted Driving Research Statistics*). There were also literature review papers (Svenson and Patten, 2005; Brace et al., 2007).

Svenson et al. categorized influences of cellular phone use on driving on human factors into Information Input, Central Processing and Psychomotor Output. They further divided researches according to types of research approaches—studies in the laboratory setting, those with driving simulator, those in actual road or test track and those of epidemiology. Table 5.11 summarizes the results of this review. The reaction time became longer and the smoothness of driving operation reduced by the use of cellular phone. The complexity of conversation was the important factor for them. From the epidemiological studies, the risk factor could be estimated as two to four times higher than without the cellular phone use. It was observed in a real road environment.

Table 5.11: Influences of Cellular Phone Use on Driving.

Research Approach	Human Factors	Driving Task	Cellular Phone Use Task	Measures	Findings
Laboratory Setting	Information input	Tracking task. Watching video movie of driving scene, etc	Telephone number entry (comparing with radio tuning). Mental arithmetic task. Conversation task	Glance behavior. Glancing patter	Longer reaction time for 50 ~ 200 ms with cellular phone use
	Central Processing			Stimulus-response time. EEG (ERP)	
	Psychomotor Output			Tracking task performance	
Driving Simulator Experiment	Information input	Driving in constant speed, vehicle following, detection of traffic signal, etc.	Telephone number entry (effect of auditory feedback). Mental arithmetic task. Conversation task	Visual information processing (memorize information on billboard)	Longer reaction time for 30 ~ 600 ms with cellular phone use. Longer glance time to inside of vehicle cabin. Increase number of large steering maneuver
	Central Processing			Heart Rate Variability. Mental workload	
	Psychomotor Output			Total task time	
On-road or test track Experiment	Information input	Driving on test track, highway, urban area. With peripheral detection task, etc	Mental arithmetic task. Memory task. Conversation task	Glance behavior. Allocation of visual attention	Longer reaction time for 130 ms with cellular phone use. Large vehicle fluctuation (03–0.4 m)
	Central Processing			Mental workload. Perceived risk	
	Psychomotor Output			Brake reaction time, TTC (Time-to-collision), vehicle behavior	
Epidemiological study	Accident, incident, risk	Traffic accident statistics		Number of accident/incident with/without cellular phone use	Twice to four times higher risk of accident/incident with cellular phone use

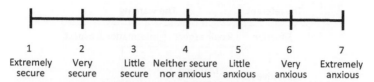

Fig. 5.52: Example of an anxiety scale.

5.4 Interaction with Advanced Driver Assistance Systems

5.4.1 Presentation and Management of Information

5.4.1.1 Design of Warning Signal

5.4.1.1.1 Warning

A warning is the presentation of message to a person when there is a possibility of a person or property getting damaged if the person does not take a proper action or does not terminate a current action to avoid it. Warnings are artificially-created messages that do not essentially exist in the environment. They are the message that the designer of the system decides to present to the user based on information obtained from the environment/object and attaches a meaning to it. They should be carefully designed in order to prompt a necessary behavioral change in the user by conveying the situation to the user clearly and accurately.

5.4.1.1.2 Warning Compliance

Cues as changes in the situation of the environment or the state of the object play an important role in our perception of risk. When there are these cues that are able to be perceived by the user, the warning can be an effective signal for the user to understand them better. When there are no such 'perceived cues' from the environment, the risk assessment becomes less accurate. In the traffic environment, we can perceive cues that allow us to understand the risk of a situation most of the times (e.g., the distance from the vehicle in front becomes shorter, the vehicle in front is waving along the road, etc.). However, in some cases, as in a traffic jam at the end of a blind curve, it is difficult to obtain a cue from the road environment. In other cases, we fail to perceive the cues because we are looking aside or our level of vigilance is low. We can perceive some sensory cues from mechanical vehicle sub-system, as strange smells or noises, but we cannot perceive any cue from electrical devices.

Warnings can be conveyed through the senses of vision, hearing, touch, taste and smell. We use physical characteristics as colors, sizes, sound frequencies, tones or vibrations for designing warning. We may also use linguistic information for visual and auditory signals. The former is an 'iconic' signal that we associate with a certain situation or condition, and the latter is expressed through clear and concrete information. Changes in behaviors caused by warnings are the result of a combination of three factors, namely, cues from the environment, iconic aspect and informational aspect of warning compliance (Fig. 5.53) (Edworthy and Adams, 1996). Generally, perception of cues from the environment has a strong influence. If it smells burnt and we hear an alarm sound, we quickly understand that a fire has broken out. On the

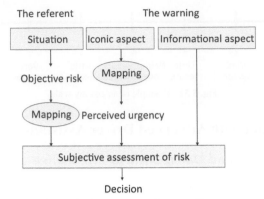

Fig. 5.53: Factors of warning compliance.

other hand, if there is no smell and we hear an alarm, we may take it to be a false alert. When a fire has broken out in a tunnel, we tend to follow the red light in the entrance of tunnel and stop the vehicle if we see smoke coming out from the tunnel; if there is no smoke, we tend to ignore the red light and go ahead into the tunnel. This example suggests that we need to make a lot of effort to develop effective warnings when there are no other cues from the environment.

Behaviors of other people, which are an external factor, also influence the behavior to avoid a risk. For the tunnel fire case, drivers are more likely to stop the vehicle if other people have stopped.

5.4.1.1.3 Expected Driver's Response

Warning from collision warning system and malfunctioning warning of vehicles: Before collision warning systems as an advanced driver assistance systems (ADAS) came into use, vehicles' warning signals informed drivers of equipment malfunctioning (Merker, 1966), as low cooling water, low oil pressure, malfunctioning of the brake system, etc. This sort of warning signal intended to prompt the driver to do something about the malfunctioning, as stopping or repairing the vehicle, in order to avoid further damage to the equipment. Although a failure itself may not harm the person, if unnoticed, failure of the brake system have a high probability of causing accidents. On the other hand, warnings of collision prompt proper behavior to prevent human damage directory.

Malfunctioning warnings and warnings of collision warning systems expect different actions from the driver. Malfunctioning warnings expect that the driver will take action,like taking the vehicle to an auto-repair shop, giving up on driving the vehicle or parking it at somewhere safe. Collision warning systems expect that the driver will take action to avoid a collision, like braking hard, changing the direction, giving up trying to change lanes and going back to where he/she was (Table 5.12) (ISO TR12204, 2012). If the driver ignores a malfunctioning warning, the car may be irreparably damaged. If the driver ignores a warning from the collision warning system, the risk of collision with other vehicles or pedestrians increases. In this way, the same term, i.e., 'warning', is used for both cases though they have different purposes.

Table 5.12: Expectation of Driver's Response by Warnings (Modified from ISO TR12204).

Expected Driver Behavior	
Preparation	
Understanding	Direct one's attention to a certain object or event (assumed to be a hazard/ threat) and to recognize the situation
Action preparation	Prepares to respond by deciding on an action
Change plan	Terminate intended action (giving up lane changing)
Response	
Braking	Hard braking is a driver's response that is intended to slow or stop the vehicle movement as quickly as possible
Acceleration	Rapid acceleration is used by drivers to avoid an imminent collision, such as when a vehicle or object is approaching the subject vehicle from the side or the rear
Steering maneuver	Emergency steering maneuver is intended to steer a vehicle around an object to avoid a collision
Retake control	The driver is required to re-take control of the vehicle when a vehicle control system becomes inactive or exceeds its range-of-control

Warnings by collision warning systems, which expect different responses from the driver, may conflict with malfunctioning warnings by using the same color (e.g., red), etc., because the warning design of the latter has already established a long history as to the use of colors and lamps. Therefore, we must design so that the driver may not misunderstand the meaning of the warning and for this we have to the different area of display, etc.

5.4.1.1.4 Warning Level and Warning Design

(1) *Criticality and urgency*: The behavior change that needs to be encouraged depends on level of criticality and urgency to avoid damage. In some cases it is required of the driver to take immediate action as slowing down when there is an imminent danger of a considerable damage; there are also some cases where there is time for the driver to think about the situation, whether he/she should take a certain action or not. There are also cases where the systems leave the decision to the driver either to follow the warning or not. These differences are called 'warning levels'.

The warning level takes both the criticality (severity of the damage) and the urgency (how close is the danger or the imminence of the danger) into consideration. The priority of the warning is determined by a combination of these two factors. When there are multiple warning systems including malfunctioning warning systems, we have to define the priority of each warning and manage the information by displaying a low priority warning only after the high priority warning has ended; for example, see 5.4.1.3. Otherwise, when more than one warning is to be presented at the same time, the driver may get confused, causing him/her to disregard the warning with a higher level of priority.

(2) *Warning level*: Warning signals are categorized into 'Warning' in the narrow sense and 'Caution', the former is made when there is a high risk or a risk of a great danger; and the latter is made when the danger or risk is low (even if the expected damage is large) or the expected damage is small. The MIL-STD-411F (Aircrew Station Alerting

Systems) standard of the US Army, which handles more general warning systems, defines 'Warning' as a signal which alerts the user to a dangerous condition requiring immediate action (MIL-STD-1472G, 2012), and 'Caution' as a signal which alerts the user to an impending dangerous condition requiring attention, but not necessarily immediate action. A signal to indicate safe or normal condition or to attract attention, including a signal to indicate a change in system status which, though important, does not require immediate action by the user is called 'Advisory Signal'.

In this way, we generally use warning signals dividing them into three levels. Although it is possible to categorize them in more detail, it would make it hard to distinguish between them. As we will show in the next section, the number of colors that can be distinguished in visual warnings is about three or five at the maximum.

5.4.1.1.5 Basic Requirements for Warning Designs

(1) *Visual presentation of warnings*: The warning must be recognized as such immediately after being activated (according to the MILSTD-1472G, in 0.5 seconds). In designing visual displays, it is possible to convey meanings through colors, blinking lamps, icons/symbols and letters.

(a) *Colors and blinking*: Below, we refer to requirements specified in MIL-STDs (MIL-STD-1472G, 2012) and (MIL-STD-411F, 1997). The color of warning signals is red and the blink frequency is of 2 to 10 Hz (MIL-STD-1472G or 3 to 5 Hz, according to MIL-STD-411F). The duty ratio of the lights (lighting time/blinking period = ratio of time that the lamp is on while the light is blinking) is 0.5 (MIL-STD-1472G). The size of the warning symbol should be over 10 mm and the font size from 30 to 60 degrees (MIL-STD-1472G), or 20 degrees at the minimum (MIL-STD-411F). The color of caution signals is yellow and the blink frequency is 2Hz (MIL-STD-1472G) with a duty ratio of 70 per cent (MIL-STD-1472G).

(b) *Symbols and letters*: For symbols and icons, we use the image of the purpose (as a cross mark: ×), abstraction of the event (broken glass) and new synthetic symbols (e.g., the inverted triangle for stop). Symbols in forward collision warning systems (FCW or FVCW) and lane change decision aid systems (LCDA) are specified in ISO 2757 (ISO 2575, 2010). Note, however, that these symbols are symbols of the system itself and not necessarily express the situation which the warning intends to convey. As mentioned in 5.4.1.1.2, it is desirable that the system uses symbols that can be easily associated with future collision for enhancing the understanding of the driver, which is improved when the symbol matches the situation to be indicated.

When displaying the warning with words, we need to select the word considering its legibility, descriptiveness and length. It should be expressed using short words (words with less than eight letters or four syllables in case of English) and less than two words if possible. MIL-STD-1472G uses 'Danger', 'Warning', 'Caution' and 'Notice' as warning labels; 'Danger' is used for the most dangerous situations and 'Notice' for use when there is no direct risk of damage. When using sentences, double negatives are not used because it may delay the understanding of the meaning.

(c) *Position of display*: The warning has to be displayed at a location that is easy to see while driving. The location needs to be determined taking into consideration the visual behavior that the driver needs to perform while driving. Visual warnings of forward collision warning system that are supposed to be presented when the driver is looking at the road ahead are displayed in HUD or the meter cluster that is within 15 degrees down. In case of lateral warning systems (that inform of vehicles approaching from the sides, etc.) and lane change decision aid systems (that inform of the existence of vehicles moving parallel when initiating lane change), it is expected that the driver will be looking at the door mirrors, as it is easy to be perceived if the warning signal is presented somewhere close to the mirror or inside it.

(d) *Auditory presentation of warning signals*: For example, when a forward collision warning appears during driving, there is a possibility that the driver is not looking forward, e.g., looking at something else inside or outside the vehicle, or his/her vigilance level is low, and fails to perceive a visual warning signal. Therefore, using auditory signals that use sound is useful to make sure that the information is being conveyed without fail (on a Warning or Caution level).

Since warning sounds need to be heard in the environmental noise while driving, their sound pressure should be 15 dB higher than the background noise (10 dB at the minimum). The frequency has to be between 250 and 8000 Hz, if possible between 500 and 2000 Hz. The sound pressure at the ear has to be less than 115 dB because excessively high sound pressures may cause a 'startle' reflex. 'Startle' reflex is also caused by sudden increase in pressure. This should not be raised more than 30 dB in the first 0.5 second (MIL-STD-1472G, 2012).

We also have to make sure that the driver will not confuse it with auditory signals from other systems. Tonal sounds (with fixed frequency or change frequency), 'earcons' (symbolic sounds that are associated with or simulate the sounds of the real environment) or verbal messages can be used as auditory warning signals.

Sounds are often used as pre-cue (or cue) signals for visual and auditory presentations. Cue signals are called 'annouciator', which are sounds given for a brief period of time before the message of the warning is conveyed and that, although not loud, can be distinguished from other sounds. Cue sounds are useful to attract attention of the driver to the warning displayed when he/she is distracted with something else.

(2) *Impression given by the design of warning signals*: The requirements regarding the use of colors and blinking lights according to the warning level are the result of empirical research. The relation between the degree of 'urgency' and 'criticality' and the impression given by a certain color or blinking pattern was experimentally investigated through an experiment with 50 participants (Uno et al., 1999).

As visual signals, the participants were given a rectangle on the screen of VDT with different colors (red, yellow, green, blue and white), brightness (from 50 cd/m^2 to 150 cd/m^2), sizes and blinking patterns (frequency and duty ratio). Participants were asked to judge according to the perceived level of 'criticality' and 'urgency' from the signal. An interval scale was obtained, using Thurstone's method of paired comparisons based on the ordered scale. The same process was used for auditory signals; i.e., formulating an interval scale according to the perceived level of the 'criticality' and 'urgency', after presenting sounds with different pressures (from 65 to 80 dB), frequencies (from 500 Hz to 2 kHz) and pattern of intermittency (cycle and duty ratio).

Regarding colors, there was a minor difference in the perceived criticality of green, blue and white, but there was a big difference for yellow and red. Blinking lights with cycles of 1 Hz were not much different from continuous lights and the perceived criticality becomes stronger at 2 or 5 Hz. If we compare duty ratios of

30 per cent and 70 per cent, we see that the perceived criticality becomes stronger at 30 per cent. The size also has a considerable effect on criticality (Fig. 5.54). As for the effect of sound frequency, the higher that the frequency (from 500 Hz to 1 or 2 kHz) is, the stronger is the impression. Although the impression is stronger when there are harmonic tones rather than pure tones, the influence of harmonic tones depends on the basic frequency and does not show a consistent effect. The higher the sound pressure, the stronger is the impression of criticality. As to the effect of the sound pattern, the perceived criticality increases with continuous sounds rather than with sounds at intervals of 1 or 3 seconds. When using intermittent sounds, the perceived criticality is weaker than continuous sounds with cycles of 1 Hz, but it gets stronger as the cycle gets shorter (2Hz, 5Hz, etc.). Also, the higher the duty rate, the stronger is the impression of criticality (Fig. 5.55). These tendencies were almost the same for the 'urgency'.

Although the results of this experiment are practically in conformity with the MIL-STD standards, the results regarding the duty ratio show that it has different effects on visual and auditory displays. While in visual displays, smaller duty ratios increase the criticality, in auditory warnings, bigger duty ratios increase it. When the

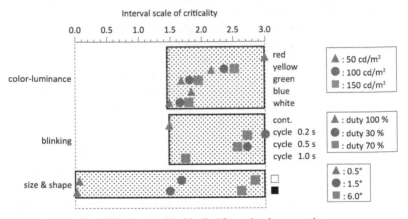

Fig. 5.54: Perceived 'criticality' from visual presentation.

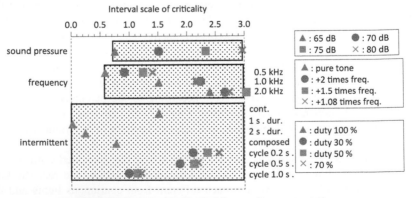

Fig. 5.55: Perceived 'criticality' from auditory presentation.

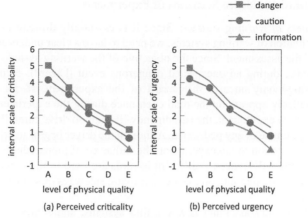

Fig. 5.56: Effect of the wording of warnings.

lights blink 'flash-flash-flash' and the sound is a 'beep-beep-beep', the impression of criticality becomes stronger.

By investigating the combined effects of visual and auditory presentation at the same time, it was shown that the perceived criticality/urgency increases almost in an additional manner. The effects of labels using words ('Danger', 'Caution' and 'Notice') with combined visual and auditory presentation were also investigated. According to the results, the impression becomes stronger when using labels but its effect in the interval scale was about 25 per cent (Fig. 5.56). When estimating the impression through magnitude estimation in actual vehicles, the effect of the label was 10 per cent. Therefore, the influence of physical properties of the sensory presentation is stronger than the linguistic expression.

5.4.1.2 Influence of Warning Signal on Driver Behavior

5.4.1.2.1 Assessment of Effectiveness of the Warning System on the Avoidance of Danger

There are studies that assess the effectiveness of ADASs in advance; however, it is essentially difficult to empirically assess their effectiveness. Situations where we need to avoid an accident are rare in the real world. Warning systems are particularly effective when a completely unexpected hazard occurs and not when the driver has perceived the danger and is driving carefully. Warning systems are expected to induce the driver to take proper avoidance action when the driver is not ready and does not expect it.

In order to empirically test the effectiveness regarding driver's behavior to avoid hazards, driving simulators or test tracks are used for maintaining the safety of the driver. The experiment is conducted through driving simulators if it is not possible to ensure the safety, i.e., when conducting an experiment on accident avoidance at high speeds. In the case of experiments at low speeds (as systems for avoiding collision with pedestrians at intersections, or for avoiding accidents at parking speeds, etc.), the security may be guaranteed even at test tracks.

5.4.1.2.2 Hazard Avoidance Scenarios of Experiments

(1) *Effectiveness of warning systems*: Since it is essentially difficult to conduct an empirical evaluation of warning systems, we need to have a clear understanding of the limitations of the assessment. Since the objective of the warning systems assessment is to be effective during an unexpected dangerous event (hazard), each participant can be submitted only once to the hazard of the experiment. If, for example, a pedestrian suddenly appears on the road even once during the experiment, after that and throughout the experiment, the participant will begin to drive carefully, expecting the sudden appearance of any pedestrian. Although the driver begins to drive carefully paying careful attention to the experiment, the influence of preparedness in avoiding unexpected events will weaken if different hazards are used in the same experiment (pedestrians suddenly jumping out, vehicles moving sideways or suddenly braking in the front, etc.).

When empirically assessing new warning systems, sometimes it is necessary to have the participants learn the meaning of the new warning design in advance; otherwise, they may not understand it. The experiment may also be conducted for investigating the effectiveness under a situation where the driver has no prior knowledge of the system. However, in real life, it is possible to assume that drivers are instructed at the moment of delivery of the vehicle or have read its manual in advance. Notwithstanding, if the participant in the experiment is informed of the warning signal that will appear at the beginning of the experiment, it is likely that he/she will be prepared for a hazard related to that warning signal. A solution to this problem is making it hard to notice the relationship by explaining that there are two experiments— one about understanding the warning signals, and another, a driving experiment. For example, by explaining to the participants that the former is an experiment about the understanding of symbols other than warnings by using a questionnaire or a computer display in the laboratory, it will be more difficult for them to notice that it is relevant to the second experiment on warning systems, using the driving simulator.

The effectiveness is assessed on the basis of the behavioral response of participants when they face a hazard in the experiment. Avoidance when the system is activated is the basic index of effectiveness (whether the participant was able to avoid it more than when without a warning system or not). Note, however, that the crash rate is an unstable probabilistic parameter because the participant may or may not avoid the collision for a slight difference. That is why sometimes it is assessed through the reaction time from the moment of the warning signal or the presence of hazard to the moment of the onset of evasive manoevers. Subjective evaluations using questionnaires are also used where participants are asked how difficult it was to avoid it, if they were able to keep calm without getting panic, etc.

In this way, there is an inherent difficulty in empirically assessing the effectiveness of the warning systems and it is not possible to estimate the effectiveness in real situations only through results of the experiments. For effectiveness in real roads, Field Operational Tests (FOT) are conducted (see 2.3.2.3). Furthermore, it is also necessary to estimate the effectiveness of the system from the actual use of the system after the product is released into the market. In some cases, traffic simulations or combined use of simulator experiments and simulations are utilized.

(2) *Assessment of the warning signal*: Although it is difficult to empirically assess the effectiveness of the warning system, we may assume that it has validity in empirical

experiments where the design of different warning signals is compared. If we want to know only the rating of the effectiveness of the signals of different designs, i.e., the relative relation between the designs, we can assess it through differences in the reaction time to the warning signal, and not through the rate of successful avoidance. Note, however, that when the same hazard is repeated, the participant starts to take avoidance actions regardless of the warning signal based on the prediction of the hazard; therefore, we need to design the experiment so as not to repeat the same hazards.

5.4.1.2.3 Assessment of the Compliance with Warning/alerting Systems

(1) *Compliance*: There are systems that present the regulatory information, such as speed limit. This kind of system encourages a driver who failed to detect or forgot it to comply with the regulation by displaying it. In this case, the index of evaluation is whether the driver took appropriate action following the signal or not. Taking action following the warning is called 'compliance' (see 5.4.1.1).

(2) Example of assessment of effectiveness of seat belts reminders:

(a) *Passenger seat-belt reminders*: The effectiveness of seat belts in reducing damage of collisions on the human body is well recognized and its use has been encouraged through regulations and campaigns. As a result, the rate of use of seat belts of the driver seat and the front passenger seat in Japan is respectively about 98 per cent and 93 per cent, which, although not 100 per cent, is quite high (Japan Automobile Federation survey, 2012). On the other hand, although it was already a legal obligation and the accident statistics showed the mortality rate increased three times when the rear-seat passenger was not using the seat belt, the rate of use of seat belts on rear seats was still low, around 30 per cent. A solution for encouraging the use of seat belts in rear seats is to introduce the seat-belt reminder (unfastened seat belt warning).

For this reason, the Japan New Car Assessment Program (JNCAP) included the installation of passenger seat belt reminders (hereinafter, PSBR) as an item of the assessment of the safety of occupants. As the effectiveness of the PSBR may differ according to the design of the warning signal, it is necessary to estimate the influence of PSBR on the use of seat belts. The effectiveness of seat-belt reminders for driver seats may be verified by field observation because they are already in the market (Freedman et al., 2007). However, since assessments are supposed to be conducted in advance, i.e., before the product is released in the market, methods as FOT cannot be applied. Therefore, the effectiveness of PSBR was empirically estimated by applying the model of compliance with the use of seat belts on rear seats as shown below.

(b) *Model of compliance with the use of seat belts of rear seats through seat-belt reminders*: In typical automobile warning signals, 'compliance' means the extent to which the driver follows the warning. Similarly, since the targets of PSBR are passengers, a simple model is the change of the rate of use by presenting the PSBR to passengers (Freedman et al., 2009). However, in Japan, the person liable to be punished for non-compliance with the rear seat-belt regulations is the driver. Therefore, displaying the PSBR will probably facilitate the driver to ask the passengers to fasten their seat belts. In other words, the effect of PSBR consists of a combination of a direct and indirect effect (the driver's request) on passengers.

If we define P as the ratio of passengers that use the seat belt even without the PSBR, $1 - P$ will be the ratio of passengers to be encouraged by PSBR to use it. Next, if we define p as the ratio of passengers that fasten the seat belt only with the PSBR (i.e., without being asked by the driver), then $(1 - P) \times p$ is the ratio of passengers that use seat belts with the PSBR and $(1 - P) \times (1 - p)$ is the ratio of passengers that do not fasten it even after the PSBR. The latter are the passengers to whom the driver requests. Finally, if the ratio of drivers that ask the passengers is q and the ratio of passengers that use the seat belt is a at the driver's request, then the total number of passengers that fastened the seat belt because they were encouraged to, i.e., the whole effectiveness of the PSBR, can be estimated by $(1 - P) \times p + (1 - P) \times (1 - p) \times q \times a$ (Fig. 5.57) (Akamatsu et al., 2011).

(c) *Estimate of the effectiveness of seat-belt reminders*: Following the model above, the values that we need to estimate are P, p, q and a. For instance, it is possible to conduct an experiment by preparing some cars equipped with PSBR, gathering a large number of people, having them ride the car and investigating the ratio of participants that fastened the seat belt on being encouraged by the PSBR. However, we may easily assume that the participants will tend to fasten their seat belts just because they are being observed. This means that in empirical experiments on the compliance with legal obligations, there is a bias as to the expected social behavior. In such cases, it is possible to remove the bias by having them evaluate not his/her own but the behavior of other people.

A subjective assessment using 100 participants was conducted for estimating the effect of the design elements of PSBR, the position of visual symbol, use of words or icons, use of blinking lights, use of beep sounds or voice messages, etc. First, experiment participants were asked to choose from 5 per cent, 25 per cent, 50 per cent, 75 per cent and 95 per cent as to how many people would use seat belts even

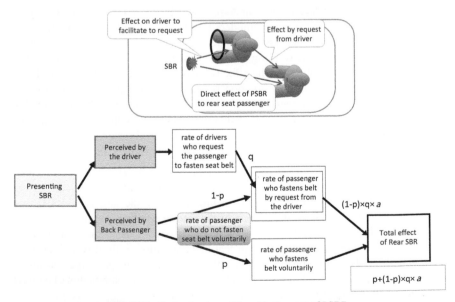

Fig. 5.57: Basic structure of the effectiveness of PSBR.

without PSBR. Next, they were asked to choose the ratio of drivers that would ask passengers to fasten the seat belt after the PSBR while sitting in the driver's seat besides choosing the ratio of people that would fasten the seat belt when prompted by the PSBR. Finally, they were asked how many people they think that would fasten the seat belt when asked by the driver. After calculating the weighted average of the number of participants that chose each alternative, the ratio for each of them was calculated.

By estimating the effect of design factors based on these results, it was noticed that the strength of the effect may be roughly grouped as follows: whether it was displayed to both the driver and the passenger or only to one of them, and whether an auditory warning was used or not.

5.4.1.3 *Priority and Management of In-vehicle Information*

5.4.1.3.1 Need for Information Management

In-vehicle information equipment and driving support systems provide many types of information—from driving-related information to non-driving-related information. Because there is a limit to the capacity of human information processing (see 1.7.4, 5.1.2 and 5.3.2), we cannot process all this information at the same time. As a general rule, human beings cannot process more than one type of information at the same time and process multiple information through time-sharing. This means that the driver cannot process information for driving while he/she is processing non-driving-related information.

Although many types of in-vehicle systems to provide information to drivers have been developed, they are usually developed separately. For this reason, each system presents information to the driver whenever necessary and requires that the driver registers it when it needs a reaction from the driver. When the vehicle is equipped with various systems, different systems may present information independently, and thus there is a possibility that various information is presented at the same time. The driver will be confused if different in-vehicle systems require interaction at the same time during driving (Fig. 5.58). Therefore, when providing two types of information or requiring interaction at the same time, as a general rule, the information that is more important should be prioritized.

Fig. 5.58: Unrestricted provision of information to the driver (DRIVE-I Project STAMMI, 1991).

5.4.1.3.2 Information Importance

The priority of the information provided (message) is determined by its 'criticality' and 'urgency' (ISO TS16951) (ISO TS16951, 2004).

(i) *Criticality*: Criticality is severity of the impact of an accident or malfunctioning that can occur when the message is not received or ignored by the driver. Table 5.13 indicates the four levels of criticality used by ISO TS16951.

(ii) *Urgency*: Urgency refers to the time within which the driver has to take action or decision if the benefit intended by the system is to be derived from the message. ISO TS16951 divides it into four levels as presented in Table 5.14.

When quantifying the priority of messages, we have to seek the function between the 'criticality' and 'urgency'. There are many possible functions, but the simplest one is the addition of 'criticality' and 'urgency'.

Because criticality and urgency depend on the situation under which the vehicle is placed, it is necessary to identify that situation before assessing the criticality and

Table 5.13: Different Levels of 'Criticality' (ISO TS1695).

Rating	Risk to vehicle, occupants and/or pedestrians	Examples
3	Severe or fatal injury	Ignoring speed warning when driving significantly above the speed limit.
		Collision as a result of loss of braking due to ignoring the brake failure warning.
		Departing roadway due to ignoring lane departure warning.
		Collicion at high speed.
		Leaving the roadway, head-on collision and collision with structures at intermediate speed.
		Following vehicle ahead too closely at high speed.
2	Injury or possible injury	Risk of collision due to following a vehicle ahead too closely at intermediate speed.
		Vehicle(side)-to-vehicle(side) collision due to ignoring collision warning at intermediate or low speed, vehicle leaving the road, head-on collision and collision with structures at intermediate or low speed.
1	No injury (vehicle damaged)	Vehicle-to-vehicle collision except head-on collision at low speed.
		Following vehicle ahead too closely at low speed.
		Collision with structures at low speed.
0	No injury (no vehicle damage)	Vehicle-to-vehicle contact at very low speed.
		Collision with structures at very low speed.

Table 5.14: Different Levels of 'Urgency' (ISO TS1695).

Rating	Description	Examples
3	**Respond immediately** Take immediate action or dicision (within zero to three seconds) according to the displayed information.	Obstacle immediately in the vehicle path. Brake immediately. Steer to avoid dangerous situations. ACC malfunctioning.
2	**Respond within a few seconds** Take action or decision according to the information within 3 to 10 seconds	Obstacle within a few seconds in the vehicle path. Brake in a few seconds. Steer away from danger as required.
1	**Response preparation** Prepare to take action or dicision according to the information within 10 seconds to 2 minutes.	Onset of detection of an obstacle
0	**Information only** No direct action or decision required by driver	System on.

urgency of a message. Some of the factors that should be take into account in such situations are the trip situation (if the vehicle is close to a turning point, etc.), road environment (expressways, urban areas, slippery roads, etc.), weather conditions (rain, fog, etc.), traffic conditions (distance to the lead vehicle, other vehicles on the sides, etc.), vehicle conditions (passenger car/large-sized car, speed, lane, etc.) and others as the arousal level of the driver, etc.

When scoring the criticality and urgency, it is necessary to clarify these conditions and have an expert to grade it. Because it is a subjective assessment, the score may differ according to expert assessment. The cause to these differences may be either the subjectivity of assessment or the large diversity of assumed conditions. After having it graded by multiple experts, we need to take an average of it as a representative value. According to Annexure C of ISO TS16951, when calculating the priority through the addition of 'criticality' to the 'urgency', more than seven experts are necessary for keeping the probability of inversion (the higher the true value, the lower is the calculated value) of the value of the 'priority' obtained under 5 per cent for two messages where the difference of the true value of the 'priority' is 1.0. Therefore, it is desirable that more than 10 experts are used for the assessment.

If we wish to score just the 'priority', there is no need to divide it into 'criticality' and 'urgency'; it is possible to apply the paired comparison method (assessing which is the most important message for each combination of messages). Note, however, that if a new system is added to an existing in-vehicle system, it is necessary to conduct another paired comparison for all the combinations again.

5.4.1.3.3 Message Management

(1) *Selection and integration of the message to be presented (priority management)*: Although the interaction must be designed taking other systems equipped in the car into account, we do not necessarily know with what systems the vehicle is actually equipped in advance. Therefore, it is necessary to devise a message management system (or message controller) for avoiding confusion in the driver by controlling multiple messages from different systems (Fig. 5.59).

If requests for displaying messages coming from each system and information on the priority or criticality are sent together to the controller, when a message with a high priority (e.g., forward collision warning) arrives, then other messages with

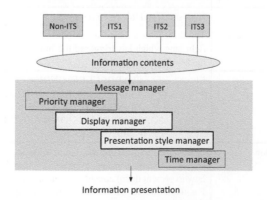

Fig. 5.59: Message management system.

lower priority are postponed (e.g., e-mails) in order to prevent high priority messages from being disturbed by low-priority messages (priority management). In principle, it is necessary to avoid messages with low priority are displayed when displaying messages with high priority.

As described above, the priority may change according to the situation. Consequently, we must not follow only the priority value calculated by the average. For example, if messages from the forward collision warning system and the side obstacle warning system are requested to be displayed at the same time, the priority of the former is assumed to be a little higher because the damage (i.e., criticality) caused by a collision is expected to be worse. Notwithstanding this, if the warning from the side obstacle warning system is not displayed just because its priority is lower, the driver may not notice the other vehicle approaching from the side and change lanes to avoid forward collision. This would dramatically increase the probability of collision with the side vehicle (Sato and Akamatsu, 2008). Since the situation may change in a second, messages with low priority or messages whose priority may suddenly increase must be displayed at the same time without being postponed (ISO TR12204, 2012). According to ISO TR12204, the policy for integration must be examined by describing expected situations when designing the integration of multiple types of warning information (Table 5.15).

Table 5.15: Example of Scenario of Displaying Warnings (ISO TR12204 Table A.16).

Road Context	Road: • Speed range > 40 km/h, • Road structure = 2 or more lanes • … Traffic description: • One vehicle forward with a low speed • One vehicle in the left lane with a high speed • … Weather: • Dry road • …
Vehicle dynamic	High speed and low tyre pressure…
Driver's current manoeuvre	The driver has initiated an avoiding manoeuvre to return in the right lane which has triggered the blind spot warning signal and the lane departure warning
Warning signals	S1 SEAT-BELT REMINDER S7b TYRE PRESSURE MONITORING (urgent) S8r BLIND SPOT DETECTION right S11a FORWARD COLLISION WARNING (CAUTION) S12r LANE DEPARTURE right
Possible integration solution	In this situation, the human factor experts will put the warning strategies (which warnings to be presented to the driver, when, and how to present them) considering the input of technical experts. For this example, *S1* and *S12r* are less important than the other ones.

(2) *Design consistency between messages from different systems*: Messages from warning systems use certain colors and blinking patterns of signals to inform the user that it is a warning (see 5.4.1.1). If the systems are developed separately, each system may present their most important messages through red blinking lights. For that reason, signals that give similar impressions by different systems, do not necessarily mean that they have the same priority level. The message management system selects an appropriate expression of the signals to avoid inconsistencies regarding the perceived priority of the message.

(3) *Display management*: Information management systems control which message is presented on which display when there are multiple in-vehicle displays (e.g., multi-function display in the meter cluster panel, display of the center console, head up display (HUD), display in the mirrors, etc.). It is important to select the message to display according to the priority level of the situation when the message is displayed at a place that is easy for the driver to see (as HUD) or on small areas.

As a general rule, messages with a high priority level are displayed at a place that is easy to see. Notwithstanding the place of display changing from place to place, the driver gets confused about what to look at. When using HUD as an integrated display, the message displayed according to the priority level may be changed as necessary, and if we display the messages at fixed places as the indicator inside the meter cluster or the center display, the driver will be able to verify the display whenever necessary.

Warnings of mechanical systems, as oil pressure warnings, use signals that give the impression of a high level of priority as red blinking lights. Although the level of priority of collision warnings is higher, it is difficult to change the signal design of oil-pressure warning lights as it is an age-old message design. In these cases, we have to design the display clearly by dividing the areas of display.

(4) *Presentation style management*: In some cases, there is no need to suppress the message with a lower priority level, particularly if there are messages that have a similarly low level of priority. In these cases, the difference in the priority level may be expressed by making changes in the design by making the size of the message with the lower priority smaller or reducing its brightness.

(5) *Time management*: Among the messages displayed, some of them are displayed for a continuous period of time (a few seconds or dozen of seconds). For instance, messages informing that the fuel is low are displayed for a certain period of time. Regarding this type of display, there is no need to display it when it is necessary to display other messages with a shorter period of time. It is enough to display it again after the display of the message with a high priority level has ended.

5.4.1.4 Estimation of Driving Demand or Workload for Message Management

5.4.1.4.1 Workload Manager in Information Management

Driving demands depend on the traffic conditions of the road (see 6.1.2.2). The condition changes from a low demand condition to a high demand condition, and the driving task is performed using mental resources according to the task demand. In case of limited capacity of mental resources (or attentional resources, or capacity of human information processing; see 5.1.2 and 5.3.2), if the driving task has high demand, the mental resources increase, reducing the spare capacity. If there is not much capacity

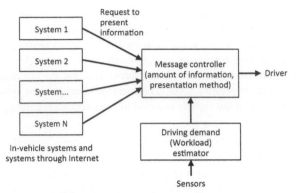

Fig. 5.60: Conceptual diagram of message controllers with workload estimator.

left for providing the driver with information from in-vehicle systems, it is dangerous to display a large amount of information because the driver will be distracted by it. The purpose of message controllers is to control the amount of information taking the driving demand into consideration when displaying messages (Fig. 5.60).

This concept was discussed in the European GIDS (Generic Intelligent Driver Support System) and proposed a message management system with a driver workload estimator (Piersma, 1993; Michon and Smiley, 1993). The purpose of the workload estimator is to avoid distractions while driving; in other words, maintain the situation awareness of the driver while driving (see 6.1.2.1.4 and 6.2.1.4). Therefore, the technical report on FG Distraction of the ITU-T, which is one of the standardization activities conducted by the ITU (International Telecommunication Union), introduces a concept similar to message management using workload estimators, i.e., Situation Awareness Management (ITU-T Focus Group on Driver Distraction).

The workload estimation methods described in Section 5.1.2 are introduced as technologies to measure workload of non-driving tasks in laboratories or controlled environments. In workload estimators, it is necessary to estimate the driving workload or task demand in real vehicles driven on actual roads.

5.4.1.4.2 Estimation of the Driving Demand based on the Road Traffic Environment

Verwey selected driving courses with a variety of road environments (road situations), had the drivers perform secondary tasks and measured the workload at each situation of the road (Verwey, 2000). Assuming that in-vehicle systems use visual information and/or auditory presentation to provide information, they used two types of secondary tasks—the first was a visual detection task (orally answering 'yes' when a two-digit number is displayed on the dashboard), and the second, a memorizing task where the participants orally answer the result of the addition of 12 to numbers given by a recorded voice.

The road environment was categorized into straight ahead at outer-city roads and that at inner-city roads, straight ahead at intersections, turning left or right at intersections, roundabouts, etc. The results showed that drivers perform the worst when turning left at intersections and passing through roundabouts, and the next worst performance was seen when turning right or at roads with curves. As to the

Table 5.16: Driving Demand of Road Situation Categorized According to the Performance of Secondary Tasks.

	Demand level	Road situation
Visual loading	Low	Standing still
	Medium	Straight ahead at outer-city and inner-city, Straight ahead at controlled intersection, Motorway
	High	Straight ahead on secondary inner-city road, Turning right at uncontrolled intersection, Curve driving
	Very high	Turning left at uncontrolled intersection, Roundabout
Auditory loading	Low	Standing still, Straight ahead at outer-city and inner-city, Straight ahead at controlled intersection, Motorway, Curve driving
	High	Straight ahead on secondary inner-city road, Turning right or left at uncontrolled intersection, Roundabout

performance of the auditory memorizing test, although it was not much different from the performance of visual tasks, the performance when turning left and at roundabouts was particularly bad. The demand level of each road situation for visual information and auditory information was determined based on these results (Table 5.16). By applying this knowledge, it is possible to manage messages according to the road environment using map databases. Since no difference was seen in the measurement of workload by the secondary tasks method when driving along the same course at two different times of day, there is no need to take the traffic volume into account. Note that this knowledge is based on the traffic environment in the Netherlands.

5.4.1.4.3 Estimation based on Automotive Sensor Signals of Driving Demand in Road Traffic Environment

There are many factors in the driving demand. Akamatsu et al. used four factors from the task demands related to the road-traffic environment, namely, 'understanding of the road structure', 'understanding of traffic situations', 'disturbance of own pace' and 'driving operations' based on the sensor signals received from the vehicle (Kurahashi et al., 2003).

They have 16 participants drive vehicles equipped with several sensors on a variety of roads (bypass roads, old roads, roads in front of stations, rural roads, etc.) for five days (five times). The courses were divided in 10 sections according to their characteristics (see photo in Fig. 5.61). Each time after the trip, the drivers were asked to answer about their subjective demand for 'understanding road structure', 'Understanding traffic condition', 'disturbance of the own pace' and 'driving operations' using a 7-point scale. The average of the results is shown in Fig. 5.61. Given that individual differences in perceived demand were removed by taking an average of the results of all participants, it can be regarded as the amount of each factor of driving demand on each road section.

It is possible to estimate the amount of each factor of the driving demand by using the variables obtained from sensors of the vehicle. They are the accelerator/

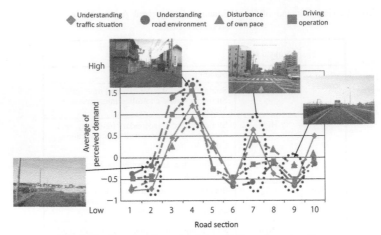

Fig. 5.61: Average of perceived demand for each road section.

Table 5.17: Coefficient of Correlation Between Factors of Driving Demand and Variables Obtained from Sensors (over 0.6 is Shaded).

	Frequency of acceleration operation	Time of braking operation	Time being foot above acceleration pedal	Time being foot above brake pedal	Frequency of steering operation	Inverse of average speed	Frequency of detection of a lead vehicle	Complexity of front view
Understanding traffic situation	0.69	0.71	0.66	0.36	0.43	0.45	0.46	0.71
Understanding road environment	0.53	0.42	0.70	0.24	0.92	-0.04	-0.22	0.57
Disturbance of own pace	0.74	0.56	0.66	0.02	0.57	0.25	0.47	0.58
Driving operations	0.59	0.47	0.79	-0.21	0.86	0.00	-0.01	0.61

brake pedal strokes, frequency of the operation of each pedal per unit of time, duration of time when the driver's foot was not stepping on the brake pedal, frequency of steering operations (number of times that the driver turned the wheel more than 10 degrees), headway distance to the lead vehicle, speed of the vehicle, complexity of the front view, etc. The complexity of the front view was calculated by an algorithm for detecting the segmentation and red color, based on the level of brightness in the image of the scene. Table 5.17 indicates the correlation between the average of each variable and the score of each factor of demand obtained for each of the 10 road sections.

A correlation is noticed between the factor of driving demand regarding the road traffic and the readiness to step on the pedals, the steering operations and the complexity of the front view, Also there is sensor information regarding the characteristics of each of them. A multiple regression analysis model was applied by using frequency of operations by the accelerator, time with the foot on the accelerator pedal, frequency of steering operations and the complexity of the front view as independent variables and each demand as a dependent variable. Figure 5.62 shows the results of the estimation from each sensor signal of each demand by using this regression method. We can see that it is possible to estimate the driving demand from the signals of sensors with a relatively high precision.

Since the values of each factor of driving demand can be estimated, we can estimate the whole workload of the driver in order to apply it to message management.

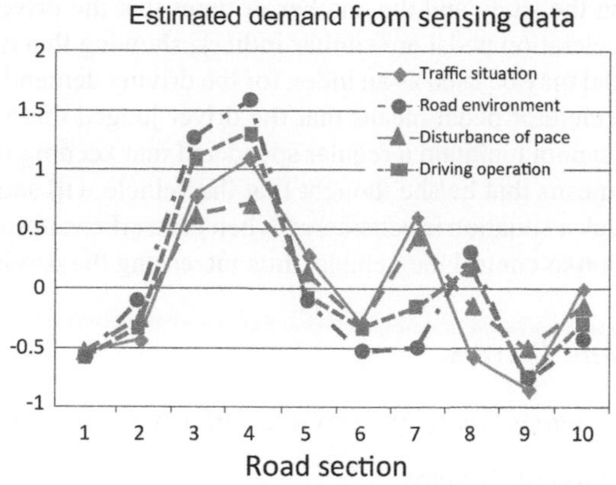

Estimated demand from sensing data

Fig. 5.62: Estimation values of demand for each road section.

When we take the individual workload sensitivity into account, it is possible to estimate the total amount of the workload by using the sensitivity of each driver obtained by a workload sensitivity questionnaire (WSQ) (Ishibashi et al., 2002), weighting the scores for each of the driving demands and adding them up. If we do not take individual differences into account, we can estimate the amount of workload by adding up the value of the each factor of driving demand. The message manager controls the amount of information on road sections where this value is high.

5.4.1.4.4 Estimation of the Driving Workload in Real-time based on Sensor Signals

Uchiyama et al. measured the driving workload in real roads by using the secondary task method, as Verway et al. (Uchiyama et al., 2002). The secondary task they used was a phonetic memory task where drivers were told five words of a voice message from the in-vehicle system and immediately asked to orally answer them. The memory tasks of about 20 seconds were repeated during the drive along a suburban road of about 7 kms (a drive of about 20 min). They had nine drivers drive five times, obtaining a total of about 2,000 secondary tasks.

They investigated the correlation between the performance of the secondary tasks and the variables obtained from the sensor signals during the drives. The variables were the speed of the vehicle, the accelerator pedal strike, number of times that the driver moved the foot away from the accelerator, the brake pedal stroke, number of times the driver moved the foot away from the brake and angle of the street and headway distance to the lead vehicle. Results showed that although the correlation coefficient was as low as 0.1, the number of times that the driver moved the foot away from the accelerator pedal had a negative correlation with the performance of the task and the workload can be estimated based on the number of times. Based on these results, they suggested that 6 seconds just after moving the foot away from the accelerator pedal should be assumed as time with a heavy workload. With this rule, it is possible to manage messages more precisely.

Although their results were obtained only from almost straight roads, the time being when the foot was above the accelerator pedal. In Table 5.17, the highly correlated driving demands in the study and the number of times that the driver moved the foot away from the acceleration pedal are similar indices, showing that operation times of the accelerator pedal may be used as an index for the driving demand. Moving the foot away from the accelerator pedal means that the driver judged that the vehicle is in a situation where it cannot maintain a regular speed and that keeping the foot above the accelerator pedal means that he/she thought that the vehicle will soon be able to pick up speed again. Such a situation is considered when the traffic is unsteady and requires the driver's attention to control the vehicle, thus increasing the driving demand.

5.4.2 Systems and Drivers

5.4.2.1 Levels of Automation of Systems and Drivers

5.4.2.1.1 Automation of Systems

The process to achieve a goal can be adopted by four-stage model for information processing: information acquisition, information analysis, decision selection and action execution (Parasuraman et al., 2000). It is important to consider to what extent the steps should be automated for designing human-machine systems.

The 'perception', which continuously captures information from the environment, and the 'cognition', which tries to understand the situation by integrating and analyzing a variety of information, are often automated as much as possible (Parasuraman et al., 2000). That is because both require continuous effort and people have difficulty in performing these functions for long periods of time. Also the fact that the number of situations where machines are more capable than humans is increasing with every progress of technology.

On the other hand, we need to think carefully as to what extent we should automate the "selection" and the 'execution'. In fact, in this field where the 'ultimate responsibility for safety is borne by humans', we adopt the view of human-centered automation in which 'the ultimate decision must be made by humans (Billings, 1991)'. In other words, there is room left for humans to actively participate in the automation system.

5.4.2.1.2 Levels of Automation

As for the 'selection' and the 'execution', there are many ways of putting humans and machines to work together towards a goal. A concept useful for expressing them in a concrete manner is the 'Levels of Automation' concept as shown in Table 5.18 (Sheridan, 1999). For a long time, there were only 10 levels in the scale of automation. However, because a new level, important from a theoretical and practical perspective, was discovered between Levels 6 and 7 (Level 6.5 of Automation) (Inagaki et al., 1998), the scale in the table consists of 11 levels. We will use both actual and hypothetical examples for explaining what each automation levels signifies.

5.4.2.1.3 Examples of Levels 1 to 3

A system that informs the driver of the existence of pedestrians uses a night-vision camera to not only project an image on the display but also to identify and set a frame around the pedestrians in the image. This sort of system, i.e., system that assists in

Table 5.18: Levels of Automation.

Level	Definition
1	The computer offers no assistance: human must take all decisions and actions
2	The computer offers a complete set of decisions/action alternatives
3	The computer narrows down the selection to a few
4	The computer suggests one alternative
5	The computer executes that suggestion if the human approves
6	The computer allows a human a restricted time to veto before automatic execution
6.5	The computer executes automatically after telling the human what it will do. No veto is allowed
7	The computer executes automatically and if necessary, informs the human
8	The computer informs the human if asked
9	The computer informs the human only if it, the computer, decides to
10	The computer decides everything, acts autonomously, ignoring the human

'perception' and 'cognition' but does not assist with 'selection' and 'execution', is considered as Level 1 of automation.

If the system mentioned above also assists the driver in the selection by indicating all the alternatives of operation that the driver may choose for avoiding the collision with a pedestrian (using the brake to decelerate or stop/ turning the car to the left or to the right to avoid the pedestrian), the level of automation rises to Level 2.

If the system also chooses or indicates the best choice among the alternatives, the automation level rises to 3.

5.4.2.1.4 Examples of Level 4 to 6

The systems commonly known as warning systems inform people that 'the best alternative for securing safety is to immediately take a counter measure' and encourage the people to take action. A classic example is the system that encourages the driver to take the necessary action for avoiding collision by warning him/her when the system detects the danger of collision. The automation level of this sort of system is Level 4.

If the system applies the brakes for avoiding collision with an obstacle by interpreting the action of the driver, who was warned by the system, putting his/her foot on the brake pedal with the intention of 'avoiding the collision with an obstacle ahead by braking as 'an approval of the system's idea (execution of the braking)', the automation level of the system is 5.

If the system applies the brakes for avoiding the collision following its own decision when the driver does not take a counter measure (including commands for stopping the execution as 'the braking operation is unnecessary in this case'), a few seconds after being warned, the level of automation is 6.

5.4.2.1.5 Examples of Level 6.5

In automation Level 6, there is a delay in the time when the system suggests an operation to the time of execution, and a delay of 1 or 2 seconds may have a crucial meaning for automobiles. The purpose of the automation Level 6.5 is to resolve this problem. A classic example of automation Level 6.5 is a system that tries to avoid lane departure by warning the driver and applying a steering torque (or applying different forces of brake to each wheel) to correct the direction of the vehicle when the vehicle begins to move out of its lane.

5.4.2.1.6 Examples of Level 7

There is a system that maintains the car in the right position by immediately applying an appropriate braking force to each wheel when the driver makes a sudden movement for avoiding an obstacle or when it detects that the car is slipping while running on a slippery road. If the system reports to the driver only after it has already taken an action for preventing the car from slipping, its automation level is 7. If there was no report at all after the event, the driver would probably not even notice that a situation with which he/she could not deal with his/her driving skills had occurred. In that sense, we may say that the report after the event has occurred as an effective means for preventing the driver from being overconfident in his/her skills.

In automation Level 6, the system executes a variety of operations following its own decision, i.e., it does not wait for a command or approval from the driver. This means that the right to make an ultimate decision by the driver is not guaranteed. This may cause a lack of understanding of the situation by the driver, automation surprise (when the driver is surprised by an unexpected action of the system) and other problems. However, as indicated in 5.4.2.1.3–5.4.2.1.5, we may not deny that systems with an automation level below 6 are essential for making designs where the machine (system) tries to save people by taking action in place of the driver if he/she is unable to respond to a situation.

Sometimes when the human-machine interface is ill-designed, the automation level rises to 8 or above, despite the intention of the designer. However, this level of automation is not necessary for vehicles where the driver and the system are supposed to work together as teammates towards a common goal. Systems that execute an operation without warning the driver on the decision and do not inform the driver of what they have done until they are asked, or systems that do not answer to the driver, saying only 'there is no need to inform' when asked about what they have done, are not accepted by drivers and the society.

5.4.2.2 Over-trust and Overdependence

When revolutionary new technologies are commercialized, people often tend to place too much trust in them. In fact, counter measures against this over-trust are already being taken. Airbags are a good example of how the over-trust in new technologies may create actual problems (Itoh, 2014). In the early nineties, when SRS airbags began to become common, people got the false idea that they were a substitute for seatbelts or that they would inflate every time the car suffered a collision. From this sort of wrong belief, people thought that the airbag would absorb the impact of the collision even if the seatbelt was not fastened and thus began to drive without the seatbelt.

The problem of over-trust or overdependence is also emphasized in the Advanced Safety Vehicle (ASV) Project by the Ministry of Land, Infrastructure, Transport and Tourism of Japan. In fact, the 'guiding principles' in the ASV clearly advocate the need for 'avoiding over-trust and overreliance' (Fig. 5.63). For that reason, at least in Japan,

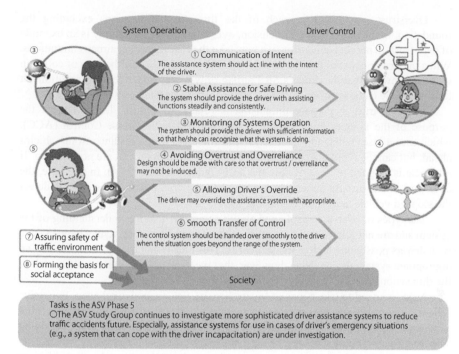

Fig. 5.63: Guiding principles of ASV technology development (Pamphlet of ASV Promotion Project Phase 5).

it is almost a fact that preventing over-trust will become a matter to discuss regarding the development of driving assistance systems and automated driving systems.

Unfortunately, the methodology for resolving the problem of over-trust is not established yet. The problem is that the concept of over-trust and overreliance has not been established among researchers and experts in the world. Certainly, the over-trust issue is discussed in other countries as well. Nevertheless, the fact that a variety of terms are used to express the same or similar phenomenon indicates that concepts have not been well established. Some examples of concepts related to over-trust are: complacency, (inappropriate) mental model, lack of situation awareness, etc. Perhaps, the choice of the term used to express this problem depends on the mother tongue of the speaker. Although from a Japanese perspective it may seem natural to use the term *kashin* (it means "overbelief"), this may not apply to other countries.

Literature with a clear focus on the ambiguity of the term 'over-trust' and 'overreliance' is limited (Inagaki and Itoh, 2013). In order to understand the practical meaning of 'over-trust', we need to take another look at the concept of 'trust'. According to Lee and Moray (Lee and Moray, 1992), trust may be divided in four following dimensions:

(1) *Purpose*: the intention or motivation of the system is convincing.
(2) *Process*: the method, algorithm and rule to govern behavior are understandable.
(3) *Performance*: we may expect desired behavior with stability and consistency.
(4) *Foundation*: it is in accordance with the natural and social order.

Over-trust may occur in each of the three first dimensions excluding the foundation. In the performance dimension, overestimation of reliability is an example of over-trust. In process dimension, we may take the case regarding airbags limitations. When the driver believes that the airbag will inflate regardless of the type of collision without knowing that the driver's SRS airbag only works under certain conditions, it falls under the category of over-trust in the process dimension.

The over-trust in purpose dimension is related to a misunderstanding of the purpose of the system. For example, traditional Adaptive Cruise Control (ACC), which uses a laser radar, adjusts the speed to maintain a safe distance from vehicles ahead, but cannot avoid the collision with objects on the road that are not moving. If a vehicle is stopped in front of you, the ACC ignores it (Fig. 5.64). In this case, the over-trust in the purpose dimension occurs when the driver expects the ACC to detect the stopped vehicle and adjust the speed.

The types of over-trust explained above occur due to a misunderstanding of the system and are not related to the morality of the driver. Therefore, it may happen even with drivers possessing high moral standards. This means that it may be avoided by an appropriate system design. Note, however, that we need counter measures for each of the dimensions of trust for avoiding the over-trust.

Next, we need to take a look at 'overreliance', i.e., excessive reliance. Excessive reliance is divided into two categories: relying too much on someone or something and not doing what one is supposed to do (monitoring the system, etc.), and doing something you are not supposed to do due to lack of tension caused by reliance on someone or something. Neglecting monitoring tasks is an example of the former category. The so-called 'risk compensation' behavior is a classic example of the latter category. It is equivalent to an error of commission that the person does something wrong and unnecessary.

Preventing over-trust is a design issue. Regarding over-trust in the performance dimension, as long as the operation is within the range of the system, enhancing the reliability of the system is the duty of the designer. As for over-trust in the process dimension, it is important to enhance the driver's understanding of the operational conditions, mechanisms and limitations of the system. However, since most drivers do

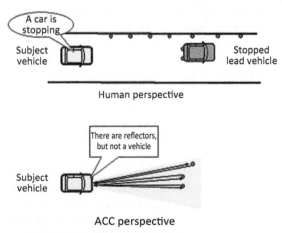

Fig. 5.64: ACC that does not detect stopped vehicles.

not know much about this sort of technology, it is not easy to make them understand details of the logic of system operations. A solution for that is to select the limited number of important points that are usually the cause of confusion in the mental model of drivers and explain clearly to clarify the operational conditions, mechanism and limitations of the system to the drivers. With respect to the over-trust in the purpose dimension, it is necessary to find proper ways to make drivers understand what the system is made for, particularly what the system CANNOT do. If possible, it is desirable to insert these methods into the system design.

For preventing overreliance, it is necessary to devise methods to make drivers do what they are expected to do by making the monitoring amusing and easy to perform. Also, it is essential to devise methods to avoid risk-compensation behavior, as lowering the target level of the risk.

5.4.2.3 Monitoring of the System Status by the Driver

5.4.2.3.1 Supervisory Control

Even when the driver is leaving all or part of the driving operations for the ADAS or driving automation system, it does not mean the he/she can take a nap or surf the internet while driving. Similarly there is a limit to human capacity and a limit in the capacity of the system to understand correctly every traffic situation under different weather conditions, select the appropriate operation or properly execute it. For that reason, the driver must always supervise whether the system is controlling the vehicle doing what he/she wants it to do (or if it is following the commands) and needs to intervene immediately for taking appropriate measures when there is something wrong with the control of the system. This form of control is called 'supervisory control' (Sheridan, 1992). Although supervisory control reduces the physical workload of the driver in controlling the car, the reduction of the mental workload depends on the HMI provided to the driver.

5.4.2.3.2 HMI in Driving Supporting Systems Using V2X Communication

What type of HMI can reduce the mental workload on the driver through supervisory control? The "HMI guidance in systems for supporting safe driving through infrastructure cooperation (Report of Advanced Safety Vehicle, 2011)" is a good reference for answering this question. Now, "systems for supporting safe driving through infrastructure cooperation" are "systems that provide information to or alert the driver when it is necessary to inform him/her of traffic situations that cannot be directly seen from inside the vehicle by collecting information from infrastructure devices (including devices mounted on other vehicles or roadside equipment or devices carried by pedestrians) through wireless communication".

As showed in Table 5.19, the 'HMI guidance' include the following five factors: (1) verification of the operating status; (2) easy communication; (3) communication reliability; (4) easy understanding of emergencies; and (5) prevention of over-trust or distrust.

(1) *Verification of operating status*: This section requires that the system allows the driver to be aware of the operating status of the system (if the system is functioning or not), for what event the system is providing support, etc.

Table 5.19: HMI Guidance in Systems for Supporting Safe Driving through Infrastructure Cooperation.

Consideration Points	Overview
(1) Verification of the operating status	Making sure that the driver can verify the operating status of the system and the type of support
(Example)	- Displaying in a way that the driver can understand if the system is operating or not - Displaying which event (action category) is being supported by the system
(2) Easy to understand	Making sure that the driver can use the system without any worry, making it easy to understand and use
(Example)	- Displaying simple information that the driver can quickly understand - Assuring the consistency of the communication when using a combination of vehicle communication and vehicle-road communication
(3) Communication certainty	Making sure of the stability of the communication
(Example)	- Using visual, auditory, tactile sensory modalities for communication when displaying warnings or alerts
(4) Easy understanding of criticality	Making sure that the driver can easily understand the level of alert (warning, alert or information provision)
(Example)	- If color display is possible, using different colors according to the level of criticality (for example, red for warnings, yellow for alerts and other colors for other information) - Adopting different frequencies, intervals, sound pressures, etc. according to the level of criticality when using sounds
(5) Prevention of over-trust or distrust	Making sure that the driver does not over-trust or distrust the system assuring an appropriate level of trust
(Example)	- Providing information to the driver at the right time taking into account the level of criticality, the driver's reaction time, time of the system delay, etc. - Displaying it when there is a failure in the system - Informing drivers of the system's limitations through manuals, etc.

First, let us take a look at the example of systems using V2X communication that avoid crossing collisions at intersections, where the view is limited, by informing the driver of the existence of other crossing vehicles (vehicles approaching the intersection) through messages as 'car coming from the right', for understanding why systems need to display the operating status. Let us suppose that the system does not display any message when the driver is waiting to move on at an intersection with a stop sign where the view is limited. In this situation, if the system does not display its status (if it is activated or not activated), it is impossible for the driver to know if there is another vehicle at the intersection or not, whether the system is not displaying anything because it is not activated or because it did not detect another vehicle. Then, a driver under this situation does not know that the system is turned off and may cause a collision accident by moving on with the vehicle thinking, 'it is alright because there is no message about another crossing vehicle'.

When automobiles equipped with the systems approach an intersection with the stop sign, the system first displays the message 'intersection with stop sign ahead'. The purpose of this message is to inform the driver that he/she needs to reduce the speed and stop the vehicle at an appropriate location. The purpose of the message 'car coming from the right' that is displayed after the driver has stopped (at the moment he/she is waiting for the right time to move on), is to avoid crossing collisions by encouraging the driver to pay attention to his/her right side. In this way, system messages to the driver must tell in a clear manner what the driver should pay attention to, what action he/she should take, etc.

(2) *Easy to understand*: Messages displayed to drivers need to be easily understood immediately. It is necessary to express them in an accurate and simple manner by taking into account the number of letters, etc., referring to the JAMA Guidelines version 3.0 (JAMA in-vehicle display guideline version 3.0). When using displays for multiple sensory modalities, it is essential to combine visual, auditory, tactile information in an appropriate manner.

Furthermore, when the system is trying to provide support based on information that it obtained through a variety of methods, there must not be any inconsistencies between each type of support. For instance, it is necessary to maintain the consistency in information when combining vehicle-to-vehicle communication system and vehicle-to-road communication system.

(3) *Communication certainty*: Failures are not allowed in the communication of urgent information. When there are multiple sensory communication modalities, it is necessary to make sure that the information is being conveyed to the driver under each type of situation by using not only one, but multiple modes of information (visual, auditory, tactile, etc.).

(4) *Easy understanding of criticality*: It is necessary to use colors according to the criticality level of information for drivers (e.g., using red colors for 'warnings' that require the driver to take immediate action against a specific event for safety; or using yellow colors for 'cautions' that inform the driver of possible obstacles and other security dangers, etc.). As for sounds, it is required to use different frequencies, intervals, sound pressures, etc. Furthermore, we need to allow drivers to easily understand the changes in information when it is necessary to change the level of information according to the urgency (e.g., changing from a 'warning' to an 'caution', etc.) (see 5.4.1.1).

(5) *Prevention of over-trust and distrust*: Since there are four levels of trust that human beings place in machines, there are also certain types of over-trust when humans overestimate the potential of machines. Therefore, it is necessary to take prevention or control measures according to the type (see 5.4.2.2). When humans experience a failure in the system in case over-trusted the system, the over-trust (or trust) may turn into distrust, where humans underestimate the potential of the system. In this sense, over-trust and distrust are different sides of the same coin.

The HMI consideration points shown in Table 5.18 indicate provision of information at the right time (it is not in Table 5.18, but note that providing information always at the same time under similar situations is one of the keys for preventing over-trust/distrust of the system) by taking into account the time delay in the system (the time it takes for the data to be sent from road infrastructure or other vehicles to process the information) and the reaction time of the driver (the time the driver takes to take action after receiving the information from the system. It is generally 3.2 seconds for cautions and 0.8 second for warnings). Although displaying it clearly and at the right time in case of error in the system is related to HMI consideration points, it is also important for preventing over-reliance.

Furthermore, excessive dependence may also be seen when the driver does not have a precise understanding of the limitations of the system and thinks that the system can handle everything. This is why the HMI guidance requires notification of information on system limitations through manuals, etc. (Note, however, that since we cannot expect the drivers to carefully read the manual, it is necessary to develop other means of informing the drivers of the limitations through HMI).

5.4.2.3.3 More General HMI in Driving Support/automated Driving Systems

With the development of various automated driving systems, there are other aspects of HMI, apart from those shown in Table 5.18, that need to be addressed. For example, the display of (a) cue information for understanding system limitations, (b) information for understanding the reasons of a system decision, (c) cue information for understanding the intention of the machine, (d) information for helping to establish a common understanding of situations between humans and machines, etc. Even if the machine has the ability to understand the situation, choose the most appropriate solution and execute it, it does not necessarily mean that it has the same way of thinking or taking action as the driver. If the driver cannot know how the machine has understood the driving conditions or what the machine is trying to do, he/she will probably shout, "What are you doing!?" This is called 'automation surprise'. The above-mentioned aspects, (a)–(d), are essential for establishing a smart human-machine relationship.

5.4.2.4 Changes in Driver's Behavior Caused by Introduction of the System

Driving support systems for providing information about dangers that are difficult to notice by drivers are being developed and one of these attempts is Driving Safety Support System (DSSS) that was tested on public roads. The effectiveness of systems that provide warning information was evaluated by using specific intersections on public roads (site-based Field Operational Tests, see 2.3.2.3). Below, we will give an outline of the evaluation methods and show the results of analyzing user-behavioral changes regarding a 'supporting system for preventing rear-end collisions with vehicles waiting at traffic lights' in the site-based FOT conducted in Hiroshima (Iwashita et al., 2012).

5.4.2.4.1 Driving Behavior Induced by the System

The purpose of warning systems is to enhance safety. Human errors from the safety point of view can be defined as 'deviant behaviors that are beyond the acceptable range of the system' or 'inappropriate decisions or behaviors that interfere in the efficiency, safety and performance of the system', presuming the existence of expected behavior to be performed by users. Based on this idea, there can be two categories of behavior—'normative behavior' that consists of actions that the system requires the driver to take and is considered to be the 'minimum requirements'; 'subsidiary desired behavior' consists of prudent actions for reducing risk and may consist of 'smoothness of action' and 'increase in the safety margin'. They are actions that 'enhance the value of the system'.

5.4.2.4.2 Definition of Road/traffic Factors Influencing Driving Behavior

Even if the driver reduces the speed, it may be an expected behavior, but there is the possibility that it may be caused by traffic jams or other vehicles cutting in. In addition, traffic signals and the speed of the traffic flow may also have influence. Therefore, it is necessary to identify factors other than the warning information that may influence the driver's behavior (hereinafter, 'noise-factors') and remove the 'contaminated' data from the analysis. In order to do that, it is necessary to identify the noise-factors, such as vehicles cutting in, by observing the traffic conditions of the test site, to specify the evaluation criteria in advance, and determine what data should by adopted by analyzing the image data (including from roadside cameras), data of the vehicle, etc.

5.4.2.4.3 Example of Analysis of Behavioral Changes Caused by the System

'Support system for preventing rear-end collisions with vehicles waiting at traffic lights' in order to alert the driver to the existence of stopped vehicles waiting at a traffic light in the intersection ahead, where the visibility of the road is poor (hereinafter, 'rear-end collision prevention support' is shown in Fig. 5.65).

Experiment participants were 30 employees (average age: 42.7 years) of the R&D department of a car manufacturing company not engaged in the development of the system being tested by the experiment. The warning signal consisted of a beep followed by a voice and visual message. One trip of the running course in the experiment site took around 7 minutes. After mastering the course, they ran three trips without the warning system and four to seven trips with the system, in that order.

In order to define the expected behavior, the following situation was assumed. The information about traffic at the intersection at the bottom of hill is provided to a driver climbing the hill. The driver makes a decision based on information about a place that he cannot see with his own eyes. When it is difficult to know the situation ahead, it is desirable for the driver to drive carefully. In this situation, the driver is expected to foresee the danger and slow down even if he is not able to see it. The normative behavior was defined as 'slowing down after the information was provided and before reaching the top of the slope (in the space where the driver cannot see ahead)' and seven other actions were set as subsidiary desired behavior (making slower speed changes in the space between the warning and the top of the slope, keeping a larger distance to the lead vehicle, smaller maximum deceleration when stopping behind a vehicle waiting for a traffic light, etc.).

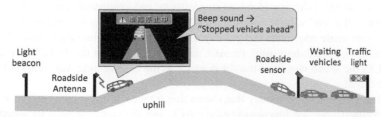

Fig. 5.65: Support system for preventing rear-end collisions with vehicles waiting at traffic lights.

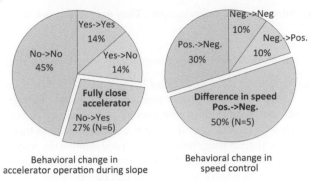

Fig. 5.66: Change in the normative behavior.

The following points were regarded as noise-factors and the data with these factors were removed from analysis. Speeding and sudden deceleration due to reduction in the headway were detected by comparing their statistical distribution in the test site. Deceleration behavior caused by vehicles cutting in from the side roads, reacting with brake lights of vehicles ahead, vehicles cutting in from adjacent lanes and motorcycles ahead were also detected by observing the recorded videos.

Normative behaviors were measured as 'time duration of fully closing the accelerator' and 'difference in the speed of the subject vehicle at the time the information was provided and at the top of the slope'. If 'difference in the speed' is positive, it indicates that the vehicle was accelerating towards the top of the slope, and if negative, then the vehicle was slowing down. When we saw the behavioral changes from driving 'without the system' to driving 'with the system', 27 per cent of the drivers changed their behavior to fully close the accelerator and 50 per cent of the drivers changed from a positive to a negative value when driving with the system (Fig. 5.66).

5.4.2.5 Compatibility of the System with Divers' Behavior

In this section, we show the user's behavioral adaptation when using the driving assistance system. The headway distance to the lead vehicle before using the system was compared with the distance when the driver is using the Adaptive Cruise Control (ACC) system.

5.4.2.5.1 Distance without the System and Distance with the ACC System

Sato et al. collected real-world data on driving behavior from 18 drivers (nine males and nine females) driving along an expressway and compared the speed and the distance to the lead vehicle with the ACC to the ones without it (Sato et al., 2007). The drivers made one round trip per day of 35 km one way at an expressway (Joban Expressway); for 20 days using the ACC and 10 days without the ACC. When using the ACC, the speed was fixed but the distance to the lead vehicle was freely selected by the driver.

Based on results of a study that shows that it takes around one week for drivers to understand both the functioning of the ACC and the characteristics of speed and distance control by the ACC (Sato et al., 2005), the data of the days after the eighth day of driving with the ACC was used. In this ACC, the driver was able to select three levels of distances (short, middle and long) and the ratio of time drivers used at each level of distance was analyzed. The results were that eight drivers selected a 'short' distance for 95 per cent of the driving time, another eight drivers selected 'long' or 'middle' distance for more than 70 per cent of the driving time and the other two drivers selected 'short' and 'long' for about the same amount of time (around 50 per cent each).

The driver has to drive by himself/herself when not using the ACC. Based on a previous research that shows that drivers need about five days to get used to the vehicle (Akamatsu, 2004), data of the days after the 6th day driving without ACC was used. By analyzing the distance selected by the eight drivers that chose a 'short' distance in the ACC (short distance group) and the distance selected by the eight drivers that mainly chose a 'long' or 'middle' distance (long distance group), it was shown that the short distance group kept a significantly shorter distance in comparison to the long distance group.

In this way, it was demonstrated that while drivers who usually kept short distances from vehicles in front, when driving on expressways, also select short distances in the ACC; drivers who usually kept long distances also select long distances in the ACC. Therefore, it is possible to say that drivers use the system for maintaining their habitual driving performance.

5.4.2.5.2 Relation Between Drivers' Characteristics, Driving Behavior and the Distance Selected in the ACC

The driving characteristics of the participants in the experiment of item (a) were investigated by conducting a questionnaire survey after the end of the experiment. A driving style questionnaire (DSQ) (Ishibashi et al., 2007), a workload sensitivity questionnaire (WSQ) (Ishibashi et al., 2002; Sato and Akamatsu, 2014) and an expressway driving style questionnaire (Sato et al., 2007) were used to examine the relationship with the ratio of time of a selected distance when using the ACC were investigated. The characteristics below are the ones that have a strong correlation with the ration of time of a selected distance when using the ACC.

- *Preparatory driving behavior for traffic lights*: Drivers who select short distances tend to prepare more for traffic lights while driving.
- *Confidence in driving skills*: Driving skills of drivers who select short distances are higher.
- *Impatient driving*: Drivers who select short distances tend to drive in a more impatient manner.
- *Overtaking tendency*: Drivers who select short distances tend to overtake other vehicles more often.
- *Attitude towards other vehicles*: Drivers who select short distances pay more attention to the behavior of other vehicles and are more aware of the traffic situation surrounding them.

A path analysis was conducted for modeling the relation between the driving characteristics of drivers, the driving behavior without ACC (speed and distance when not using ACC) and the tendency regarding the selection of distance in the ACC (ratio of time using a 'short' distance). Figure 5.67 shows the most suitable model

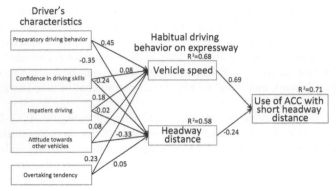

Fig. 5.67: Relation between the distance selected in the ACC, habitual driving behavior on expressways and drivers' characteristics.

for the collected data. It indicates that the habitual driving style is determined by the characteristics of the driver and the driver selects a distance in the ACC that will allow maintaining one's habitual style.

5.4.2.6 Human Factors in Automated Driving Systems

5.4.2.6.1 Intersection Between Automated Driving Systems and Humans

The issues regarding the human factors in automated driving systems are in the interactions between the automated vehicle and the human being. 'Humans' can be 'Drivers', 'Other traffic participants' (drivers of non-automated vehicles, pedestrians, etc.) and the 'society'. Therefore, it is possible to say that there are three types of interactions, namely, the one between the driver and the automated driving system, the one between the automated driving vehicle and the other traffic participants and the one between automated driving vehicles and the society (Fig. 5.68). Here, we discuss the human factor issues in the interaction between the system and the driver and in the interaction between automated vehicles and other traffic participants according to each level of automation as defined by SAE (SAE J3016, September 2016).

At Level 1, when driving an automation system, the driver has to control the steering or the pedals by himself/herself and supervise part of the control tasks, the surrounding environment and the operation status of the system (monitoring task).

Fig. 5.68: Three interactions between automated driving vehicles and humans.

Interaction between vehicle and driver/surrounding road users/society		Level of automation (SAE, 2016) and research questions			
		Level 1	Level 2	Level 3	Level 4
Vehicle - Driver	**System use**				
	A-1 Understanding system functions	How to avoid over trust, over reliance, misunderstanding of functional limitations?			
	A-2 Understanding system states	How to avoid misunderstandings of system's current state and future actions.			
	A-3 Understanding system operations	How to improve usability of complicated HMI (switches)?			
	A-4 Understanding system behavior	How to avoid worries and discomfort for system's driving manner differing from driver's manner?			
	Driver's state				
	B-1 Driver state with automation		How to maintain required driver's state with automation?		
	B-2 Transition from automation to fully manual		How to avoid degraded response action of the driver unready to take over the vehicle control?		
	B-3 User benefits of automation		How to overcome the negative benefit of fight against drowsiness /boredom?	How to overcome the negative benefit of interruption of relax time?	How to compensate for the decreased value of homogenized car performance?
Vehicle - Surrounding road users	C-1 Communication between the autonomous vehicles and surrounding drivers		How to enable communication with crossing pedestrians, pedestrians in parking, communication in shared space and other situations?		
	C-2 Communication between the autonomous vehicle and surrounding vulnerable road users		How to enable communication at intersections, merging, lane change and others?		
	C-3 Mediation between formal rules and traffic efficiency			How to mediate yielding with priority, difference between speed limit and traffic speed, and other conflicts?	
Vehicle - Society	D-1 Social value and acceptance of the autonomous vehicles			Liability for crashes and police tickets caused by the system.	
	D-2 Liability			Licensing requirements for autonomous vehicles.	
	D-3 Licensing			How to design functional deployment over time to raise social acceptance?	

Fig. 5.69: Human factor issue in driving automation systems.

At Level 2, the driver is not responsible for any control task, but only for monitoring tasks and has to be prepared to intervene in controls tasks whenever necessary. At Level 3, the driver can leave both control and monitoring tasks to the system under certain conditions (ODD: operational design domain (SAE J3016, September 2016)), but when conditions change, he/she has to replace the system for conducting all tasks. In this case, it is assumed that there is sufficient time for the driver to take over these tasks from the system. At Level 4, the driver does not need to get involved in the driving to any extent. There are human factor issues for each level of driving automation (Fig. 5.69).

5.4.2.6.2 Understanding of the System

(1) *Understanding of system's functions*: When intending to use a system, humans need a prior knowledge on the purpose of the system. Users interpret the system behavior based on this knowledge: the HMI design and the using experience and construct a mental model of it.

If the system and its HMI design are not appropriate, he/she creates an incorrect mental model, what leads to incorrect interpretations of the presented information or improper operations. Incorrect understanding of the functional limit of the system becomes the cause of over-trust or overdependence (see 5.4.2.2). Establishing an incorrect mental model is not only the cause of improper operations, but also the reason for leading to dangerous situations where the driver is surprised at unexpected system behaviors (Mode Confusion or Automation Surprise) (see 5.4.2.3).

(2) *Understanding of the system status*: Driving automation systems consist of a variety of subsystems and each of these systems has multiple modes, according to operating conditions. Since the whole system is a combination of these modes, the system has a large number of modes. Therefore, the driver will have difficulty in understanding the differences of all the modes. For that reason, it is necessary to reduce the number of modes of the system as a whole as much as possible by integrating a

few of these modes. It is also necessary to design a system display that is consistent with the system behavior and easy for the user to understand.

(3) *Understanding of the system operation*: The operation of current ACCs have been developed based on cruise controls, which were devices for fixing the speed of the vehicle. Therefore, the system architecture and its operation needs to be re-designed when evolving from ACC-like system to 'automated' system in accordance with the mental model of automated driving systems.

(4) *Understanding of the behavior of the system*: Each driver drives in a different way; for instance, there are differences in the timing of onset of the braking, in the way of the braking, the way of accelerating, etc. The driver feels uncomfortable or uneasy when the control of the vehicle by the system is different from when he/she is controlling it by himself/herself (see 5.4.2.5). System control at the moment of entering a curve or changing lanes needs to be designed taking that into account.

5.4.2.6.3 State of the Driver

(1) *State of the driver when using automated driving systems*: At Level 2, the driver does not need to perform control tasks but needs to keep attention on the traffic conditions of the road. When the traffic conditions are stable and there is nothing particular, keeping attention (vigilance, see 4.3.2) is difficult for the human. Drivers' attention to the environment outside will easily decline. Therefore, it is necessary to develop technologies that guide their attention to certain targets, when necessary.

There is a risk that the situation awareness of the driver will decrease and he will not perceive or anticipate actions of other close vehicles when he/she stops paying continuous attention to the traffic environment of the road (see 6.1.2.1.4). Therefore, it is desirable that the system proactively provides information on the situation awareness to the driver. As to what sort of expression should be used to communicate with the driver, that is an HMI issue.

When there are no significant changes in the outside environment and the driver is not engaged in any driving operation for a long time, the arousal level decreases with it. Under this situation, the driver may not be able to take immediate action when the system suddenly requests him/her to do. In order to prevent it, it is necessary that the system monitors the arousal level of the driver and takes necessary measures for avoiding its decrease or for enhancing it (see 4.3.3).

(2) *Gap in the transition to the state where the driver is able to execute driving tasks*: At Level 3, there is a large gap with the state where the driver is able to execute driving tasks because the he/she does not have to execute any driving task while using the automated system. When changing to the manual mode, there has to be a transition to the state where the driver is able to execute driving tasks for filling is this gap. Because the driver is not prepared in terms of attention to the outside/ situation awareness/operation actions while using the automatic mode, the HMI must be designed to arouse driver's 'readiness/availability, ISO/TR 21959-1, 2018' to take over driving by calling the attention of the driver, enhancing his/her awareness and preparing him/her for taking operation actions. Furthermore, technologies will be needed to make the transition from a sleeping state to an awakened state when he/she is asleep (see 4.3.3).

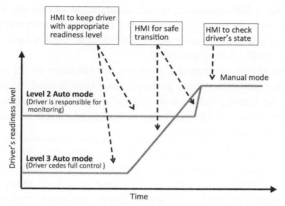

Fig. 5.70: Mode transition of Level 2 and Level 3 driving automation systems and HMI.

The gap between the state of the driver when driving in the automatic mode of Level 3 and the state where the driver is able to perform driving tasks is larger than the one at Level 2. Therefore, we should think that the time it takes to make the transition becomes longer (Fig. 5.70). In an experiment using DS, it took about 30 to 40 seconds for the vehicle to stabilize after the automatic mode was turned off and the driver took over the control (Merat et al., 2014).

5.4.2.6.4 Value of Automated Driving Systems for Humans

The general image concerning automated driving systems is that they will allow us to reach our destination without much effort. A survey conducted by Payre et al. shows that around 70 per cent of people are willing to use automated driving systems, and that there are cases where people would use it for being bored in a traffic jam and cases where people are ill or drunk (Payre et al., 2014). You will still be bored after changing to the automatic mode if you are not allowed to do any non-driving related activity. A Level 2 system is a system that does not allow drivers to sleep when feeling sleepy and a Level 3 system is a system where even if the driver is allowed to be asleep, it is maintained at a shallow level to prepare for taking over driving when going out from ODD.

We have impression that a car is good or bad by feeling the smoothness of the acceleration, the comfort, etc., when we are driving it. We have this sort of feelings, particularly when we are actively operating the steering, the pedals, i.e., operating the vehicle in an active manner. Therefore, it will be difficult to see any difference, if we stop performing driving operations by ourselves. In other words, it will be difficult to distinguish vehicles according to their automotive technology. Moreover, there is a concern that our driving skills may decline as we increase the use of automated driving systems.

5.4.2.6.5 Interaction Between the Car and other Traffic Participants

(1) *Communication between traffic participants*: Although vehicles have turn signals, stop lamps and hazard lamps for conveying the intention or action of the driver to other people, in addition to that, the driver utilizes informal communication signals as

the behavior of the vehicle (position, speed, acceleration and deceleration, etc.), eye contact, hand gestures for conveying one's own intention to other traffic participants or interpreting the intention of other traffic participants. That is how a coordinated, effective and safe traffic flow is generated.

Kitazaki and Myhre conducted a study on the influence of the communication signals as vehicle behavior and hand gestures that a vehicle sends when two vehicles are approaching an intersection without the yield sign at the same time and found that the behavior of the vehicle and hand gestures influenced the decision and its confidence (Kitazaki and Myhre, 2015).

(2) *Communication functions that automated vehicles must have*: As described above, informal communication signals to other traffic participants play an important role. If that is so, automated driving systems must be able to understand informal communication signals from other traffic participants and send communication signals for conveying the intention of the system to them.

In order to understand informal communication signals, one must be capable of understanding the intention of the behavior of other vehicles, pedestrians, bicycles, etc. According to Renge et al. there are cases where communication between traffic participants differs by region (Renge et al., 2001). Therefore, automated vehicles need to learn communication features of each region. As for sending communication signals, we need to control the vehicle's behavior to clearly express the future state of the vehicle, such as stopping at a crosswalk. Furthermore, there have been discussions to develop the external HMIs as communication signals from automated driving vehicles; for instance, devices to communicate with other traffic participants through the LED arrays (ISO/TR 23049, 2018). It may be also possible to smoothen the interaction by displaying the identity of automated driving vehicles (informing that the vehicle is an automated driving vehicle).

References

Akamatsu, M. (2004). Evaluating driver while driving in real road environment using equipped vehicle. Journal of Society of Automotive Engineers of Japan, 58(12): 53–58 (in Japanese with English abstract).

Akamatsu, M. (2008). Japanese approaches to principles, codes, guidelines, and checklists for in-vehicle HMI. pp. 425–443. *In*: M.A. Regan, J.D. Lee and K.L. Young (eds.). Driver Distraction, Theory, Effects, and Mitigation, CRC Press, London, UK.

Akamatsu, M., H. Hashimoto and S. Shimaoka. (2011). Assessment method of effectiveness of passenger seat belt reminder. SAE Technical Paper, 2012-01-0050.

Alliance of Automobile Manufacturers: Statement of principles on human-machine interfaces (HMI) for in-vehicle information and communication systems. Version 2.0, Washington, D.C, Retrieved July 15, 2007.

Asoh, T. (2002). Study of a static test method on the evaluation of the total glance time for car navigation systems while driving. In: Symposium on Mobile Interactions and Navigation (in Japanese).

Billings, C. (1991). Human-centered aircraft automation. NASA Technical Memorandum 103885. http://hdl.handle.net/2060/19910022821.

Brace, C.L., K.L. Young and M.A. Regan. (2007). Analysis of the literature: The use of mobile phones while driving. Monash University Accident Research Centre.

Broadbent, D.E. (1958). Perception and Communication, London, Pergamon Press.

Brookhuis, K., D.D. Waard and B.E.N. Mulder. (1994). Measuring driving performance by car-following in traffic. Ergonomics, 37(3): 427–434.

Brown, I.D. and E.C. Poulton. (1961). Measuring the spare 'mental capacity' of car drivers by a subsidiary task. Ergonomics, 4(1): 35–40.

Brown, I.D., A.H. Tickner and D.C. Simmonds. (1969). Interference between concurrent tasks of driving and telephoning. Journal of Applied Psychology, 53: 419–426.

Bruyas, M.P. and L. Dumont. (2013). Sensitivity of detecting response task (DRT) to the driving demand and task difficulty. Proceedings of 7th International Driving Symposium of Human Factors in Driver Assessment, Training and Vehicle Design, 64–70.

Card, S.K., Thomas P. Moran and A. Newell. (1983). The Psychology of Human-Computer Interaction, Lawrence Erlbaum Associates, Hillsdale, NJ.

Charlton, J.L., M. Catchlove, M. Scully, S. Koppel and S. Newstead. (2013). Older driver distraction: A naturalistic study of behavior at intersections. Accident Analysis and Prevention, 58: 271–278.

Cooper, G.E. and R.P. Harper. (1969). The use of pilot rating in the evaluation of aircraft handling qualities. Report TN-D-5153, NASA, Ames Research Center, Moffett Field, California.

De Waard, D. (1996). The Measurement of Drivers' Mental Workload, The Traffic Research Centre VSC, University of Groningen, The Netherlands.

Dingus, T.A., M.C. Hulse, J.F. Antin and W.W. Wierwille. (1989). Attentional demand requirements of an automobile moving-map navigation system. Transportation Research A, 23(4): 301–315.

Dingus, T.A., S. Klauer, V. Neale, A. Petersen, S. Lee, J. Sudweeks, M. Perez, J. Hankey, D. Ramsey and S. Gupta. (2006). The 100-car Naturalistic Driving Study, Phase II-Results of the 100-car Field Experiment (No. HS-810 593).

Donders, F.C. (1969). On the speed of mental processes. Acta Psychologia, 30: 412–431.

Driver workload metrics project, DOT HS 810 635 (2006). Department of Transportation, National Highway Traffic Safety Administration, NHTSA.

Edworthy, J. and A. Adams. (1996). Warning Design—A Research Prospective, Taylor & Francis.

Fairclough, S.H., M.C. Ashby and A.M. Parkes. (1993). In-vehicle displays, visual workload, and usability evaluation. pp. 245–254. In: A.G. Gale, C.M. Haslegrave, H.W. Kruysse and S.P. Taylor (eds.). Vision in Vehicle IV, Amsterdam, North Holland.

Freedman, M., N. Lerner, P. Zador, J. Lopdell and E. Bergeron. (2007). The effectiveness of enhanced seat belt reminder systems—Observation field data collection methodology and findings. DOT HS 810 844.

Freedman, M., N. Lerner, P. Zador, J. Singer and S. Levi. (2009). Effectiveness and acceptance of enhanced seat belt remainder systems: Characteristics of optimal reminder system. DOT HS 811 097.

French, R.L. (1990). In-vehicle navigation—Status and safety impacts. Technical Papers from ITE's 1990, 1989, and 1988 Conference: 226–235, Institute of Transportation Engineers.

Freundschuh, S. (1989). Does anybody really want (or need) vehicle navigation aids? Proceedings of the 1st Vehicle Navigation and Information Systems Conference, Toronto, 439–442.

Fuiji, T., H. Naito, K. Shinohara, T. Matsuoka, K. Ishida, S. Enokida, T. Asao, S. Suzuki and K. Kotani. (2014). Validity of multi-modal stimulus detection task for measuring driver's mental load for operation of IVIS during actual driving. Transaction of Society of Automotive Engineers of Japan, 45(4): 723–728 (in Japanese with English abstract).

Green, P. (2004). Driver distraction, telematics design, and workload manager: Safety issues and solutions. SAE Paper 2004-21-0022.

Harms, L. and C. Patten. (2003). Peripheral detection as a measure of driver distraction: A study of memory-based versus system-based navigation in a built-up area. Transportation Research, Part F, 6(1): 23–36.

Hart, S.G. and L.E. Staveland. (1988). Development of NASA-TLX (Task Load Index): Results of empirical and theoretical research. pp. 139–183. In: P.A. Hancock and N. Meshkati (eds.). Human Mental Workload, Amsterdam, North Holland.

Hatanaka, S., N. Sanma and K. Nakagawa. (2012). Remote control device. Proceedings of JSAE Spring Meeting, 20125239. 36-12: 5–8.

HMI guidance for in systems for supporting safe driving through infrastructure cooperation. Report of Advanced Safety Vehicle, Ministry of Land, Infrastructure and Transportation, Japan (2011) (in Japanese).

Hoel, J., M. Jaffard and Van Elslande Pierre. (2010). Attentional competition between tasks and its implications, paper presented at the European Conference on Human Centre Design for Intelligent Transport Systems, 575–581.

http://www.ghsa.org/html/stateinfo/laws/cellphone_laws.html. [accessed 2014.09].

Inagaki, T., N. Moray and M. Itoh. (1998). Trust, self-confidence and authority in human-machine systems. Proceedings of IFAC HMS, 31(26): 431–436.

Inagaki, T. and M. Itoh. (2013). Human's overtrust in and overreliance on advanced driver assistance systems: A theoretical framework. International Journal of Vehicular Technology, Vol. 2013, Article ID 951762: 8 pages.

Ishibashi, M., M. Okuwa and M. Akamatsu. (2002). Development of driving style questionnaire and workload sensitivity questionnaire for drivers characteristic identification. Proceedings of JSAE Spring Conference, 55-02: 9–12 (in Japanese; English abstract is available as SAE Technical Paper 2002-08-0243).

Ishibashi, M., M. Okuwa, S. Doi and M. Akamatsu. (2007). Indices for characterizing driving style and their relevance to car following behavior. In: SICE, 2007 Annual Conference, IEEE: 1132–1137.

ISO TS16951 (2004). Ergonomic aspects of transport information and control systems (TICS)— Procedures for determining priority of on-board messages presented to drivers.

ISO 9241-110 (2006). Ergonomics of human-system interaction—Part 110: Dialogue principles.

ISO 16673 (2007). Ergonomics aspects of transport information and control systems—Occlusion method to assess visual demand due to the use of in-vehicle systems.

ISO 2575 (2010). Road vehicles—Symbols for controls, indicators and tell-tales.

ISO 26022 (2010). Ergonomics aspects of transport information and control systems—Simulated lane change test to assess in-vehicle secondary task demand.

ISO TR12204 (2012). Ergonomic aspects of transport information and control systems—Introduction to integrating safety critical and time critical warning signals.

ISO TS 14198 (2012). Ergonomics aspects of transport information and control systems—Calibration tasks for methods which assess driver demand due to the use of in-vehicle systems.

ISO 15007-1 (2014). Measurement of driver visual behavior with respect to transport information and control systems. Part 1, Definition and Parameters.

ISO 17488 (2016). Road vehicles—Transport information and control systems. Detection-response task (DRT) for assessing attentional effects of cognitive load in driving.

ISO TR 21959-1 (2018). Road vehicles—Human performance and state in the context of automated driving. Part 1: Common underlying concepts.

ISO TR 23049 (2018). Road Vehicles—Ergonomic aspects of external visual communication from automated vehicles to other road users.

Itoh, M. (2011). A model of trust in automation: Why humans over-trust? In: SICE Annual Conference (SICE), 2011 Proceedings of IEEE, 2011: 198–201.

ITU-T Focus Group on Driver Distraction. Report on Situation Awareness Management, Version 1.0, 2013.

Iwai, A. (2013). Design and evaluation of an automotive service integration platform. DENSO Technical Review, 18: 172–180 (in Japanese with English abstract).

Iwashita, Y., M. Ishibashi, Y. Miura and M. Yamamoto. (2012). Changes of driver behavior by rear-end collision prevention support system in poor visibility. International Journal of Automotive Engineering, 3(3): 89–95.

Jahn, G., A. Oehme, J.F. Krems and C. Gelau. (2005). Peripheral detection as a workload measure in driving: Effects of traffic complexity and route guidance system use in a driving study. Transportation Research Part F, 8(3): 255–275.

JAMA in-vehicle display guideline version 3.0. http://www.jama-english.jp/release/release/2005/In-vehicle_Display_GuidelineVer3.pdf.

Kahneman, D. (1973). Attention and Effort, Prentice-Hall, New York, USA.

Kakihara, M. and T. Asoh. (2005). JAMA guideline for in-vehicle display systems. In: Proceedings of 12th World Congress on ITS, 3231.

Kirchner, W.K. (1958). Age differences in short-term retention of rapidly changing information. Journal of Experimental Psychology, 55(4): 352–358.

Kitajima, M., H. Tahira and H. Saito. (2006). A method for evaluating and revising hierarchical menus—In the case of menu system for destination setting in a car navigation system. The Japanese Journal of Ergonomics, 42(2): 87–97 (in Japanese with English abstract).

Kitazaki, S. and N.J. Myhre. (2015). Effects of non-verbal communication cues on decisions and confidence of drivers at an uncontrolled intersection. Proceedings of the Eighth International Driving Symposium on Human Factors in Driver Assessment, Training and Vehicle Design, 113–119.

Klauer, S.G., T.A. Dingus, V.L. Neale, J.D. Sudweeks and D.J. Ramsey. (2006). The impact of driver inattention on near-crash/crash risk: An analysis using the 100-car naturalistic driving study data. DOT HS 810 594, National Highway Traffic Safety Administration (NHTSA).

Knowles, W.B. (1963). Operator loading tasks. Human Factors, 5(2): 155–161.

Kondoh, T., T. Yamamura, N. Kuge, P. Miguel and T. Sunda. (2015). Development of a real-time steering entropy method for quantifying driver's workload. Transactions of Society of Automotive Engineers of Japan, 46(1): 167–172 (in Japanese with English abstract).

Kurahashi, T., M. Ishibashi and M. Akamatsu. (2003). Objective measures to assess workload for car driving. In: SICE 2003 Annual Conference, IEEE, 2003: 270–275.

Lee, D.H. and K.S. Park. (1990). Multivariate analysis of mental and physical load components in sinus arrhythmia scores. Ergonomics, 33(1): 35–47.

Lee, J. and N. Moray. (1992). Trust, control strategies and allocation of function in human-machine systems. Ergonomics, 35(10): 1243–1270.

Lei, S. and M. Roetting. (2011). Influence of task combination on EEG spectrum modulation for driving workload estimation. Human Factors, 53(2): 168–179.

Lynch, K. (1960). The Image of the City, MIT Press., Cambridge, MA, USA.

Marie-Pierre, B. and D. Laëtitia. (2013). Sensitivity of detection response task (DRT) to the driving demand and task difficulty. Proceedings of the seventh International Driving Symposium on Human Factors in Driver Assessment, Training, and Vehicle Design, 64–70.

McLean, J.R. and R. Hoffmann. (1975). Steering reversals as a measure of driver performance and steering task difficulty. Human Factors, 17(3): 248–256.

Mehler, B., B. Reimer, J.F. Coughlin and J.A. Dusek. (2009). The impact of incremental increases in cognitive workload on physiological arousal and performance in young adult drivers. Transportation Research Record, 2138: 6–12.

Merat, N., A.H. Jamson, F.C. Lai, M. Daly and O.M. Carsten. (2014). Transition to manual: Driver behavior when resuming control from a highly automated vehicle. Transportation Research Part F, 27: 274–282.

Merker, H.M. (1966). Imperial sentry signal warning system. SAE Paper 660045.

Michon, J.A. and A. Smiley. (1993). Generic Intelligent Driver Support, Taylor & Francis, London, UK.

Miller, D.P. (1981, October). The depth/breadth tradeoff in hierarchical computer menus. In: Proceedings of the Human Factors Society, 25th Annual Meeting, 25(1): 296–300, SAGE Publications CA: Los Angeles, CA, USA.

MIL-STD-411F (1997). Department of Defense, Design Criteria Standard, Aircrew Station Alerting Systems.

MIL-STD-1472G (2012). Department of Defense, Design Criteria Standard, Human Engineering.

Nakamura, Y. (2008). JAMA guideline for in-vehicle display systems (Human- machine interface). SAE paper 20083934 No.2008-21-0003.

Nakayama, O., T. Futami, T. Nakamura and E.R. Boer. (1999). Development of a steering entropy method for evaluating driver workload, SAE Technical Paper 1999-01-0892.

National Highway Traffic Safety Administration (NHTSA) (2010). Driver Distraction Program, DOT HS 811 299.

National Highway Traffic Safety Administration (2013). Visual-manual NHTSA driver distraction guidelines for in-vehicle electronic devices. No. NHTSA-2010-0053, Department of Transportation, NHTSA.

Neisser, U. (1976). Cognition and Reality, Freeman, San Francisco, USA.

Newman , W.M. and M.G. Lamming. (1995). Interactive System Design, Addison-Wesley Pub. Co.

NSC: Distracted Driving Research Statistics, http://www.nsc.org/learn/NSC-Initiatives/Pages/distracted-driving-research-studies.aspx.

Obinata, G., T. Usui and N. Shibata. (2008). On-line method for evaluating driver distraction of memory-decision workload based on dynamics of vestibular-ocular reflex. Review of Automotive Engineering, 29(4): 627–632.

O'Donnell, R.D. and F.T. Eggemeier. (1986). Workload assessment methodology. pp. 1–49. *In*: K.R. Boff, L. Kaufman and J.P. Thomas (eds.). Handbook of Perception and Human Performance, Vol. II, Cognitive Processes and Performance, Wiley, New York, USA.

Ogden, G.D., J.M. Levine and E.J. Eisner. (1979). Measurement of workload by secondary tasks. Human Factors, 21(5): 529–548.

Ohtani, A., H. Uno, T. Asoh and A. Iihoshi. (2007). Dual task method to estimate task demand caused by operating in-vehicle system—Discussion on lane change test. Transaction of Society of Automobile Engineers of Japan, 38(6): 235–240 (in Japanese; English abstract is available as SAE Technical Paper, 2007-08-0263).

Ohtani, A., H. Uno, T. Asoh, Y. Nakamura and K. Marunaka. (2007). Sensitivities of lane change test (LCT) to auditory tasks: Devised measurement for lane change initiation to make LCT efficient. JARI Research Journal, 29(11): 587–590 (in Japanese).

Ohtani, A., H. Uno, T. Asoh, A. Iihoshi and K. Narunaka. (2008). Dual task method to estimate task demand caused by operating in-vehicle systems—Discussion on lane change test. Review of Automotive Engineering, 29(2): 165–172.

Olssson, S. and P.C. Burns. (2000). Measuring driver distraction with a peripheral detection task. Driver Distraction Internet Forum, NHTSA.

Pamphlet of ASV Promotion Project Phase 5. http://www.mlit.go.jp/jidosha/anzen/01asv/data/asvpanphlet-e.html (accessed 2018 July 13th).

Parasuraman, R., T.B. Sheridan and C.D. Wickens. (2000). A model for types and levels of human interaction with automation. IEEE Trans. SMC—Part A, 30(3): 286–297.

Pauzié, A. (2008a). A method to assess the driver mental workload: The driving activity load index (DALI). IET Intelligent Transport Systems, 2(4): 315–322.

Pauzié, A. (2008b). Evaluating driver mental workload using the driving activity load index (DALI). In: Proc. of European Conference on Human Interface Design for Intelligent Transport Systems, 67–77.

Payre, W., J. Cestac and P. Delhomme. (2014). Intention to use a fully automated car: Attitudes and a priori acceptability. Transportation Research Part F, 27: 252–263.

Pettitt, M., G. Burnett and A. Stevens. (2005). Defining driver distraction. In: Proceedings of the 12th ITS World Congress, San Francisco, USA, ITS America, 1–12.

Petzoldt, T., N. Bär, C. Ihle and J.F. Krems. (2011). Learning effects on the lane change task (LCT)—Evidence from two experimental studies. Transportation Research Part F, 14(1): 1–12.

Petzoldt, T., S. Bruggemann and J.F. Krems. (2014). Learning effects in the lane change task (LCT)—Realistic secondary tasks and transfer of learning. Applied Ergonomics, 45(3): 639–46.

Piersma, E.H. (1993). Adaptive interface and support systems in future vehicles. pp. 321–332. *In*: A.M. Parkes and S. Franzen (eds.). Driving Future Vehicles, Taylor & Francis.

Posner, M.I. (1980). Orienting of attention. Quarterly Journal of Experimental Psychology, 32(1): 3–25.

Rakauskas, M.E., N.J. Ward, E. Bernat, M. Cadwallader, C. Patrick and D. De Waard. (2005). Psycho-physiological measures of driver distraction and workload while intoxicated. Proceedings of the Third International Symposium on Human Factors in Driver Assessment, Training and Vehicle Design, 150–157.

Ranney, T.A., J.L. Harbluk and Y.I. Noy. (2005). Effects of voice technology on test track driving performance: Implications for driver distraction. Human Factors, 47(2): 439–454.

Ranney, T.A. (2008). Driver distraction: A review of the current state-of-knowledge. DOT HS 810 787, National Highway Traffic Safety Administration (NHTSA).

Regan, M.A., C. Hallett and C.P. Gordon. (2011). Driver distraction and driver inattention: Definition, relationship and taxonomy. Accident Analysis and Prevention, 43(5): 1771–1781.

Reid, G.B. and T.E. Nygren. (1988). The subjective workload assessment technique: A scaling procedure for measuring mental workload. pp. 185–218. *In*: P.A. Hancock and N. Meshkati (eds.). Human Mental Workload, Amsterdam, North Holland.

Reid, G.B., S.S. Potter and J.R. Bressler. (1989). Subjective Workload Assessment Technique (SWAT): A User's Guide, AAMRL-TR-89-023, Armstrong Aerospace Medical Research Laboratory, Ohio.

Renge, K., G. Weller, B. Schlag, M. Peraaho and E. Keskinen. (2001). Comprehension and evaluation of road users' signaling—An international comparison between Finland, Germany and Japan. Proceedings of International Conference on Traffic and Transport Psychology-ICTTP, 2000: 91–100.

Rockwell, T.H. (1988). Spare visual capacity in driving-revisited—New empirical results for an old idea. pp. 317–324. *In*: M.H. Freeman, P. Smith, A.G. Gale, S.P. Taylor and C.M. Haslegrave (eds.). Vision in Vehicle II: Amsterdam, North Holland.

Ross, V., E.M. Jongen, W. Wang, T. Brijs, K. Brijs, R.A. Ruiter and G. Wets. (2014). Investigating the influence of working memory capacity when driving behavior is combined with cognitive load: An LCT study of young novice drivers. Accident Analysis and Prevention, 62: 377–387.

SAE J3016 (September 2016). Taxonomy and definitions for terms related to driving automation systems for on-road motor vehicles. SAE International.

Sanders, A.F. (1997). A summary of resource theories from a behavioral perspective. Biological Psychology, 45(1-3): 5–18.

Sato, T., M. Akamatsu, A. Takahashi, K. Yoshimura, Y. Shiraishi, T. Watanabe and T. Sugano. (2005). Analysis of driver behavior when overtaking with adaptive cruise control. Review of Automotive Engineering, 26(4): 481–488.

Sato, T., M. Akamatsu, A. Takahashi, Y. Takae, N. Kuge and T. Yamamura. (2007). Influence of naturalistic driving behavior on selection of pre-set headway distance while driving with ACC. Proceedings of JSAE Spring Conference, 38-07: 9–12 (in Japanese; English abstract is available as SAE Technical Paper 2007-08-0167).

Sato, T. and M. Akamatsu. (2008). Preliminary study on driver acceptance of multiple warnings while driving on highway. Proceedings of SICE Annual Conference, 872–877.

Sato, T. and M. Akamatsu. (2014). Development of an inference system for drivers' driving style and workload sensitivity from their demographic characteristics. Advances in Human Aspects of Transportation: Part III, 9: 199.

Senders, J.W. (1970).The estimation of operator workload in complex systems. pp. 207–216. *In*: K.B. DeGreene (ed.). Systems Psychology, McGraw-Hill, New York, USA.

Sheridan, T.B. (1992). Telerobotics, Automation, and Human Supervisory Control, MIT Press, USA.

Sheridan, T.B. (1999). Human supervisory control. pp. 591–628. *In*: A.P. Sage and W.B. Rouse (eds.). Handbook of Systems Engineering and Management, John Wiley & Sons, NJ, USA.

Shinohara, K., A. Shimada, T. Kimura, M. Ohsuga and M. Wakamatsu. (2012). Assessing driver's cognitive load by stimulus detection task when a driver operates an in-vehicle information devise. Proceedings of 2012 JSAE Spring Conference No. 81-12: 13–18 (in Japanese; English abstract is available as SAE Technical Paper 2012-08-0362).

Shneiderman, B. (1992). Designing the User Interface: Strategies for Effective Human-Computer Interaction 2nd edition, Addison-Wesley, MA, USA.

Sirevaag, E., A.F. Kramer, R. DeJong and A. Mecklinger. (1988). A psycho-physiological analysis of multi-task processing demands. Psycho-physiology, 25(4): 482.

Skipper, J.H., C.A. Rieger and W.W. Wierwille. (1986). Evaluation of decision-tree rating scales for mental workload estimation. Ergonomics, 29(4): 585–599.

Sternberg, S. (1969). Memory scanning: Mental processes revealed by reaction time experiment. American Scientist, 57(4): 421.–457.

Svenson, O. and C.J. Patten. (2005). Mobile phones and driving: A review of contemporary research. Cognition, Technology & Work, 7(3): 182–197.

Takeda, Y., N. Yoshitsugu, K. Itoh and N. Kanamori. (2012). Assessment of attention workload while driving by eye-fixation-related potentials. Kansei Engineering International Journal, 11(3): 121–126.

Tijerina, I. (2006). Report of the road safety committee on the inquiry into driver distraction. Report No. 209, Melbourne, Victoria, Australia, Road Safety Committee, Parliament of Victoria.

Treisman, A.M. and G. Gelade. (1980). A feature-integration theory of attention. Cognitive Psychology, 12(1): 97–136.

Tsang, P.S. and V.L. Velazquez. (1996). Diagnosticity and multi-dimensional subjective workload ratings. Ergonomics, 39(3): 358–381.

Uchiyama, Y., S. Kojima, T. Hongo, R. Terashima and T. Wakita. (2002). Voice information system adapted to driver's mental workload. Proceedings of HFES Annual Meeting, 1871–1875.

Uno, H., K. Hiramatsu, H. Ito, B. Atsumi and M. Akamatsu. (1999). Communication of criticality and urgency by assignment of visual and auditory qualities for in-vehicle display. *In*: Proceedings of 6th World Congresson Intelligent Transport Systems (ITS), Toronto, Canada, November 8–12.

Verwey, W.B. (2000). On-line driver workload estimation—Effects of road situation and age on secondary task measures. Ergonomics, 43(2): 187–209.

Victor, T.W. (2008). Distraction assessment methods based on visual behavior and event detection. pp. 135–165. *In*: M.A. Regan, J. Lee and K. Young (eds.). Driver Distraction, Theory, Effects, and Mitigation, CRC Press Taylor & Francis Group, Boca Raton, FL, USA.

Waard, D. De (1996). The Measurement of Drivers' Mental Workload, The Traffic Research Centre VSC, University of Groningen, The Netherlands.

Warner, W.H., M. Ljung Aust, J. Sandin, E. Johansson and G. Björklund. (2008). Manual for DREAM 3.0, Driving Reliability and Error Analysis Method, paper presented at the Deliverable D5.6 of the EU FP6 project Safety Net, TREN-04-FP6TR-SI2.395465/506723.

Wickens, C.D. and J.G. Hollands. (1999). Engineering Psychology and Human Performance (Third Edition), Prentice Hall, New Jersey, USA.

Widyanti, A., A. Johnson and De Ward, D. (2013). Adaptation of the rating scale mental effort (RSME) for use in Indonesia. International Journal of Industrial Ergonomics, 43: 70–76.

Wierwille, W.W. and F.T. Eggemeier. (1993). Recommendations for mental workload measurement in a test and evaluation environment. Human Factors, 35(2): 263–281.

Wierwille, W.W., M. Rahimi and J.G. Casali. (1985). Evaluation of 16 measures of mental workload using a simulated flight task emphasizing meditational activity. Human Factors, 27(5): 489–502.

Yerks, R.M. and J.D. Dodson. (1908). The relation of strength of stimulus to rapidity of habit-formation. Journal of Comparative Neurology and Psychology, 18(5): 459–482.

Young, K.L., M.G. Lenné and A.R. Williamson. (2011). Sensitivity of the lane change test as a measure of in-vehicle system demand. Applied Ergonomics, 42(4): 611–618.

Zaidel, D.M. (1992). Quality of driving with route guidance assistance: An evaluation methodology. Transport Canada Report TME 9201.

Zijlstra, F.R.H. (1993). Efficiency in Work Behavior: A Design Approach for Modern Tools, doctoral thesis, Delft University of Technology, Delft, The Netherlands, Retrieved from: http://repository.tudelft.nl/view/ir/uuid%3Ad97a028b-c3dc-4930-b2ab-a7877993a17f/. [Accessed 2015.09.25].

Zwahlen, H.T. and D.P. De Bald. (1986). Safety aspects of sophisticated in-vehicle information displays and controls. Proceedings of the Human Factors Society 30th Annual Meeting, 30(3): 256–260.

6

Driver Behavior

6.1 Human Characteristics Related to Driver Behavior

6.1.1 Visual Cognitive Functions

6.1.1.1 Visual Attention and Its Psychological Measurements

Driving an automobile is human behavior which relies heavily on visual information, and is critical to obtain information in a timely manner for proper driving. If the information is obtained inadequately, dangerous situations, such as oversight and delayed perception may arise and cause an accident. The key function of visual attention in driving is to distinguish the critical information out of the vast amount of visual information outside of the automobile. In this section, three major functions for driving are described: shift of attention, selection of visual information from a fixation point, and useful field of view.

6.1.1.1.1 Shift of Attention

A feature of human visual function is that people have good eyesight only in a very narrow range of their central field of view. To obtain visual information from a wide visual field, it is necessary to constantly move your viewpoint. We can study how drivers obtain visual information by analyzing patterns in the eye movement. This eye movement is closely related to the shift of visual attention.

There are two control mechanisms in shift of attention (Posner, 1980). One type is endogenously-controlled attention, which is based on conscious thought, and the other is exogenously-controlled attention, which is triggered automatically by an external stimulus. The former is top-down attentional control as it is driven by one's intention while the latter is bottom-up attentional control as it is driven by external stimuli. In psychology, the spatial cueing method (Fig. 6.1) is often used to study attention.

In this method, a subject is first given a cue indicating where to pay attention, and is then presented with a stimulus to detect. During this process, if an arrow is displayed as a cue to indicate where to pay attention (central cueing), the subject is expected to exert endogenously-controlled attention; and if a stimulus is presented to move the subject's attention to a certain location (peripheral cueing), the subject is expected to exert exogenously-controlled attention. These spatial cueing studies demonstrate that

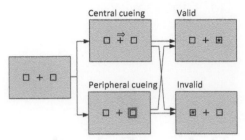

Fig. 6.1: Spatial cueing method.

the subject's attention is oriented more quickly by means of exogenously-controlled attention (100–300 ms) than endogenously-controlled attention (approx. 300 ms). In addition, these studies also indicate that when a time gap between the appearance of a cue and a stimulus is set to be longer than 300 ms, a phenomenon called 'inhibition of return' may take place, whereby the subject's attention to the location indicated by a cue is inhibited. Also, it can estimate the time required to move attention due to a cue by comparing the reaction time between the cases where the location at which a stimulus is presented matches the location indicated by a cue (valid) and the cases where these locations do not match (invalid).

Endogenously-controlled attention relates to the performance of visual search in front of and around the automobile while driving. Drivers intentionally direct their attention to where they should—that is, based on traffic signs, previous experience, and visual information indicating possible danger. By investigating how endogenous attention works, it can evaluate the driver's cognitive skills related to information acquisition for safe driving. Also, when new visual information-display systems for automobiles are introduced, it can evaluate how the information displayed by the system affects the driver's attention to the forward direction. For example, experiments have been conducted in which drivers first looked at a map on an in-vehicle display and then directed their attention to the forward direction. The study revealed that it takes a certain amount of time for drivers to completely shift their attention (Miura and Shinohara, 1998).

Exogenously-controlled attention plays an important role when drivers are in situations where they must notice and recognize a sudden appearance of danger or warning signs issued by a device. When drivers must swiftly respond to an emergency situation, it is vital for them to exert exogenously-controlled attention. Accordingly, when designing the appearance of automobiles that is required to attract attention of other people in the traffic, such as police cars, or designing warning signs that drivers can notice readily under any circumstances, the exogenously-controlled attention must be considered.

6.1.1.1.2 Selection of Visual Information at a Fixation Point

To select and perceive visual information about an object we are gazing at, we must pay attention to the object. According to a feature-integration theory of attention—a basic theory of visual cognition—we automatically process the features, such as colors and line segments in parallel without giving attention to them (Treisman and Gelade, 1980). For us to integrate these features into a whole object and perceive it, we need

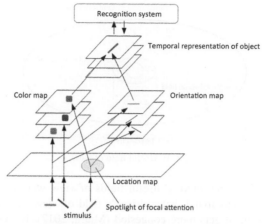

Fig. 6.2: Feature-integration theory of attention.

to pay attention to the location at which the object is present (Fig. 6.2). In other words, we do not need to pay attention just to notice the presence of an object that is different from other objects in terms of perceptual features associated with them. However, to genuinely know what the object really is, we need to pay attention to it.

Phenomena, such as inattentional blindness, where we fail to perceive sufficiently large objects or changes as visual stimuli as we do not pay attention to them, and change blindness have become well known. The following phenomenon is commonly observed—subjects are asked to view a video in which a mixture of two teams, one in white clothes and the other in black, throws passes. When the subjects count the number of passes made by the team in white, they tend to overlook the sudden appearance of a person in a black bear-suit in the video.

Similar phenomena occur during driving. Studies were conducted in which subjects were asked to make a phone call using a hand-free telephone while driving in a driving simulator (Strayer et al., 2003). After driving, the subjects were asked to remember the signs they saw alongside the driving route. It was found that the probability of the subjects remembering the signs decreased if they had made a phone call despite the fact that they carefully looked at the signs. These results indicate that the subjects' attention toward visual information coming into their view was insufficient when their attention was directed toward phone conversations.

6.1.1.1.3 Useful Field of View

People obtain visual information not only from high-resolution central vision, but also from the periphery of a fixation point. The visually perceivable area surrounding the part of the periphery of the fixation point beyond the central vision area (2°) is called the useful field of vision (Fig. 6.3) (Muira, 2012).

The useful field of view ranges between 4 to 20° and its size is influenced by various factors including factors related to visual stimuli, such as the visual load of central vision, spatial density of and similarity in stimulus background, and the context of anticipation—and unrelated factors, such as the motivation for tasks, temperature of the work environment (high temperature), and a feeling of danger.

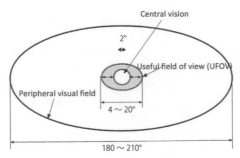

Fig. 6.3: Useful field of view.

In researches in which drivers' useful fields of view were measured in actual driving situations, it was found that a driver's field of view becomes narrower as the driving environment gets more congested (Muira, 2012). It was concluded that the narrowing of the driver's useful fields of view is not caused by deterioration of attention due to a complicated traffic situation, but rather that the driver's attempt to deal with complex situations through in-depth processing of visual information was obtained from various fixation points, yet there was a limit to how much attention the driver could give. Consequently, the drivers actively controlled their attention, which led to the narrowing of their useful fields of view (Fig. 6.4).

At present, there is no standardized/normalized method to measure a useful field of view. As such, the current situation is that individual researchers invent and apply their own measurement techniques based on the definition of the useful field of view. In general, researchers let subjects perform some kinds of visual processes involving central vision (central vision tasks). At the same time, the subjects are asked to detect visual stimuli in their peripheral vision (peripheral vision tasks). When the stimuli are detectable, they are assumed to occur in the useful field of view by definition operationally. Based on this definition, the researchers determine the area in which stimuli are detectable.

In the most basic method for measuring a useful field of view, the head of the subject is kept in a fixed position in a laboratory in such a manner that measurements can be made while the subject fixes attention at a still target (Fig. 6.5). This experimental setup is very different in nature from the driving condition. But this method can be useful in pursuing such questions as: When a driver is looking at information displayed on a device in the automobile, how large is the driver's area of perceivable peripheral vision?

Kitamura et al. conducted an experiment to see if the area of useful field of view in the forward direction differs, depending on whether both the eyes are used or only one eye is used to read information displayed on a head-up display (HUD) (Kitamura et al., 2014). In this study, the HUD was considered as a central vision task, and stimuli were presented behind the HUD as peripheral vision tasks. It was reported that the area of useful field of view was greater when the driver used only one eye than when both the eyes were used.

The measurement of the useful field of view in a laboratory setting can also be an effective method for evaluating a difference in visual attention function among individuals. In particular, the so-called Useful Field of View test (UFOV) was

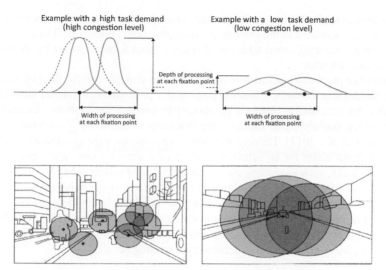

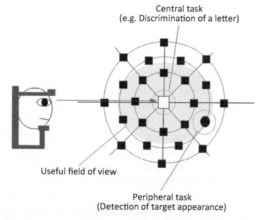

Fig. 6.4: Relationship between traffic conditions and useful field of view.

Fig. 6.5: Measurement of useful field of view in a laboratory setting.

developed as a method to evaluate cognitive functions of elderly drivers, and ample knowledge was gained using this method (Ball and Owsley, 1993).

The subsidiary task method has been applied for measuring useful fields of view while driving a vehicle (see 5.1.2). Several small lights are set on the windshield as visual stimuli. These lights are turned on at random, and as soon as a driver detects that a light is turned on, the driver orally responds to it (Muira, 2012). The driver in the experiment wears an eye-tracker so that the following data can be obtained: positions of fixation points and visual stimuli at the moment stimuli are presented, and the areas of useful fields of view that are estimated based on the results of whether or not each stimulus was detected by the driver. Also, the data on how fast the drivers detect stimuli can be used as an index for the estimation of their useful fields of view. In the useful field of view, a driver can pay visual attention to detectable levels of visual stimuli applied at any location. Assuming that the time it takes for drivers to detect stimuli is related to the amount of visual attentional resources, it can assess the distribution pattern of visual attention in areas around fixation points by measuring drivers' reaction time.

Another proposed method was to estimate the size of a useful field of view by using the distance between a fixation point and a position of peripheral stimulus as stimulus strength, and by fitting a psychometric function between the distance and the detection probability to estimate the detection threshold for peripheral stimuli (Yamanaka et al., 2014). Driving simulator experiments were conducted in which subjects drove either on an urban area route or a circuit route, while performing subsidiary tasks, such as reading out text or performing additional task. Under these conditions in which drivers were allowed to perform free visual behavior, their useful fields of view were measured. As a result, it was found that useful fields of view were estimated to be smaller for subjects who drove on the urban area route, which was more congested than the circuit route, and for those who performed the additional task, which had greater cognitive demand than reading out text.

6.1.1.2 *Physiological Measurement of Attention*

6.1.1.2.1 Attentional Resource Allocation and Event-related Potentials

For safety driving, the following two points from the viewpoint of drivers' attention are critical: (i) A driver has a sufficient amount of attentional resources to process the information on events occurring outside of the automobile, and (ii) a driver directs attention to important information and processes selectively (allocation of attentional resources). The amount of attentional resources mentioned in the first point is strongly related to the arousal level, and thus is considered to be reduced by mental fatigue and drowsiness (Kahneman, 1973). It is known that when people experience mental fatigue or drowsiness, a neuro-physiological change occurs, namely that the low-frequency component of electroencephalogram (EEG) (< 13 Hz) increases. Thus, such a change has been used as an index to evaluate mental fatigue and drowsiness of drivers.

Even when a driver's arousal level is high and attentional resources are sufficient, safety-related issues still may arise unless the driver allocates attentional resources to the appropriate information. If drivers are allocating their attentional resources to an event unrelated to driving, they cannot process visual information necessary for safe driving. Sometimes it is called 'inattentive driving'.

An event-related potential (ERP) technique is a neuro physiological method to evaluate what type of information is processed and to what extent (see 4.4.2.1). The ERP, except for a part of its components, changes its amplitude according to attentional resources allocated to a visual event. The amount of allocated attentional resources influences the visual information processing from the early stage. It is known that directing one's attention to an event causes an increase in amplitudes of early ERP components (e.g., *P1* and *N1* components) (Hillyard et al., 1998).

While the ERP is a reliable index to represent cognitive states, such as attentional resource allocation, it is necessary to repeatedly collect EEG data on similar events and perform additional processes. We also need to identify when drivers begin to process perceived information in relation to events occurring outside of the vehicle within an accuracy of milliseconds. Because of these restrictions, various efforts, other than measurements in the laboratory setting, need to be taken to apply the ERPs in the real world setting.

6.1.1.2.2 Evaluation of Visual Attentional Resource Allocation using Eye-fixation-related Potentials

The eye-fixation-related potential is an ERP which enables researchers to estimate the amount of attentional resources allocated to visual information without disturbing the participants from performing in certain field activities. Under the condition in which people are performing visual tasks, the eye-fixation-related potential, an ERP, can be obtained by performing asynchronous addition of EEG data triggered by the timing of end of saccadic eye movement (Fig. 6.6). It is generally believed that eye-fixation-related potentials are associated with the acquisition of new visual information taking place when eyes are fixed in position at the end of eye movement. The amplitude of the positive component (*P100* component), recorded in occipital region of head 100 ms after the end of eye movement, is known to attenuate when attentional resources allocated to visual information diminish due to mental fatigue (Takeda et al., 2001). The amplitude of the *P100* component of eye-fixation-related potentials is used as an index to investigate attentional resource allocation of drivers. For example, Takeda et al. showed that the amplitude of the *P100* component attenuates if the drivers simultaneously conduct tasks while driving that impose spatial working memory load (Takeda et al., 2012).

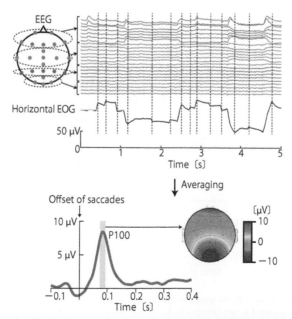

Fig. 6.6: An example of analysis on eye-fixation-related potentials (EFRP). An electroencephalogram (EEG) and horizontal electro-oculogram (EOG) are simultaneously recorded, and identify the time at which saccadic eye movement ended (dotted line) based on the EOG. EFRP is obtained by performing addition averaging processes on the EEG based on the identified times. A prominent positive component (*P100* component) appears in the EFRP recorded in the occipital region of the head, approximately 100 ms after the end of eye movement.

Drivers often perform saccadic eye movement, and their eye-fixation-related potentials can be obtained by using data collected for a few minutes, and the eye-fixation-related potential is thought to be an effective index for evaluating drivers' attentional resources allocated to visual information present outside of the automobile. However, the amplitude of the *P100* component of eye-fixation-related potentials changes not only due to attentional resource allocation, but also due to visual environmental factors (e.g., brightness and contrast). For this reason, it should be noted that measurements taken from very different visual environments cannot be quantitatively compared, for example, attentional resource allocation measured during daytime and night-time cannot be compared.

6.1.1.2.3 Evaluation of Attentional Resource Allocation Using Probe Methods

When it is required to compare attentional resources allocated to visual information measured in very different visual environments, the probe methods can be applied. In these methods, auditory and tactile stimuli (probe stimuli) are given to a participant at a time, independent of visual events of primary interest, and ERPs induced by the probe stimuli are measured (Fig. 6.7). Assuming that the total amount of attentional resources is constant, when a subject allocates a large amount of attentional resources

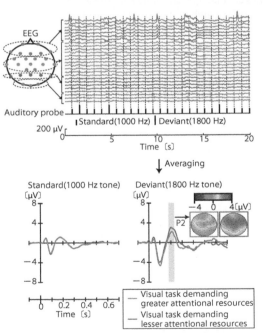

Fig. 6.7: An example of analysis on event-related potentials using probe stimuli (oddball auditory stimuli) that are unrelated to the primary tasks. Along with visual tasks, pure-tone stimuli at 1,000 Hz (standard) and 1,800 Hz (deviant) are presented frequently and infrequently, respectively. Event-related potentials induced by auditory stimuli can be obtained by performing addition averaging processes on an electroencephalogram (EEG), based on the timing of the onset of auditory stimulus presentation (dotted line). In this particular case, when subjects performed visual tasks, to which a large amount of attentional resources was allocated, attenuation of amplitude was observed in the *P2* component generated in the frontal region and the parietal region of head.

to visual information of primary interest, only a small amount of attentional resources is left unused (see 1.7.4). Regarding potentials that are related to events to which a subject allocates only a small amount of attentional resources, the amplitude of the potentials attenuates. As such, when the amplitude of ERPs induced by the probe stimuli is small, it is presumed that a large amount of attentional resources is allocated to primary visual tasks (Kramer et al., 1995).

Within probe methods, there are task-relevant methods, in which subjects are instructed to perform tasks related to probe stimuli, such as detecting deviant sound in addition to conducting primary tasks, and the task-irrelevant methods, in which subjects are instructed to ignore probe stimuli. In theory, the amount of attentional resources allocated to primary tasks is expected to be estimated more accurately using the task-relevant probe method than the task-irrelevant one. However, because an addition of secondary tasks (tasks related to the probe stimuli) may have an impact on subjects performing primary tasks (see 5.3.2), it is recommended to use the task-irrelevant probe method for evaluations under a field environment.

There are probe methods to measure ERPs induced by a sequence of stimuli consisting of one type of stimulus repeatedly presented, and those to measure EPRs induced by infrequent deviant stimuli using a sequence of stimuli consisting of frequent standard stimuli and infrequent deviant stimuli (e.g., a random sequence consisting of pure-tone stimuli at 1,000 Hz and 1,800 Hz, with probabilities of these frequencies appearing being 90 and 10 per cent, respectively. This sort of sequence is called an 'oddball sequence'). Both the probe methods can estimate attentional resource allocation, showing that when participants allocated more attentional resources to visual information, their ERPs, induced by probe stimuli had attenuated amplitude.

There was a study in which probe stimuli not related to a task were presented to subjects in the form of oddball sequences when they were engaged in driving games (visual tasks). In the study, the amount of attentional resources that the participants allocated to the visual tasks was manipulated by adjusting the difficulty level of the driving games. As the difficulty level increased, the participants' ERPs, which were induced by tactile probe stimuli, had attenuated amplitude based on the *P2* component measurement (Sugimoto and Katayama, 2013). However, to evaluate the amount of attentional resources allocated to visual information using probe methods, it is necessary to meet the assumption that the total amount of attentional resources is constant between different conditions to be compared. Accordingly, it should be noted that some studies may be invalid if the participants have different arousal levels between different conditions to be compared, due to mental fatigue and drowsiness.

6.1.1.3 *Visual Attentional Models*

In the human visual system, resolution is high in the central field of view while it is low in the periphery. This characteristic is attributed to the structure of eyes and the brain. In order to obtain high-resolution visual information from an object, we need to look straight at it. Visual attention plays an important role when we shift our eyes toward the object in a precise manner. Before we shift our eyes, we first direct our attention to the position where the object is (Rizzolatti et al., 1987).

The objective of developing visual attentional models is to explain and predict the sequence of subjects' attentional shift while looking at an image, among locations and objects in the image. Visual attentional models are useful in understanding drivers' visual behavior, safety issues as the question of what sort of objects drivers tend to overlook, and issues related to the evaluation of HMIs regarding as to what extent do in-vehicle devices distract drivers from obtaining information needed for driving.

Visual attentional models are developed based on psychological phenomena related to human visual behavior and knowledge on visual information processing in physiology and brain sciences. A number of such models, including conceptual and computer models, have been proposed from the perspective of psychology, physiology, brain sciences and computer science. Here, we primarily explain models based on the concept of the saliency map as they have broad application. We start with the model of Itti and Koch, a major visual attentional model (Itti and Koch, 2001).

6.1.1.3.1 Saliency Model of Itti and Koch

Feature integration theory (Fig. 6.2) constitutes the foundation of visual attentional models (Treisman and Gelade, 1980). Consistent with this theory, the model of Itti and Koch also assumes that it consists of a stage in which each basic feature, such as color, brightness, and the orientation of line segments, is separately processed to create feature maps and the stage at which these features are integrated into a saliency map. Feature maps reflect stimulus selectivity and receptive field functions, which are commonly observed in the neurons of the visual system in the brains of mammals, such as humans, monkeys and cats.

The initial stage of visual information processing in the brain (visual cortex) involves neurons that respond to basic visual features, such as brightness, color, and the orientation of line segments. These neurons have a relatively narrow receptive field. Each neuron responds to a stimulus only when the stimulus is given within the receptive field of vision. Also, neurons of the visual cortex in their early stages have a structure in which adjacent neurons have receptive fields at similar visual field. Consequently, locations of visual fields and locations of neurons on the visual cortex almost perfectly match on a one-on-one basis (retinotopy). As such, neurons of the visual system in their early stage play a role of coding both features and locations of visual stimuli. In attentional models, features of the visual system are expressed in terms of feature maps. The forms of feature maps differ depending on the models used. In general, models that extract basic visual features, such as brightness, color and the orientation of edges, are commonly used.

Feature maps are used to compare feature values at different locations for each feature type. This process is called 'feature contrast' computation. When a feature at a certain location is similar to that in the surrounding area, the location is assigned a low contrast value. On the other hand, when the feature at a certain location is very different from that in the surrounding area, the location is given a high contrast value. There are different ways to perform these calculations. Feature contrast values from different feature maps are added up at each corresponding location. The resulting product is the saliency map. To add feature contrast values from different feature maps, several methods have been proposed to normalize the output from each map.

A saliency map only shows relative saliency among positions, and not saliency values. An attentional spotlight shifts in the saliency map from locations of high values to those of lower values. A so-called 'winner-take-all' algorithm has been proposed as a method to calculate the locations of the highest saliency (Koch and Ullman, 1985). In this algorithm, local saliency values are compared, and locations with high values win over locations with lower values in a manner similar to competition in a tournament. Because this process is performed simultaneously across the entire visual field, the location with the highest saliency value can be identified quickly.

A computer program that implements the model of Itti and Koch has been published, and it enables the prediction of positions and objects to which attention is likely to be drawn, based on the saliency of features in an input image. For example, this model has been used to determine which part of a webpage attracts the greatest attention (Rosenholtz et al., 2011).

The model of Itti and Koch is known to enable the explanation of visual search behavior of humans to a certain extent. However, this model has limitations. The biggest limitation is that it assumes that observers' attentional behavior is determined solely by the saliency of visual information. In reality, top-down factors, such as observers' expectations and knowledge, are also relevant. The model does not adequately reflect these factors. Another limitation is that this model is only compatible with still images and cannot be applied to moving images, such as driving scenes.

6.1.1.3.2 Models that take Account of Top-down Factors

Observers' attention is greatly influenced by not only bottom-up factors but also top-down factors (see 6.1.1.1). The three major top-down factors are explained below.

The first factor is observers' knowledge of the characteristics and category of the object. For example, when a subject looks for a specific shop on a street, they would rely on certain features, such as color and the shape of the shop's logo. In such a case, their knowledge of a specific feature can be taken into account in attentional models by weighing the signal sent from the relevant feature map to the saliency map (Navalpakkam and Itti, 2006). On the other hand, when a person looks for an object with a specific shape out of a myriad of objects with different shapes, they need to make a sequential search. Therefore, models used to search for objects with complex shapes are almost equivalent to the models of recognizing one object after another in their field of view.

The second top-down factor is the ability to understand the comprehensive meaning and context behind a scene of interest, and the ability to use that information to direct attention. For example, when a driver intends to turn at an upcoming intersection, he or she would make sure, as a top priority, to check in advance whether any pedestrian is crossing the road into which the driver is turning. In this way, people understand the comprehensive meaning behind a scene of interest, such as the intersections and crosswalks, and the location of certain objects. They use this type of information to direct their attention. Rensink demonstrated that objects in a scene context could be extracted quickly (Rensink, 2000). He proposed a model that can be used to extract the arrangement and gist of the scene without requiring the users to focus their attention. Another study reported that the gist can be calculated based on the distribution of statistics related to low-order features (Oliva and Torralba, 2006).

The third top-down factor is the demands of the task. For example, a person performing a driving task is expected to allocate attention according to the driving situation. When turning at an intersection, the driver takes in a series of visual behavior, such as checking the traffic light, checking if any car is approaching from the opposite direction, and checking for any pedestrians crossing the road which the driver is turning into. Numerous studies have demonstrated that people take visual behavior to meet the demands of the task (Hayhoe and Ballard, 2005).

When people perform dual tasks, the impact of task demands on their visual behavior becomes more evident. For example, the operating a cell phone or car navigation system during driving influences driving because the driver takes visual behavior to meet the demands of non-driving tasks: like operating the device. Regarding the allocation of attention between different tasks, models related to information processing time, rather than image processing models presented in this section, are applicable to handle such a case (Wickens et al., 2003; Wortelen et al., 2013).

6.1.1.3.3 Application of Models to Moving Images

Under moving conditions, the subject's attention is attracted by a bottom-up factor, namely a drastic change of features or positions with time, in addition to various features that are notable in still images. A model that calculates spatio-temporal saliency was also proposed (Marat et al., 2009).

6.1.2 Information Processing and Cognitive Models for Humans

6.1.2.1 Driver Information Processing Models

In the field of ergonomics, human models are commonly used as a basic framework for understanding the objects of interest. Because human systems are very complex, some means of simplification are often used to address ergonomic issues. In this section, human models that are typically used in ergonomic studies are described. It should be noted that such models are merely used as a framework for understanding humans, and that an actual human system and its model are not the same.

6.1.2.1.1 Basic Three-stage Information Processing Models for Humans

The concept of information processing was developed with the emergence of the computer. Information processing refers to the acquisition of new information by processing (e.g., applying computation) existing information. Information processing consists of several stages to follow. While myriad information processes are simultaneously occurring in the complex human systems, a very simple information-processing model consists of three components: sensory system, central nervous system, and motor system. In this model, an individual receives information from the outside world through the sensory organs. Then, the individual extracts information on the states of the outside world and objects, interprets the meaning of these states, and selects the type of responses to the external states. Finally, the individual takes actions using the motor organs. Accordingly, the three-stage model can be regarded as a model that represents sensory organs, the central nervous system, and motor organs. In an analogy with this model, devices/systems also can be considered to consist of three components: display system, central processing unit, and operation system. The relationships among these components are illustrated in Fig. 6.8. This model has been widely used as a basic model for humans in ergonomics.

The middle stage of the model corresponds to the central nervous system. The brain, as an organ of the central nervous system, has a very complex structure and has

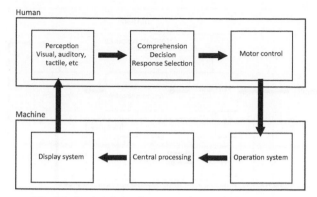

Fig. 6.8: Basic information-processing model for humans.

various sub-regions with feedback and reciprocal neural connections. Although it is complex, in order to understand human information-processing functions, the function stages along information processing within the central nervous system have been examined experimentally in the reaction time studies. Historically, Donders developed the subtraction method to discover stages in information processing (Donders, 1868). The reaction time of two different tasks was measured where one required some stages and the other required those plus an additional stage. The difference in the reaction times was considered as the time required to process the additional stage. This method was criticized because it was difficult to add stages without affecting other stages. Sternberg developed the additive factors methods where the non-interacting experimental factors were investigated on the choice reaction time (Sternberg, 1969). They prepared two different conditions for each stage and conducted reaction time experiments for all the combination of the conditions. If statistical tests for the reaction times revealed that there was no interaction effect between the factors, the factors that corresponded to the stage were considered as independent. As the times required to process the stages were independent and additive, the stages were regarded as different stages in information processing. There have been many psychophysical studies to identify stages in human information processing. However, even if various sub-stages can be identified in the laboratory setting, relevance of sub-stages to the driving task is not clear. Therefore, simple models are preferable in the automotive field.

6.1.2.1.2 Information-processing Model taking Account of Memory and Attention

The brain's memory functions can be classified into short-term memory, long-term memory, and working memory. Short-term memory is temporary storage of information obtained via sensory organs, and is retained for only several seconds before disappearing. In contrast, long-term memory is a function to store information for long periods of time. It includes not only the type of information that people can recall but also templates and filters acquired through perceptual experience, which do not surface to consciousness but are necessary for information processing. Working memory is a temporary memory function to manipulate perceived information within the brain and judge if the perceived information matches the information that the brain is looking for. The brain is considered to perceive shapes of objects using template

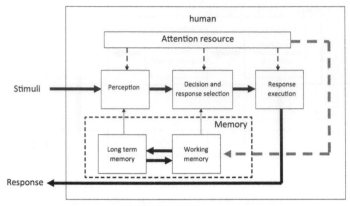

Fig. 6.9: Human information-processing model explicit on the involvement of memory and attentional resources.

that is kept in long-term memory. For the brain to decide what actions to take based on perceived information, it is necessary for it to take account of and refer to the goals in order to properly interpret the perceived information. In addition, attentional resources (see 6.1.1) must be involved to activate the perception, decision, and action selection stages using the three memory functions mentioned above during the course of performing a task. Therefore, Wickens's model includes the memory functions and attentional resources to the information processing (Fig. 6.9).

6.1.2.1.3 Norman's Seven-stage Action Model

Human information processing can be divided into any number of functional stages depending on which function he/she is interested in specifically. In his model, Donald Norman divided human information processing into seven stages: perception, interpretation of perception, evaluations of interpretations with respect to the goals and intention, goals, intention, specifying action sequence, and execution of the actions (Fig. 6.10) (Norman, 1986). This model was proposed to explain why human failed to achieve the goals by human error, like slip and lapse.

In the three-stage information-processing model, the second stage is decision. In this stage, action to be taken is selected depending on the goals. To make this

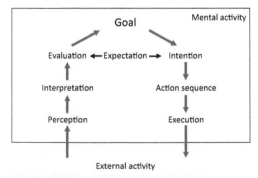

Fig. 6.10: Norman's seven-stage action model.

dependency clear, Norman's model has the 'goals' stage. Through perception, the brain obtains information on the attributes of perceived objects, such as shapes and positions. Depending on the perceiver's goals, some attributes may be ignored during the interpretation of perceived information through filtering. Then the evaluation of consistency between the goals and interpreted information takes place. If the processed information is useful in achieving the goals, the intention to take actions is triggered. Finally, a sequence of actions is selected and executed.

6.1.2.1.4 Situation Awareness Model

Situation awareness refers to the understanding of one's situation that enables appropriate actions to be taken in the situation. For example, suppose that you are in the midst of an aerial combat. You need to know the conditions of your aircraft, of your allies, and of your enemies in order to make a decision on whether to escape or go on the offensive, and take actions accordingly. You can apply similar ideas to automobile driving in the sense that, when you drive, you look for (potential) risk by observing the behavior of nearby vehicles, road conditions, etc.

Mica Endsley proposed the situation awareness (SA) model (Fig. 6.11) (Endsley, 1995). In this model, information processing generally consists of three stages: situation awareness, decision, and performance of actions. Basically, these stages correspond to the three stages of the human information-processing model mentioned above: perception, decision, and action, except that the first stage is further divided in the SA model. That is because a driver needs to perceive/recognize the outside world, not only to perceive the shapes and positions of objects (Level 1), but also to comprehend how the positions and movement of the objects relate to the driver (Level 2). Then, the driver projects the current situation into the future to anticipate what might happen (Level 3). The driver's recognition is influenced by goals and

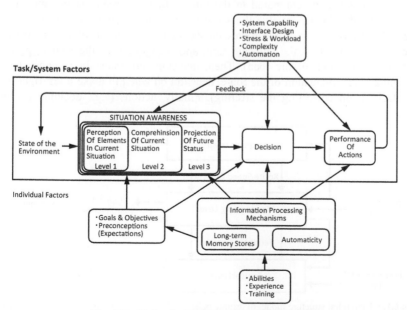

Fig. 6.11: Endsey's situation awareness (SA) model.

expectations. For example, even if the driver is exposed to the same state of the external world, his or her projection varies, depending on the mood, e.g., optimistic or pessimistic. Also, memory functions play an important role as they vary depending on the individual's capability and experience, especially for Level 2 and Level 3. The driver's situation awareness is also affected by the characteristics of the vehicles that are being driven. These relationships are illustrated in Fig. 6.11.

Although Endsley developed the SA model based on studies on airplane pilots, his model is applicable to automobile driving as well. However, it is not easy to evaluate drivers' situation awareness when they are actually driving. For driving simulator studies, SAGAT and other methods are used (see 6.2.1.4).

6.1.2.1.5 Hierarchical Model of Driving Behavior

The hierarchical model of behavior explicitly indicates the difference in temporal scale among different types of behavior. Given that the concept of combat was introduced into the model, it consists of the operational, tactic, and strategic levels (Fig. 6.12). This model is often called Michon's hierarchical model (Michon, 1985).

When the hierarchical model is applied to automobile driving, terms of 'control' and 'maneuvering' are sometimes used instead of 'operational' and 'tactic'. A driver selects a route to a destination as 'strategy', 'maneuvers' the vehicle to make left/ right turns at intersections, and 'controls' the vehicle to properly stay on the road, for example. Strategic behavior also includes deciding how far away from the edge of the road one will drive to have a wide safety margin. Then, to put this strategy into practice, one maneuvers the vehicle accordingly. Similarly, one makes a strategic plan to arrive at the destination in time, and maneuver the vehicle in terms of selecting velocity and changing lanes.

The approach of the hierarchical model is to comprehend human behavior from a different angle compared to the information-processing stages model. All three levels of behavior involve the acquisition of information through perception (e.g., identification of possible routes on the map, recognition of the intersection where one intends to make left/right turn, and recognition of the vehicle position relative to the width of the road). Then, decisions are made in response to the perception (e.g., evaluation of the route, decision on when to make a left/right turn, and decision on correcting the steering). Finally, actions (e.g., deciding a route,

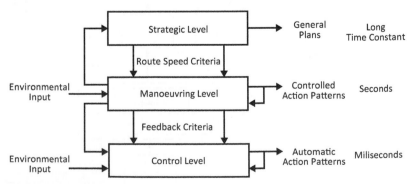

Fig. 6.12: Hierarchal structure model of driving behavior (modified from Michon (Michon, 1985)).

starting to make a left/right turn, and operating steering wheel and pedals to perform the turn) are taken.

6.1.2.1.6 Rasmussen's Skills-rules-knowledge (SRK) Model

From the perspective of human error at a large plant, Jens Rasmussen proposed an SRK model (Fig. 6.13) (The SRK model was not named by Rasmussen himself but by other researchers to make it easier for them to refer to the model) (Rasmussen, 1990). The *S*, *R*, and *K* respectively denote skill-based behavior, rule-based behavior, and knowledge-based behavior. Skill-based behavior refers to unconscious behavior acquired through everyday activities. Rule-based behavior denotes behavior that is selected according to event-condition-action (if and then) rules. A person conducting this type of behavior does not carefully examine other options of actions appropriate to the context, and follows predetermined procedures. Knowledge-based behavior means a type of behavior examined through the use of previously or newly-acquired knowledge. The concept of these types of behavior is applicable to the behavior of automobile drivers. For example, in general, drivers unconsciously correct their steering to keep the lane (skill-based behavior) and decide to stop because the upcoming traffic light is red (without considering that ignoring the red light would be dangerous; rule-based behavior), and determine the route to the destination using a map (knowledge-based behavior).

The SRK model and the hierarchical model appear similar but are actually different. For example, regarding route selection, if a route is selected to an unfamiliar place using one's knowledge, that behavior is knowledge based; if one drives following a step-by-step map drawn by someone, that behavior is rule based; and if one drives on a very familiar route to work and makes left/right turns without thinking, that behavior is skill based. Even if steering the automobile appears to be skill-based behavior, it also can be rule or knowledge based. For example, when one steers to park the vehicle, aiming to position it in a specific location (such as precisely parallel to the white line for park zone), that behavior might be knowledge based; when one starts to turn the steering wheel at the predetermined position of the vehicle relative to a specific pole, that behavior is rule based; and if one steers the vehicle without thinking about when to turn, that behavior is skill based.

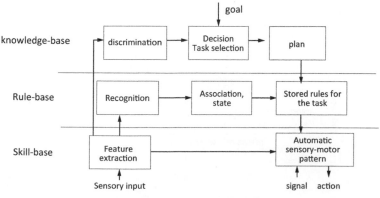

Fig. 6.13: Rasmussen's SRK model.

6.1.2.1.7 Relationship Among Different Human Information-processing Models

Here, we have introduced models commonly used in ergonomics. They were initially developed for purposes other than modeling the driving behavior. When model-users apply these models to automobile driving, they should note that these models describe human functions from different angles (Ranney, 1994). Figure 6.14 shows different viewpoints of these three models using a three-dimensional block (Theeuwes, 2001). Strictly speaking, some of the orthogonal relationships may not be necessarily true, but this may help model-users to clarify which aspect of the driver's behavior is to be discussed. It may facilitate the discussion on how to design the driver assistance systems in their development phase (Fig. 6.14). It is important to sort out and clarify the issues like which of the stages of the human information processing (perception, decision, and action) will one support; which level of the hierarchical model (control, maneuvering, and strategic) will one support and which type of behavior in the SRK model will one support.

6.1.2.1.8 Extended Contextual Control Model (E-COM)

In the information-processing models mentioned above, the processing of sensory input information is described in stages. These models were proposed to describe how humans comprehend sensory input information and take actions accordingly. In automobile driving, drivers continuously process sensory input and take actions. Therefore, it may be appropriate to assume that an input/output cycle of tasks is formed in drivers. This concept is visualized by Hollnagel's extended contextual control model (E-COM) (Fig. 6.15) (Hollnagel and Woods, 2005).

The E-COM consists of four layers of control tasks: the tracking layer, which controls the driver's actions in real time; the regulating layer, which determines task target values at the tracking control or driving trajectories; the monitoring layer, which checks whether the driver is on the right route to the destination; and the targeting layer, which determines the destination. Each layer of control is checked by the upper layer in order to meet the goal. In some cases, the goal at one layer is modified by the control tasks at the lower layer. For example, when a driver notices a hole in the road surface, he/she might change the driving trajectory. The speed of the cycle of the control differs among the layers, and at the tracking layer, the control, including

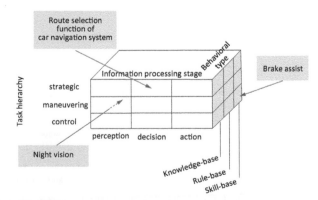

Fig. 6.14: Relationship among different human information processing models.

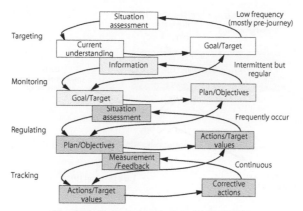

Fig. 6.15: Extended control model (E-COM).

corrective actions, performs continuously. At the regulating layer, the cycle of control, including changing lanes and making left/right turns, is completed in several tens of seconds to one minute. Also, we presume that the driver normally confirms that he/she is properly heading towards the destination every several minutes.

The lower three layers of the E-COM appear to almost perfectly match Michon's hierarchical model. Also, the situation awareness model appears to be equivalent to the information processing taking place at the regulating layer of the E-COM. The E-COM is explicit about the dynamic cycle of controls at each layer and dynamic interactions between the layers. Thus, the model can serve as a framework for understanding interactions between driving behavior and the control provided by the driving support system (Engström and Hollnagel, 2007).

6.1.2.2 Task-capability Interface Model

The task-capability interface model, one of the motivational models of driving behavior, along with the risk homeostasis model and the zero risk model (Ranney, 1994; Wilde, 1982; Summala, 1988) was proposed by R. Fuller (Fuller and Santos, 2002; Fuller, 2005). In this model, the chance of an accident occurring during driving is determined based on the relationship between the task demands and the driving performance (capability). If the task demands exceed the driving performance, the probability of having an accident increases (Fig. 6.16). The features of this model are that it specifies factors of task demands and driving capability; the task demands are affected by driving behavior itself; and the relationship between the task demand and driving capability is interpreted as the concept of task difficulty homeostasis.

Driving capability or performance is determined by the driver's comprehension and decision-making function, driving style, skill, and competency. Elderly drivers with declined comprehension and decision-making function are probably less proficient in driving performance compared to younger drivers. Aggressive driving styles probably have an impact on one's driving performance; also, the level of driving education/experience affects the driving performance. Not only these static factors but also dynamically changing factors of driving performance, such as attentional resource input and mental/physical conditions, affect the driving performance. Driving performance may temporarily decline when the drivers are tired or drowsy.

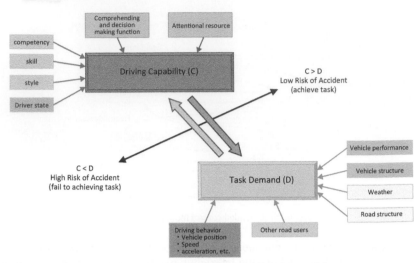

Fig. 6.16: Concept of task-capability interface model.

Task demand changes with vehicle structure/performance, road environment, traffic situations, and driving behavior. As related to road conditions and traffic situations, task demands are considered to be higher when people drive under the condition in which the lane width is narrow, or sharp curves occur, and traffic is heavy than when they drive under the condition in which lane width is wide and traffic is light. Under the high task demand condition, drivers supply large amounts of attentional resource and thereby increase their driving performance.

Even under the same road conditions and traffic situations, people who drive fast and keep only a short distance to the lead vehicle undergo higher task demands than those who drive more slowly and maintain a wide distance to the lead vehicle. The former people must respond quickly to the movement of the lead vehicle. From this viewpoint, drivers can adjust their task demands by changing their driving behavior. In reality, elderly drivers with declined mental/physical functions do not always exhibit driving behavior leading to an accident as compared to young drivers. Thus, elderly people are presumably conscious about driving properly so that they control their task demands to match their driving performance as an adaptive behavior (see 6.2.2.4.2).

In the task-capability interface model, task difficulty is defined as the difference between the driving performance and task demand, and people adjust to it during driving. When the driving task is relatively easy, people might drive in such a manner as to increase their task demands; alternatively, they might relax their mental and physical states. Since drivers can accept a certain range of task difficulty, they adjust their driving performance so as not to exceed the threshold of task difficulty. When the task difficulty exceeds a threshold, drivers feel the risk of an accident. Based on this theory, simple experiments were conducted, and the results demonstrated that people start feeling the risk of an accident when the level of their driving performance reaches the task demands (Fuller, 2005). In this regard, the task-capability interface model is different from the risk homeostasis model. The former model sets certain risk criteria, whereas the latter model assumes that task difficulty homeostasis and risk homeostasis match only when task difficulty exceeds a threshold (Fuller, 2011).

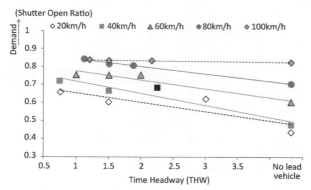

Fig. 6.17: Assessment of visual demand using the occlusion method for different speeds and headway distances.

There have been several empirical studies regarding the task capability interface model. In an on-road driving study, driving behaviors were identified to reduce task demands while approaching intersections by using the Bayesian network model (Sato et al., 2010). Some drivers expanded the headway distance when the subject vehicles were far from the intersection. Other drivers did not apply an accelerator to avoid being interfered with change in speed of the lead vehicle when the subject vehicles were closer to the intersections. Sato et al. conducted a series of driving simulator experiments in which task demands were quantified for different road structures using an occlusion method (a method by which task (visual) demands are estimated by the shutter open time with the driver wearing shutter goggles; that was used to assess visual demand when using an in-vehicle device, see 1.7.3 and 5.3.3.2) and showed that task demand was high while driving along curves and merging points and when entering tunnels (Sato et al., 2013). They also showed that task demands reduced almost linearly as the time headway increased but this relationship disappeared when driving at 100 km/h (Fig. 6.17). As the driving capability is considered to depend on the amount of mental resource, they applied the ERPs induced by the probe stimuli (see 6.1.1.2.3) and examined the relationship with the acceptable level of task demand (Soto et al., 2016).

6.2 Driving Performance

6.2.1 Driving Performance Measures

6.2.1.1 Longitudinal Driving Performance

In order to evaluate the driving performance quantitatively, various measures were used in studies. However, no common index has been established for driving evaluation so far. As a result, the same term is sometimes used for different measures, and different terms are sometimes used for practically the same measure. Therefore, we must be careful when comparing the results of different studies.

SAE has established definitions of driving performance measures and standard terminology (SAE J2944, 2015). In this section, we introduce measures for longitudinal driving performance based on SAE J2944 (Fig. 6.18).

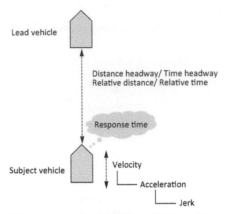

Fig. 6.18: Measures for longitudinal driving performance.

6.2.1.1.1 Velocity, Acceleration, and Jerk

Vehicle velocity is the basic measure indicating the state of a vehicle moving in the longitudinal direction. Velocity can be estimated from the speed pulse signal. As the speed pulse signal is generated due to rotation of the wheel, the accuracy of the estimated speed is low when the vehicle motion is slow. Also it becomes inaccurate when the wheel slip occurs. Difference in the tyre diameter (outer perimeter) is another factor of bias in the speed signal. Vehicle velocity can also be obtained from GSP signal. However, as velocity is calculated by differentiating the vehicle position, the accuracy is not high. Also the time resolution of the GSP speed depends on the update rate of the position data.

Acceleration is calculated by differentiating the vehicle velocity once. When the velocity data is time sample data, a difference calculation is applied instead of differential calculation, to obtain acceleration. It should be noted that the difference calculation should be carefully applied because the time resolution and the accuracy of the velocity strongly affect the accuracy of the calculated acceleration. When an accelerometer is installed in the subject vehicle, the acceleration is obtained directly from the accelerometer. In this case, the axe of the accelerometer must be aligned with the longitudinal axe of the vehicle.

If the acceleration is differentiated once (or the vehicle velocity twice), a jerk is obtained. The vehicle occupants are sensitive to the jerk of the vehicle in motion (see 3.1.3). The jerk can be used to detect the onset of acceleration change. Acceleration and jerk, and also yaw motion, can be used as measures for lateral control of the vehicle.

6.2.1.1.2 Response Time

To decelerate and stop a vehicle, the driver releases the accelerator pedal, moves his/her foot to the brake pedal, and steps on the brake pedal. To accelerate a vehicle, the driver moves his/her foot from the brake pedal or floor to the gas pedal and steps on it. Drivers' behaviors vary depending on whether they are accelerating or decelerating.

There are variations in the definitions of the onset and the end of response time among the studies. Should the measurement begin when a response-triggering event takes place or when the driver releases the gas/brake pedal? Also, should the measurement end when the driver places the foot on the gas/brake pedal or when he/she starts pressing the pedal? The operational definition depends on what kind of signal is available in the study. Thus, it is very critical to clearly define the onset and end points of the measurements (Green, 2013).

A variety of research has been conducted to measure a driver's response time. For example, driving simulator experiments indicated that the time it takes for drivers to step on the brake in response to a warning signal is 1.08 sec. (0.52–2.4 sec.) on an average (Cheng et al., 2002). Some of the other results are summarized in the reference (Gawron, 2008).

6.2.1.1.3 Headway Distance and Time

Various measures have been proposed to represent distance and time between the vehicle of interest and the lead vehicle. They include THW (time headway), calculated as headway distance ÷ velocity of the subsequent vehicle, and TTC (time to collision), calculated by headway distance ÷ relative velocity. THW denotes the time it takes for the subject vehicle to reach the current position of the lead vehicle if the former maintains the current velocity. A driving simulator study showed that when the subject vehicle followed the lead vehicle at almost a constant speed, THW strongly affected the driver's risk perception in rear-end collision to the lead vehicle (Kondoh et al., 2008). TTC indicates the time it takes for the subject vehicle to collide with the lead vehicle if the current relative velocity is maintained (velocity difference between the two vehicles). The reciprocal of TTC is equivalent to temporal change in the size of the lead vehicle as perceived by the subject driver. It is also equivalent to temporal change of the log of the headway distance. Some suggest that drivers tend to apply the brake in a manner consistent with the reciprocal of TTC (Morita et al., 2006). THW and TTC have been examined with respect to being the index of safety in a real driving environment (Vogel, 2003). Other proposed measures include the temporal differentiation of TTC (Lee, 1976), second-order projection of TTC (Barber and Clarke, 1998), and temporal change in the area of the lead vehicle as perceived by the driver of the subject vehicle in terms of decibels (performance measure for approach and alienation, KdB) (Wada et al., 2007).

Because there are different definitions for headway distance, it is important for each researcher to accurately explain how it is defined (Fig. 6.19). In traffic engineering, headway distance is often referred to as the distance between the front end of the preceding vehicle and the front end of the subsequent vehicle. This definition is also called 'headway' because traffic engineers conduct site-based observations using a sensor that detects the front end of vehicles as the reference point of the measurement. On the other hand, researchers who conduct on-road experiments using instrumented vehicles define headway distance as the distance between the tail end of the preceding vehicle and the front end of the subsequent vehicle. Because sensors installed in the front end of the subsequent vehicle detect the tail end of the preceding vehicle, their definition of headway distance is shorter than that of traffic engineering by the length of the preceding vehicle. This shorter headway distance is appropriate for studies relevant to rear-end collision. In SAE J2944, the distance between front ends of the preceding and subsequent vehicles is defined as headway, while the distance between the tail end of the preceding vehicle and the front end of the subsequent vehicle

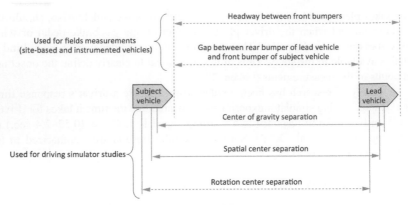

Fig. 6.19: Various headway distance measures.

is defined as a 'gap'. One should be careful about these terms as in many studies, headway is used to actually mean a gap defined in SAE J2944.

Vehicle dynamics of driving simulator are often formulated based on the vehicle's center of gravity, and therefore, headway distance commonly refers to the distance between centers of gravity of the lead and subject vehicles in driving simulators. Other definitions include the distance between the centers of the two vehicle bodies and the distance between rotation centers of the two vehicles.

As described above, the definition of headway distance varies due to different measurement methods used among different experimental environments. Also, the positions of front/tail ends of vehicles vary, depending on the shapes of the vehicles. Accordingly, it is vital for researchers to clearly define headway distance and time used in their studies.

6.2.1.2 Lateral Driving Performance

In this section, we introduce measures for the driving performance in the lateral direction (Fig. 6.20), referring to SAE J2944 which defines driving performance (SAE J2944, 2015).

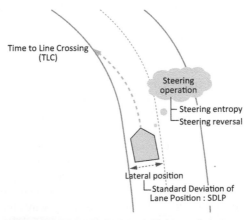

Fig. 6.20: Measures for lateral driving performance.

6.2.1.2.1 Steering Operation

Drivers steer the vehicle appropriately to travel around curves and change lanes. Regarding the maneuvering of a vehicle in response to some stimuli, for example, a driver changes lanes as he/she sees the lead vehicle turning brake lights on, the following definitions are relevant: the time it takes between the onset of stimulus and the initiation of steering operation, called the steering reaction time. The time it takes between the initiation and completion of steering is called steering movement time (in the case of changing lanes, it refers to the time it takes to complete the shift of the vehicle from its initial position to the adjacent lane). Finally, the combination of the two (steering reaction time and steering movement time) is called steering response time. Steering measures are often used to evaluate the driving performance while activating a lane-departure warning system or using a lane-keeping assistance system.

6.2.1.2.2 Steering Reversal

Steering reversal is defined as the switching of steering directions within a short time period where, the steering reversal exceeds a threshold angle (Konz and McDougal, 1968). Thresholds of 6°, 5°, and 3° have been proposed (Ostlund et al., 2004). If the threshold value is too small, the calculation of steering reversal may become too sensitive to steering vibration/noise. On the other hand, if the threshold value is set too large, the calculation of steering reversal may not reflect any performance made by the driver. Accordingly, it is critical to specify the threshold value when calculating the steering reversal.

The number of steering reversal events performed is associated with the distraction of driver's attention. The more extensively the driver is distracted, the less frequently the driver reverses steering, but the amount of steering reversal increases (McGehee et al., 2004).

6.2.1.2.3 Steering Entropy

Using time-series steering angle data, it is feasible to predict the steering angle of the next moment based on the angles of the last several moments (the number of moments used varies depending on the sampling frequency). Then, assuming that drivers perform smooth steering, the difference between predicted steering angle (calculated based on the second-order Taylor expansion) and actual steering angle can be calculated (steering angle prediction error). When drivers concentrate on steering control, it is expected that their steering will be smooth and the frequency distribution of prediction errors will be close to zero. On the other hand, when drivers do not concentrate on steering control due to distraction by non-driving tasks, corrective steering operations will increase and the frequency distribution of prediction errors will widen. The extent of frequency distribution spread can be viewed as the degree of entropy, and can be used to evaluate the smoothness of steering (see 5.1.3.2) (Nakayama et al., 1999; Boer et al., 2005).

In this method, the distribution of steering prediction errors obtained under normal driving conditions is used as a standard. Then, by comparing it with the distribution of prediction errors obtained in the driving situations under study, 'deviation from norm' can be evaluated. However, this evaluation method requires making the standard and target conditions identical in terms of driving route, road environment, and driving speed, and thus its applicability is limited. To address this issue, an enhanced evaluation method was proposed in which an identical driving environment is not required from each individual (see 5.1.3.3) (Kondoh et al., 2015).

6.2.1.2.4 Lane Position of a Vehicle

The position on the road of the vehicle under study is specified mainly in terms of the distance from a certain road feature to a certain vehicle feature. The three commonly used road features are the middle position across the road (between lane markers), average values of all vehicle trajectory, and the edge of the centerline of the road (the edge refers to the side of the white line closer to the vehicle under study). The vehicle features commonly used are the middle point of the front axle, the edge of tyres (this is used for an actual vehicle such as an instrumented vehicle), and the vehicle's center of gravity (used mostly in a driving simulator). Because there are several features referred, it is important to specify the features used to calculate the lane position of the vehicle.

The width of roads (distance across lanes) varies greatly and it is recommended to use the middle position across the road as a reference. When vehicles run in curved roads, they tend to travel along more inward trajectories than in straight roads. Accordingly, to make measures consistent between straight and curved roads, average values of all vehicle trajectory are sometimes used.

Using vehicle position data, evaluation measures, such as the number of times vehicles departed from a lane and the number of times vehicles switched lanes, can be obtained. When these measures are used, they should specify necessary details (e.g., which lane and vehicle features were used to calculate the lane position of the vehicle, and where did the vehicle start/complete departing from the lane).

6.2.1.2.5 Standard Deviation of Lane Position (SDLP)

SDLP is the most common measure used to evaluate the driving performance in the lateral direction. It can be calculated using standard deviations of lane position data, and indicates lateral wobbling of vehicles in selected road sections. Road sections may include a straight road only, a curved road only, or a combination of the two, depending on the type of analysis intended.

A study found that SDLP ranges from 0.2 to 0.3 m and varies depending on the driver's age, road environment, and driving speed (Green et al., 2003). Also, driving simulator experiments found that SDLP is 0.38 m under ordinary driving conditions but decreases to 0.3 m when a lane-departure warning system is activated (Mullen et al., 2010). SDLP can be used as an index of the driver's visual distraction as it showed a high correlation with workload when using a car navigation system (in terms of the total glance time at the system and subjective anxiety) (Akamatsu, 2008).

6.2.1.2.6 Time to Line Crossing (TLC)

TLC is comparable to TTC (time to collision), another driving performance index, but TLC concerns lateral movements while TTC concerns longitudinal movements. TLC is defined as the number of seconds it takes for a vehicle to reach a road edge if the driver keeps the current steering angle and speed. A driving simulator is often used for the calculation of TLC because accurate data need to be collected for the behavior of the vehicle (velocity and yaw rate), position of the vehicle within a lane, and road alignment coordinates. The calculation involved the division of lateral distance between the road edge and the vehicle by the speed of the vehicle in the lateral direction. An approximate calculation method was also proposed in which the lateral distance is divided by both the speed and acceleration of the vehicle (Ostlund et al., 2004). However, it should be noted that the latter calculation method must be carefully examined for its applicability and not for vehicles that are kept to stay in the same lane, change lanes, and with lateral fluctuation, as well as their expected TLC ranges (e.g., the calculation method works properly for vehicles with TLC values less than 0.5 sec., but it may not work with a TLC value of 0.5 sec. or more) (Winsum et al., 2000).

In general, the smaller the TLC is, the weaker the lateral position control of the vehicle is. Accordingly, expected average TLC values were reported for simulated highway and suburban road driving (Ostlund et al., 2006). However, very large TLC values do not necessarily mean a high level of safety. Therefore, when researchers perform statistical analysis using TLC values, they need to exclude very large TLC values (20 sec. or more) from the analysis (Ostlund et al., 2006).

6.2.1.3 Parking Maneuver

6.2.1.3.1 Cognitive Function Necessary for Parking Maneuver

Most drivers can operate vehicles intuitively when they drive forward on a wide road. However, when they back up into a narrow parking space, some drivers may feel it very difficult while others feel it easy. The performance of drivers in maneuvering a vehicle depends not only on their gender and experience but also their cognitive capability (Nishizaki et al., 2013).

A series of studies suggest that drivers' capability of parking maneuver is influenced by their cognitive function related to mapping between body coordinate system and external coordinate system. In this view, drivers who can easily associate their steering operations with resulting vehicle movement are expected to be excellent maneuvers compared to those with lesser capability in this aspect.

Normally, when people roll a vehicle forward, they can intuitively feel the forward movement of the vehicle. However, when people with poor capability for body—external coordinate system mapping roll backward, some of them may not be able to intuitively feel the consequence of steering, and get confused regarding the direction in which they should turn the steering wheel in order to move the vehicle to the intended direction. As such, they need to steer slowly. Moreover, unlike driving forward, when people drive backward, they look at the room mirror or the side mirror, or over their shoulder to confirm the backward safety, requiring them to change their body posture. Because the relationship between steering (body coordinates) and the resulting vehicle movement (external coordinates) becomes more complex in this situation, the vehicle maneuvering seems more difficult.

6.2.1.3.2 Prediction of One's Capability for Park Maneuver based on the Psycho-motor Tests

Capability to map between body coordinates and external coordinates was measured in studies where subjects were asked to perform tasks using a computer mouse in a mirror-reversed mode. The mouse was modified so that the left-to-right movement of the mouse produced right-to-left movement of the cursor on the computer screen (i.e., if a participant wanted to move the cursor to the right on the screen, he/she needed to move the mouse to the left). In this experiment, subjects were asked to work on randomly distributed letters and numbers on the screen and rearrange them in the instructed order (a -> 1 -> b -> 2 -> ...) as quickly as possible. This task required subjects to re-associate their perception of external visual coordinates with their body coordinates. This task appears to be totally unrelated to steering a vehicle. However, it has been confirmed that performance in reverse driving and reversing to park can be predicted based on performance of this task. Moreover, it was revealed that people who excel at this task are also good at maneuvering a vehicle and include reverse driving, even if they are inexperienced drivers (Nishizaki et al., 2013).

Interestingly, males performed better than females in this task of handling a mouse in a mirror-reversed mode. This result suggests that male drivers trend to do better in parking maneuver of a vehicle compared to female drivers.

For drivers to properly maneuver a vehicle, it is important for them to be able to visually imagine the paths they plan to take for the maneuver. Also, drivers' visual imaging capability is considered to relate to their ability to sense the position of their vehicle relative to the road and obstacles. It is also critical for drivers to associate between steering operation and resulting vehicle movement based on visual imaging. Drivers' visual imaging capability can be measured by the mental rotation tasks (Shepard and Metzler, 1971).

6.2.1.4 Situation Awareness Evaluation Methods

In the context of automobile driving, situation awareness (SA) refers to the comprehension of one's relationship with other traffic participants in addition to the driving condition of one's own vehicle. According to Endsley's SA model, there are three levels of SA: perception (Level 1), comprehension of current situation (Level 2), and anticipation (Level 3) (see 6.1.2.1.4).

It is important for drivers to comprehend the surrounding situation while driving for ensuring safety and deciding their next action. Also drivers' SA may be raised when warning signals are given from collision avoidance systems. On the other hand, drivers' awareness of the surrounding situation may be lowered when they conduct non-driving activities (using a cellular phone, listening radio, talking with a passenger, etc.) or when an ACC or automated driving system is activated.

The following methods to measure situation awareness have been proposed:

- Method for freezing scenes (SAGAT)
- Real-time probe technique
- Subjective rating (SART)
- Rating by observers
- Evaluation method based on task processes
- Evaluation method based on task performance

Of these methods above, 'rating by observers' is a method to evaluate the subject's SA level by a trained observer. The evaluation is conducted based on where the subject paid attention and what action he/she took subsequently (see 6.2.2.1). Thus, the reliability of the evaluated subject's SA level depends on the ability of the observer.

In the 'evaluation method based on task processes', drivers' visual behavior (eye and head movement) is measured using an eye tracker. Then, SA evaluation is conducted based on drivers' gaze direction/duration. In the 'evaluation method based on task performance', SA evaluation is conducted on the basis of whether the drivers prevented accidents and dangerous situations from occurring under complex situations. In these objective methods based on external observations and measurements, it is difficult to identify drivers' SA levels. Overviews of SAGAT, real-time probe technique, and SART are given below.

6.2.1.4.1 Situation Awareness Global Assessment Technique (SAGAT)

In this method, participants usually perform automobile driving using a driving simulator. During the drive, the display of the scene freezes and goes blank. Then, the participant is asked to answer questions on the situation when the display had gone blank (Endsley, 2000). The freezing of the screen is set to occur at random occasions. When the display goes blank, dashboard gauges are turned off so that the subject will not be given a hint that reminds him/her of the driving speed when the display went blank. Video footage can be used in the simulation. However, it should be noted that the range of viewing field is limited in the video and that video footage is inconsistent with actual driving conditions, given that the subject does not actually control a vehicle in the simulation.

After the display is frozen, the subject answers questions on the situation at the moment of freezing. If studies are to comply with the Endsley's SA model, they should use a set of questions corresponding to different SA levels. Some examples of questions are shown in Table 6.1. In some experiments, the subject is asked to draw pictures, indicating the configuration of the intersection, and the distribution, orientation, and moving speed of other traffic participants around the intersection (Fig. 6.21). To make the subject's work easier, the researcher can provide him/her with a drawing of the roads, a model car to be placed on the drawn roads, and stickers representing traffic lights and other relevant features (Franz et al., 2015).

It should be noted that in SAGAT, the subject's driving behavior is unnatural because the display gets frozen during a trial, and because the subject might rely on his/her visual memory of the images before the display went blank, and might perform memory tasks (memorizing and reproducing).

6.2.1.4.2 Real-time Probe Technique

This method does not require freezing the display during driving simulations, and subjects are asked to answer questions (like those in Table 6.1) when they are

Table 6.1: Examples of Questions Asked in the SAGAT Experiments.

Level 1 SA	Where are a pedestrian with red close, a bicycle and a bus? How many passenger cars are there? What color is a big track?
Level 2 SA	In which direction is a bicycle moving? Where is the pedestrian going to cross the road headed for?
Level 3 SA	Thinking about speed and position of your car and the surroundings, what will happen after this?

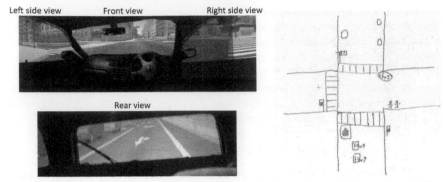

Fig. 6.21: The moment at which the display was frozen (left) and an example of an answer (right).

driving. Measurements made on the subject's answers (whether the subject correctly/ incorrectly answered, and the time it took for the subject to answer the questions) are used as indices. Given that this method does not require freezing the display, subjects can drive continuously without interruption. However, they need to answer questions during driving, which may affect their driving behavior. In addition, like the case with SAGAT, the same questions are reused in this method. So the subject's driving behavior might be affected by what he knew about the questions in advance, and therefore the state of his or her SA during a driving simulation may be different from that during actual driving.

6.2.1.4.3 Subjective Rating (SART: Situation Awareness Rating Technique)

Subjective rating is often used for SA evaluations as it is easy to perform, can be completed quickly, and does not require any equipment. Subjects answer questions with a seven-point rating scale or analogue scale (low vs high) on a sheet after they complete the driving tasks (Table 6.2) (Taylor and Selcon, 1991). Unlike SAGAT and the real-time probe technique, SART advantageously allows subjects to focus on driving tasks without interference. However, SART also has the following issues: because subjects answer questions after completing the driving tasks, they must rely

Table 6.2: Examples of Questions Asked in the SART.

Instability	Likewise of situation to change suddenly
Complexity	Degree of complication of situation
Variability	Number of variables changing in the situation
Arousal	Degree of readiness for activity
Concentration	Degree to which thoughts bear on situation
Division	Amount of division of attention in the situation
Spare capacity	Amount of attention left to spare for new variables
Information quantity	Amount of information received and understood
Information quality	Degree of goodness of the information gained
Familiarity	Degree of experience with the situation

on their memory for details and their answers may not be very accurate. Also, subjects tend to give high SA ratings to themselves if they felt good about their overall driving performance (they believe that they were satisfactorily aware of the surrounding situation if they did not cause an accident).

6.2.2 Driving Ability Evaluation for Elderly Drivers

6.2.2.1 Ability Evaluation of Driving Behavior

McKnight et al. conducted a set of psychological tests on 407 people who were at least 63-years old and some of whom had caused a car accident (McKnight and McKnight, 1999). They showed that the tests identified which of the people had caused an accident with 80 per cent accuracy. They also conducted on-road driving ability tests to evaluate the items listed in Table 6.3. They are categorized by the driving situation, like driving through an intersection, driving on a straight road, driving on a curved

Table 6.3: Items of Driving Behavior Evaluated during on-road Test.

Intersections	*Lane change*
Signal — signaling a turn in advance	Signal — signaling a lane change in advance
Search — looking both ways before entering	Search — checking mirrors, both sides before change
Entry position — correct lane and position	Speed — matching speed of traffic in lane
Stop position — stopped at the correct point	Time — making change quickly but safely
Gap acceptance — gap correctly accepted or rejected	Gap acceptance — gap properly accepted or rejected
Gap estimation— estimating last safe chance to enter path — remaining in lane through turn	Space — proper following distance after lane change
Speed — vehicle speed appropriate to turn	
	Merge
Straight stretches	Speed — vehicle speed during merge
Speed — vehicle speed close to legal limit	Lane selection — merge into closest lane
Position — vehicle positioned properly in lane	Position — properly positioned within lane
Spacing — adequate separation from other vehicles	Gap estimation — under- or over-estimating of gap
Curve	*Attention sharing*
Entry — safe speed upon entering curve	Speed — keeping proper speed
Speed — safe speed throughout curve	Position — keeping proper position
Path — proper path through curve	Navigation — turning at named streets, following sequence of turns, and finding the way 'around the block'

road, changing lanes and merging. However, the data analysis did not yield sufficient correlation in identifying drivers who had experienced a car accident.

Sommer et al. at the University of Groningen developed a simple driving-ability evaluation method called TRIP (test ride for investigating practical fitness to drive) (Sommer et al., 2003; Devos et al., 2013). This method was designed to evaluate the driving ability of elderly drivers and drivers with limited cognitive functions. This method is compatible with the concept of hierarchical levels of driving behavior described in the GADGET-Matrix (Christ et al., 2000). Each evaluation item is subjectively scored (bad, not good enough, good enough, good) by drivers. Not only driving safety but also social factors, such as securing mobility in depopulated areas without public transportation, should be taken account for the driving-ability evaluations for elderly drivers. Therefore, it is necessary to obtain the social consensus on evaluation methods.

6.2.2.2 Evaluation of Perceptual-Motor Coordination

There are four types of perceptual-motor coordination testing methods to assess the driving ability of elderly people in Japan: (i) simple reaction time tests, (ii) choice reaction time tests, (iii) steering wheel control tests, and (iv) attention allocation/ multitasking tests (National Police Agency, 2009). Performance measures of these tests include: (i) accuracy, speed, and (ii) variability of reaction, (iii) accuracy/bias and learning effect of steering operation, and (iv) accuracy, speed, and variability of focusing and diverging attention. Innes et al. reported SMCT (sensory-motor cognition testing) methods in which several psychological tests were combined, including perceptual-motor coordination tasks (Innes et al., 2007).

For example, complex attention testing is conducted as shown in Fig. 6.22. As soon as the position of the green circle in one of the two gray boxes was shifted to the other, a participant turns the steering wheel to match the horizontal position of the 'up' arrow with the green light. The time it takes for the subject to move the 'up' arrow until it passes the green line is used as an index. This index is used not only for steering operation performance but also mental and physical fatigue.

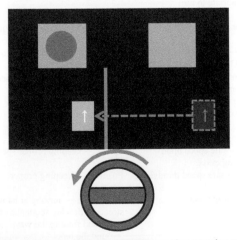

Fig. 6.22: Schematic diagram of SMTC complex attention testing.

6.2.2.3 Evaluation of Cognitive Functions

6.2.2.3.1 Neuro-psychological Tests and Driving Ability

Because cognitive functions are difficult to define and measure directly, most cognitive function evaluations are conducted based on the subject's performance on cognitive tasks that might represent essential aspects of driving behavior.

For example, Marshall et al. systematically reviewed the conceptual model of various functions related to the driver's characteristics and mapped neuro-psychological results based on these functions (Fig. 6.23) (Marshall et al., 2007). Then, they reviewed correlations between results of neuro psychological tests and on-road driving tests. This study was conducted to enable stroke patients to drive again. So these study results may not be applicable to elderly drivers in general, except that they provide some useful information. There have been also studies for establishment of testing systems and screening methods (Staplin et al., 2003; Rapoport et al., 2013; Anstey and Wood, 2011). According to Marshall et al. (Marshall et al., 2007), the trail making test (TMT) (Reitan, 1986), Ray-Osterrieth complex figure test (ROCF) (Rey, 1941), and useful field of view test (UFOV) (Owsley and Ball, 1993) are suggested to be effective screening methods for stroke patients who intend to regain their driving ability.

The trail making test (TMT) is one of the widely used tests to evaluate executive function, and there are two types: Part A and B (Reitan, 1986). In Part A, subjects are provided with a test paper showing randomly distributed numbers, and are assigned to draw lines to connect them in a sequential order; in Part B, subjects are given a paper showing numbers and letters, and are asked to connect numbers and letters alternatingly in an ascending order. The time it took a subject to complete the task is used as an evaluation index. The difference in time taken to complete Part A and B is also sometimes used as an index (Fig. 6.24).

In the ROCF tests, subjects are asked to reproduce a complicated figure that has no particular meaning and consists of several line segments and a circle with three dots in it (Fig. 6.25) (Rey, 1941). There is a set of conditions in the ROCF. Subjects first copy the figure freehand. Then, the original and copied figures are removed and the subject is asked to reproduce the figure from memory, immediately after the removal and again in three minutes. Each copy is scored for the accuracy of specific figure parts with a score of up to 2 points given for reproduction of each part (for example, 2 points are given for accurate reproduction of both shape and position, 1 point is given for accurate reproduction of either shape or position, and 0.5 point is given for inaccurate drawing completed).

The UFOV test consists of three subtests: (1) discrimination of the stimulus in the central visual field (car or track icon), (2) discrimination of stimulus in the central visual field and identification of position of stimulus presented in the peripheral visual field, (3) the stimulus discrimination in the central visual field and the stimulus position identification in the peripheral visual field where a disturbing stimuli is presented (Ball et al., 1988). It is considered to be designed to measure (1) speed of information processing, (2) ability of divided attention, and (3) ability for selective attention. The effect of age was observed in each test and it was found that the higher the age, the lower was the score. It was also observed that improvement occurred in the performances by practicing.

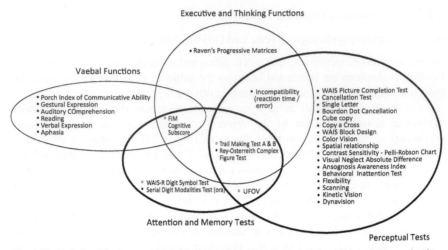

Fig. 6.23: Relationship among various neuropsychological tests used for driving ability evaluations (Marshall et al., 2007).

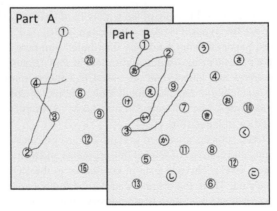

Fig. 6.24: Example of trail making test.

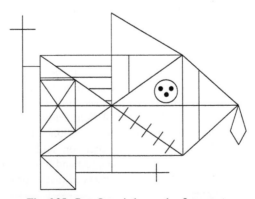

Fig. 6.25: Ray-Osterrieth complex figure test.

6.2.2.3.2 Screening Test for Elderly Drivers

Based on the meta-analysis of research on driving by dementia patients, Reger et al. reported that dementia patients' driving ability as judged by caregivers had a correlation with the result of visuo-spatial test (e.g., Kohs Block Design Test) and the score of mental status (e.g., MMSE) (Reger et al., 2004). In studies like this, one suggests that appropriate screening methods may vary depending on the severity of the subject's medical condition. To address this issue, studies were conducted for implementation of efficient screening tests. NHTSA extracted a list of tests based on discussions by specialists in various fields, and conducted the Maryland pilot older driver study (Loren et al., 2003a; Loren et al., 2003b).

6.2.2.4 Models of Driving Ability for Elderly People

6.2.2.4.1 Multifactorial Model for Enabling Driving Safety

There are individual differences in driving behavior, especially among elderly drivers. What specific ability or functions of drivers contribute to the differences? One approach is the use of the multi-factorial model for enabling driving safety, a schematic representation of the relationship between drivers' cognitive functions and their driving behavior (Anstey et al., 2005). The purpose of this model is to comprehensively understand which types of decline in mental/physical functions contribute to accident-prone driving behavior in elderly people. The model presents the relationship among three components: mental/physical functions, safe driving ability, and driving behavior. In this section, we introduce the relationship of the three components using the multi-factorial model for enabling driving safety.

The multi-factorial model for enabling driving safety was developed based on the surveys of peer-reviewed literature published between 1991 and 2002. The survey focused on the relationships among drivers' perceptual, cognitive, and mental/physical functions, and their accident records/driving behavior (e.g., correlation coefficient). People who were 60-years old or older participated in the experiments or were surveyed. The objective of the study was to characterize elderly people in general, not just those with dementia.

The multi-factorial model for enabling driving safety is shown in Fig. 6.26. According to the model, driving behavior is determined by the 'capacity to drive safely' and the 'self-monitoring and beliefs about driving capacity' components (Wood et al., 2013). The 'capacity to drive safely' consists of useful the field of view and hazard perception—an ability to notice an event or objects that might lead to an accident in traffic (e.g., a bicycle suddenly appearing from the side of the road, a vehicle in the crossroad entering an intersection with poor visibility) (Anstey et al., 2012). The 'self-monitoring and beliefs about driving capacity' category includes self-awareness of the useful field of view, hazard perception, and 'cognitive/physical abilities' that decline due to aging (Horswill et al., 2011). This model explicitly indicates that one's driving behavior changes in response to his/her awareness of age-related changes in driving ability.

The 'capacity to drive safety' is influenced by their 'cognition', 'vision', and 'physical functions'. Accordingly, drivers with sound functions are likely to drive safely. The 'self-monitoring and beliefs about driving capacity' are also related to

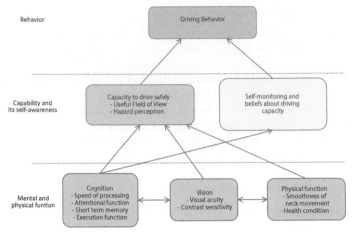

Fig. 6.26: Multi-factorial model for enabling driving safety: structure of critical safe driving factors.

their cognitive functions. Thus, if a person's cognitive functions deteriorate, his/her awareness of driving ability may weaken.

Cognitive function includes 'reaction time/speed of processing' in which a quick reaction and movement enable the driver to respond to a change occurring in the road; 'attentional function' means properly directing one's attention to the information critical for driving, and ignoring stimuli irrelevant to driving; 'short-term memory' implies retaining information for decision-making and processing information into a usable form,; 'executive function' pertains to a driver facing a complex driving situation and anticipating and coordinating sensory, cognitive, and motor reactions; 'visual functions' consist of eyesight and contrast sensitivity which enable drivers to properly perceive objects in the road (road alignments, behavior of others, hazard). When maneuvering a vehicle, 'smooth neck movement' is an important 'physical function' and include their 'health conditions', with special focus on physical condition that might lead to an increased risk of falling down, arthritis or cardiac disease (Donorfio et al., 2008).

6.2.2.4.2 Adaptive Driving Behavior of Elderly People

This model explicitly indicates that one's driving behavior may change in response to his/her awareness of age-related changes in driving ability and mental/physical functions. For example, some elderly drivers may be aware of the fact that their reaction to external stimuli and the movement of other vehicles is slow. However, they can still realize safe driving by choosing not to drive during peak hours or heavy traffic. On the other hand, elderly drivers having difficulty in looking at traffic lights/ signs and with failing night vision might continue to drive at night, thereby increasing the risk of an accident.

If elderly people change their driving behavior in response to their awareness of lowered driving ability, the change is viewed as adaptive (or compensatory) behavior in response to declined mental/physical functions with age (Smiley, 2004). Based on studies in which the types of adaptive driving behavior practiced by elderly people by monitoring their driving and interviewing them, the questionnaire study using

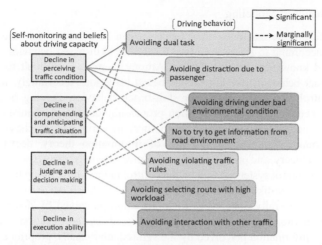

Fig. 6.27: Self-awareness of declined cognitive function and adaptive driving behavior.

100 elderly and 100 non-elderly drivers was conducted to examine the relationship between declined cognitive/physical functions of the elderly drivers and the types of driving behavior adapted (Akamatsu et al., 2006; Sato et al., 2007). The results are shown in Fig. 6.27. According to the results, after the drivers became aware of their declined ability to perceive traffic conditions, they ceased performing other tasks (e.g., listening to the radio) and having conversations with passengers while driving. They also tended to avoid driving under bad driving conditions and chose familiar routes so as not to need additional information on the road. Elderly drivers with lower ability to comprehend/anticipate traffic situations tried to look at traffic signs more carefully so as to o judge traffic situations. Elderly drivers, who feel like their judging/decision-making ability has weakened, tend to drive on wide roads, simple traffic conditions, and on familiar routes. Finally, elderly drivers, who are aware of declined execution ability, tend to drive at their own pace, avoiding paying excessive attention to other vehicles.

Accordingly, by understanding the relationship between drivers' awareness of their abilities and changes in their driving behavior, it is feasible for professionals to give proper advice to elderly drivers about adaptive behavior, specific to the type of cognitive functions they claim to have weakened.

6.3 Driver's Behavior Models

6.3.1 Driving Behavior Models

6.3.1.1 Driver Steering Control Models

In order to move a vehicle as intended, humans must operate them appropriately. As the human operator is part of the vehicle control system, studies on vehicle dynamics must take account of, not only the kinetic characteristics of automobiles, but also the characteristics of the people operating them in the context of the driver-vehicle system. Various driver models have been proposed as analysis tools.

6.3.1.1.1 Basics of Modeling

When people operate machines, their behavior can be classified into the following three types according to Rasmussen (see 6.1.2.1.6): skill-based behavior, rule-based behavior and knowledge-based behavior. Among them, it is difficult to model the highly abstract behavior, like knowledge-based behavior. Accordingly, most driver steering control models are those for skill-based behavior. Driving is considered to be performed by complex information processing and the mechanism of human information processing is unknown. As such, there are various approaches to modeling human information processing in terms of information theory, decision-making theory, control theory, and probability theory. Conventionally, drivers were considered as mechanical control systems based on the control theory, and driving behavior often resulted in black-box-like control models.

There are generally three types of information perceived by drivers from the surrounding environment: motion cues, expressed by acceleration and angular velocity, e.g., information perceived by the semicircular canals, proprioceptive cues, e.g., information perceived as reaction force and displacement of the steering wheel and pedals, and visual cues, the most important information type. There are two types of output: steering angle and torque as related to lateral movement of automobile, and pedal stepping force and pedal pressure as related to forward and backward movement.

Among all driver steering control models available, most frequently studied models are control-theory-based models, which focus on skill-based behavior related to lateral movement control, e.g., maintaining the vehicle position within the width of the lane, and assume that a driver adjusts steering angle (output) in response to visual input. During this process, three types of skill-based steering control behavior are involved as shown in Fig. 6.28 (McRuer et al., 1977).

Precognitive control refers to an open-loop control system which operates when a driver knows in advance the characteristics of the desired driving path. This system stores several steering programs developed, based on the driver's learning experience, selects an appropriate program according to the driving environment and the condition of the automobile, and steers the automobile. During this process, input to the driver triggers selection and execution of an appropriate steering program.

Pursuit control denotes a feed-forward control system which directly uses information on a desired path. As indicated in the control theory, the feed-forward control system alone does not provide robustness on modeling errors and disturbance.

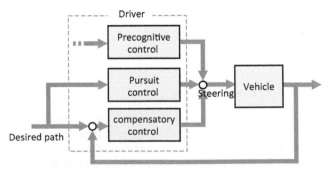

Fig. 6.28: Three types of steering control behavior.

Therefore, it operates in combination with a compensatory control, which corrects steering control in response to the vehicle's state feedback.

6.3.1.1.2 Major Examples of Driver Steering Control Models

(1) *Preview-predictive model*: When a person drives along a target path, the driver can see a certain distance ahead and use the visual information for steering control. The steering control that uses future values of target signals is called 'preview' control. The driver also can project the future state of the vehicle through extrapolation of the information on the direction of the vehicle and the path taken up to the present. This type of steering control is called 'predictive' control, where a driver predicts future situations based on the information up to the present and makes a decision on steering control. A model had been constructed assuming that the driver obtains future values of the desired path through the preview approach, determines the future vehicle position through the predictive approach, and takes a deviation between the two approaches to decide the steering angle. This model is called a 'preview-predictive' model. This was the first driver steering control model proposed and is also sometimes called a 'linear preview' model (Kond, 1958).

The concept of the preview-predictive model is illustrated in Fig. 6.29. The driver is gazing at a point, $L(m)$ ahead of the current position in the desired path. While doing so, the driver presumably predicts the future vehicle position y_p based on extrapolation, and calculates f, a lateral deviation between y_p and the gaze point y_d. Then, the driver performs compensatory steering control. The fundamental structure of the driver–vehicle system in the framework of the preview-predictive model is shown in Fig. 6.30. $P(s)$, $B(s)$, and $H(s)$ respectively denote preview function, prediction function, and driver's control characteristics. Assuming that the driver can accurately preview the desired path, his/her preview function can be expressed as a simple lead element.

$$P(s) = e^{Ts}$$

where T = preview time. The T value can be approximated by the formula, $T \approx L/V$, where V = vehicle velocity. There are several prediction functions. The simplest one can be expressed as a straight line extending in front of the vehicle (dotted line in Fig. 6.29). The predicted lateral position of the vehicle can be approximated by calculating $y + L\psi$. The more common prediction methods include the use of the first and second order Taylor expansion.

$$B(s) = 1 + Ts \qquad \text{(first-order prediction)}$$

$$B(s) = 1 + Ts + \frac{1}{2}T^2s^2 \quad \text{(second-order prediction)}$$

In general, a driver's maneuverability to control prediction error can be expressed as

$$H(s) = K\frac{T_L s + 1}{(T_I s + 1)(T_N s + 1)}e^{-Ts}$$

where K = gain due to the driver's proportional element, T_L = time constant for the driver's differential element, T_I = time constant for the driver's integral element, T_N = time constant of first-order lag element representing the driver's neuromuscular

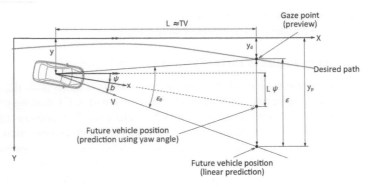

Fig. 6.29: Concept of the preview-predictive model.

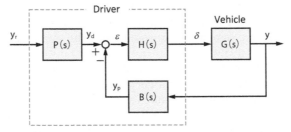

Fig. 6.30: Fundamental structure of driver-vehicle system in the framework of the preview-predictive model.

system dynamics, and τ = dead time. T_N and τ represent driver-specific lag time, and they are almost constant values of T_N= 0.1 (s) and τ = 0.1–0.3 (s). In contrast, K, T_L, and T_I vary greatly depending on the characteristics of targets that the driver controls. Accordingly, the values of these variables differ among drivers and represent their adaptability to a given target.

To simplify the model, researchers often approximate the driver's control behavior based only on proportional gain and dead time. Also, it was reported that the stability of the driver-vehicle system can be more consistently determined when the yaw angle corresponding to lateral displacement ε_θ is used as input to the driver model and not lateral displacement error ε (Fujioka, 2007). More advanced preview-predictive models have been proposed, e.g., a model with multiple gaze points and a model that applies the optimal preview control theory (Sharp et al., 2000; MacAdam, 1981).

(2) *Describing function model*: Normally, the control characteristics of a human cannot be expressed by transfer function as they include non-linear components. However, if we focus only on basic components of a human output that is linearly related to an input, the input-output relationship can be expressed using an equivalent transfer function. The equivalent transfer function in this context is called a 'describing' function. From the 1950s, numerous studies were conducted to model humans' control behavior in terms of a describing function. Subsequently, a series of research was carried out to apply the findings of these studies to automobiles.

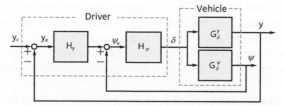

Fig. 6.31: Describing function model including the yaw angle feedback.

For example, double-loop models with yaw angle and lateral displacement feedback (Fig. 6.31) were developed (McRuer et al., 1977). In the figure, H_y and H_ψ are describing functions indicating drivers' characteristics to control lateral displacement and yaw angle, respectively. This model does not involve the use of preview information by drivers. Thus, to make the closed-loop system stable only by lateral displacement feedback, drivers must perform strong differential action. However, when humans are able to control something over a long period without any difficulty, they are presumably performing proportional control action including dead time, or very weak derivative or integral action. Drivers probably respond to not only lateral displacement feedback but also feedback with other information. Therefore, the model in Fig. 6.31 also includes yaw angle feedback in addition to the afore-mentioned lateral displacement feedback. It has been theoretically demonstrated that with this control structure, without requiring a driver to perform differential control action, the system with simple control rules, where H_y has a proportional gain only as a parameter and H_ψ has proportional gain and dead time as parameters, enables to achieve a desirable stable closed-loop control system.

(3) *Pursuit control model*: This is a feed-forward control system which uses information on a target path. This model is expected to achieve desirable feed-forward characteristics simply by using proportional action without having differential action, assuming that the driver uses input information on the curvature of a target path.

(4) *Other models*: See Plöchl et al. for many other driver models, including longitudinal vehicle control and their applications (Plöchl and Edelmann, 2007).

6.3.1.2 Model of Visual Recognition During Driving

6.3.1.2.1 Perception of Direction of Travel

Automobile driving can be viewed as a special mode of human locomotion. First, perception of the traveling direction for human locomotion is reviewed. When observers are traveling, they perceive the motion information, which is projected on the retina and is called 'optic flow' (Gibson, 1950) (sometimes it is called 'retinal flow' in this definition). When they travel forward, their optic flow takes the pattern of radial expansion from a point. This center point is called the focus of expansion (FOE). When they move straight ahead, the FOE matches their traveling direction and they basically perceive their traveling direction using the FOE (Gibson, 1966).

An FOE can be determined by identifying the intersection point of two or more extrapolated velocity vectors. Taking account of noise effect, one can perceive his/her traveling direction more accurately by using velocity vectors that are located near the FOE (Crowell and Banks, 1996). In experiments concerning the perception of traveling direction using random dots, it was found that difference threshold values (sensitivity measurement) for traveling direction were very small—the smallest value being 0.66 degrees and average, 1.2 degrees, and that even if the number of random dots is reduced to only about 10 dots, these results remained nearly unchanged (Warren et al., 1988).

When observers are moving their heads or eyes, their traveling direction does not match the FOE. For example, if one pays attention to roadside signs or traffic lights while driving, one's head or eyes move. At such a moment, an image on the retina becomes a curved flow. As a result, the motionless center point of radial flow, FOE, deviates from the direction in which one is traveling (Hildreth and Royden, 1998). In simulated eye movement studies, observations of optic flow were compared between the two conditions in which subjects either kept their eyes still or moving. When the subjects moved their eyes slowly at a rate of 1 degree/s or less, they were able to perceive their traveling direction correctly. In contrast, when they moved their eyes more quickly, their perception of traveling direction was biased by the FOE (Royden et al., 1992). However, despite these study results, when people move their eyes, they in fact can correctly perceive their traveling direction irrespective of eye movement speed, indicating that people perceive their traveling direction using eye movement information (motor commands or proprioception).

In addition, when there are independently moving objects, e.g., other vehicles, in their fields of view, it is impossible to uniquely detect an FOE. In psychological experiments, subjects have correctly perceived their traveling direction even when seeing an independently moving object in their field of view. However, it has been demonstrated that when a large object moves across the subject's field of view while the subject is moving forward, his or her perception of direction may be biased towards the object's direction (Royden and Hildreth, 1996).

6.3.1.2.2 Use of Tangent Points

There is a large difference between the perception of traveling direction in general locomotion and that of the automobile driving in terms of the following aspects: automobile driving almost always takes place on the road and requires accurate steering to stay in a lane. In particular, steering around curves is called 'curve negotiations'. There was a study in which drivers' eye movements were measured in order to identify the types of information they use during curve negotiations (Land and Lee, 1994). Drivers undergo saccade eye movement 1 to 2 seconds before entering a curve and look at tangent points on the inner side of the curve. Then, drivers spend 80 per cent of the time looking at tangent points 0.5 seconds after they entered the curve. In curves, especially narrow and winding ones, the angle between the direction of the vehicle and the direction of a tangent point viewed from the driver may be only information available with which he/she can visually estimate the curvature of the upcoming bend (Fig. 6.32). Some researchers pointed out that this is the reason why drivers look at tangent points when driving around curves.

Curvature of a bend can be calculated using the following equation:

$C = 1/(d \cos\theta) - 1/d,$

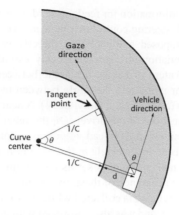

Fig. 6.32: Relationship among the direction of a tangent point from the driver, traveling direction, and the curve.

where θ = angle between the direction of a tangent point viewed from the driver and the direction of the vehicle, C = curvature of the bend (reciprocal of curve radius), and d = the distance between the driver and the inner side of the curve. The value of C can be calculated if θ is known, and with known C, an appropriate amount of steering can be determined. In addition, without calculating C, people's steering behavior, particularly when entering a curve, was adequately explained by a driving model that takes account of θ and the distance between the tangent point and the driver (Boer, 1996).

Driving simulator experiments have been conducted to address the question whether the driving performance actually improves if the drivers look at tangent points (Mars, 2008; Robertshaw and Wilkie, 2008). In any of the studies conducted, however, no improvement in driving accuracy was observed under the conditions in which the drivers were instructed to gaze at tangent points. Moreover, it was observed that drivers' steering was biased towards the direction of their gaze points (Robertshaw and Wilkie, 2008).

6.3.1.2.3 Use of Information on Near and Far Areas

A model was developed to explain the driving behavior assuming that drivers see either nearby or distant areas (Donges, 1978). Using a driving simulator, measurements were taken on subjects' driving behavior under the condition in which their fields of view were restricted to a part of a road. When the subjects were allowed to see only a distant part of the road, their steering was fluent and stable, but their driving positions greatly deviated from the centerline of the road. In contrast, when subjects were allowed to see only a near part of the road, their steering was very rapid and busy, but they drove close to the centerline of the road. On the other hand, when subjects were allowed to see only an intermediately distant part of the road, their driving behavior had intermediate characteristics between the driving behaviors observed under the first two conditions (Land and Horwood, 1995). These results indicate that drivers use the information on the distant part of the road to estimate upcoming curves, while they use the information on the near part of the road to make lateral adjustments. In other words, drivers use distant information for feed-forward trajectory planning and nearby information for a feedback steering control mechanism (see 6.3.1.1).

Some researchers proposed a model that directly calculates proper steering angles during curve negotiations by simultaneously estimating the following two angles instead of the curvature of bends: the angle between the vehicle's direction and the direction of a near point, and the angle between the vehicle's direction and the direction of a far point (Salvucci and Gray, 2004). A near point refers to the center of the road immediately in front of the driver. On the other hand, the definition of a far point depends on the situation. If a lead vehicle is running ahead, that vehicle is set as a far point. If there is no lead vehicle in the straight road, and if driving a curve without lead vehicle, the tangent point is set as a far point. The driving model was to steer proportional to the difference of these two angles. The model was compared with the aforementioned driving data collected while restricting drivers' view to either near, far, or intermediate area, and was also compared with drivers' corrective steering actions that were taken when they temporarily lost vision (Salvucci and Gray, 2004; Hildreth et al., 2000). Based on these comparisons, it was demonstrated that the model will account for driving characteristics (see 6.3.2.2.2).

Also, experiments were conducted to study the effects of visual information on road edge (curb) and environmental optic flow on drivers' performance during information derived from near and distant areas (Fig. 6.33) (Kishida et al., 2008). In driving simulator experiments, when drivers were allowed to see information from the near area only, they were able to reduce the deviation from the centerline of the road with improved driving accuracy. However, driving workload appeared to be high based on rapid steering indicated by large standard deviations of steering speed. In addition, when random dot optic-flow stimuli were added to the near-area information, standard deviations of steering speed further increased, indicating higher driving workload. When drivers were allowed to see information from the distant area only, standard deviations of steering speed decreased, indicating reduced driving workload. However, driving accuracy deteriorated as the vehicle largely departed from the centerline of the road. When optic flow stimuli were added to the far-area information, the extent of departure got reduced.

Moreover, test track studies were performed in which a board was set in the vehicle in such a way that the driver was able to see both the road edges (curbs) but was unable to see the surface in the middle of the road. After driving with this setting, it was found that standard deviations of steering speed decreased. These results indicate that optic flow stimuli from a nearby area increase driving workload while those from a distant area increase driving accuracy.

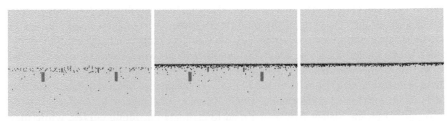

Fig. 6.33: Optic flow and other stimuli derived from near area (left), whole area (middle), and distant area (right).

6.3.1.2.4 Effect of Gaze Direction

A study was carried out to understand the effect of drivers' gaze direction on their steering behavior (Readinger et al., 2002). Using a driving simulator, subjects were asked to drive while staring at an activated gaze point among several points arranged horizontally, from left to right. As a result, their steering was biased towards the direction of the gaze point. This effect was observed irrespective of the driving speed and was generated even by gaze points located slightly to the left/right of the center (5°). Furthermore, drivers' steering also deviated towards the direction of a gaze point even when the relationship between steering wheel rotation and resulting vehicle direction change was reversed (i.e., when the steering wheel is turned counterclockwise, the vehicle turns toward right, instead of left). These results suggested that drivers' perception of traveling direction appears to be affected by their gaze direction, and not by linking eye movement and hand movement (e.g., when you look towards the left-hand side, your tendency to move your hands towards the left increases). In the experiments in which subjects were asked to gaze at certain points, including tangent points, while driving, a similar trend of steering deviation towards the gaze points was observed (Robertshaw and Wilkie, 2008). Thus, the direction in which the driver gazes has a major impact on perception and control of traveling direction (Wilkie and Wann, 2002).

6.3.2 Information-processing Models Related to Driver's Behavior

6.3.2.1 Information-processing Models for Drivers Using Car Navigation System

When people are driving to a destination, they collect and process a variety of information regarding the route to the destination and intersections to turn, while checking various objects along the way serving as landmarks. People used to refer to road maps and information signs to obtain information necessary to drive to a destination. Then, car navigation systems were introduced in the market around 1990, and the main function of early systems was only to display the driver's current location and a destination on an electronic map (Fig. 6.34). Route guidance functions of car navigation systems and mobile devices were popularized later on (see 5.2). The new technology has relieved drivers' concern about losing the way, and reduced their workload of route selection. On the other hand, drivers' information processing for route selection and route following has changed in terms of the types of information used and interactions with navigation aid.

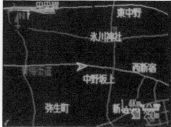

Fig. 6.34: Example of early car navigation systems.

In this section, information-processing models for drivers, who use an in-vehicle car navigation system or a navigation application for mobile device, are described.

6.3.2.1.1   Information-processing Models for Drivers using a Digital Map System with Self-localized Function

The digital map system with self-localized function displays his/her current traveling position on a digital map by providing information about traveling direction, surrounding roads, etc. The system is not equipped with voice guide or the capability to display an enlarged view of an intersection. Comparing driving with paper maps, the system allows drivers to easily find their current position and information on surrounding roads, whereas paper maps are not that handy.

A study was conducted to compare information using characteristics between drivers who choose a route to a destination using a digital map system with self-location and those who do it using paper maps. In the study, a verbal protocol method (thinking aloud method, Ericsson and Simon, 1993) was applied to the two groups of subjects and the content of information used by them for route selection was extracted (Takeda et al., 2012). Then, the information extracted was categorized by cognitive map component (see 5.2.2.2) and frequency of each category was calculated (Table 6.4).

When drivers use a digital map system with self-location, they use the path and node information obtained from the in-vehicle device, and the landmark and node information obtained externally (i.e., road environment). Another experimental study with different road environments showed that the ratio of use between internal and external information sources for landmark and path information was similar to the abovementioned study (Akamatsu et al., 1997). When drivers used paper maps, they obtained landmark and path information from in-vehicle source (i.e., map) and landmark information externally. These results suggest that drivers may use different types of information, depending on the functions available from the digital map with self-location or those from paper maps.

Based on these results, information-processing models were constructed for drivers using either the digital map systems with self-location or paper maps. Models shown in Fig. 6.35 describe drivers' use of information on their current vehicle position and routes and which is vital for choosing a route to a destination.

When drivers use digital maps with self-location function, they can track their continuously-changing vehicle position by following a cursor displayed on the digital map (Fig. 6.35(a)). In addition to the current vehicle position, drivers also gather information from the digital map about the direction to the next intersection where

Table 6.4: Types of Information Used for Route Selection (Daimon and Kawashima, 1996).

	Digital map with self-location			Paper map		
Landmark	20.7 %	(38.4%	/ 10.3%)	63.2 %	(69.8%	/ 49.4%)
Path	33.6 %	(16.8%	/ 43.5%)	19.7 %	(14.5%	/ 29.3%)
Node	25.7 %	(36.8%	/ 19.2%)	7.6 %	(9.1%	/ 4.6%)
District	8.9 %	(4.8%	/ 11.2%)	8.4 %	(5.1%	/ 14.4%)
Edge	1.8 %	(1.8%	/ 0.9%)	1.2 %	(1.5%	/ 2.3%)
Car location	9.4 %	(——	/ 15.0%)			

Information from outside of vehicle/that from in-vehicle

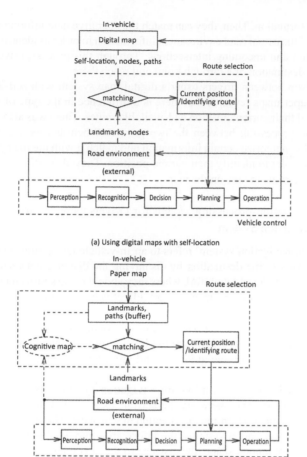

Fig. 6.35: Information-processing models for drivers to identify their current location and available routes.

they should make a turn, the road they are currently taking, and upcoming cross roads. Accordingly the drivers can decide the route to a destination or a through-point, and drive there.

On the other hand, when drivers use a paper map, they need to identify their current vehicle position on the map since the map does not show the current vehicle position (Fig. 6.35(b)). To estimate the current vehicle position on the paper map, drivers need to select objects and paths on the map that can be used as landmarks and identify them with those in the real road environment. Then, the drivers have to estimate their current vehicle position on the map by comparing the information on the map with information obtained from the road scene. However, as drivers have to look away from the road environment for a fair amount of time to read a paper map, they need to do so when the vehicle is not moving and memorize the landmarks and paths shown on the paper map by temporarily storing the information in their short-term memory. In some cases, drivers can form a simple cognitive map based on memorized information. Then, they can match the cognitive map information with the visual scene of the actual road environment information in order to identify the current vehicle position and upcoming intersections. Through this process, drivers can plan the route to a destination or a through-point and reach there.

Comparison between drivers using a digital map system with self-location and those using paper maps revealed that there is a difference in the type of information they use to find their current position and available routes. There was also a difference in information processing between the two groups. When drivers use paper maps, they temporally memorize certain information and match it with the road environment information in order to identify their current position. They also estimate their current position and the route to take on the map.

6.3.2.1.2 Information-processing Models for Drivers using a Turn-by-turn Navigation System

A 'turn-by-turn navigation system' refers to an in-vehicle navigation system capable of guiding drivers to the destination by displaying the distance between the current location and the next intersection/fork where they should make a turn and the direction to travel (Fig. 6.36). This system may work effectively in areas where intersections are relatively sparsely distributed and the current vehicle position can be estimated accurately. Navigation systems providing turn-by-turn information were introduced in the market in the 1990s, but today, their use is only supplemental in head-up displays and certain scenes during route guidance, and are often used in combination with a digital map.

As explained above, a turn-by-turn navigation system provides drivers with information on the distance between the current location and the next intersection where they should make a turn and the direction in which they should travel. Compared to other types of navigation systems, the information provided is limited. When an intersection has a common shape, the distance to the next intersection where a driver should make a turn constitutes the critical information. According to Asoh et al. the number of glances to the turn-by-turn display changed, depending on whether the system was equipped with audio function (Asoh et al., 2000). Drivers looked at the display less frequently with the voice guide. Figure 6.37 illustrates an information-processing model for drivers using a turn-by-turn navigation system, taking account of the presence/absence of the voice guidance function.

When drivers use a turn-by-turn route guidance system with audio functions, they find the location of the next intersection to make a turn based primarily on the voice guide. They visually or auditorily gather information from the in-vehicle system about the distance to the intersection and the direction to travel at the intersection. They also collect external information about the distance to the upcoming intersection and spatial characteristics along the way to the intersection. By comparing information from the in-vehicle device and outside, they determine whether to make a turn at the upcoming intersection or not.

When the turn-by-turn navigation system is not equipped with audio functions, drivers judge the upcoming intersection mostly by perceiving the road environment. Then, they visually gather information from the in-vehicle display about the distance to the intersection where they should make a turn, and collect external information about the distance to the upcoming intersection and spatial characteristics along the

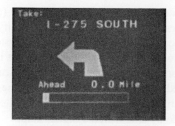
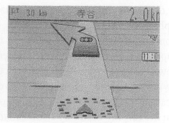

Fig. 6.36: Examples of turn-by-turn navigation systems.

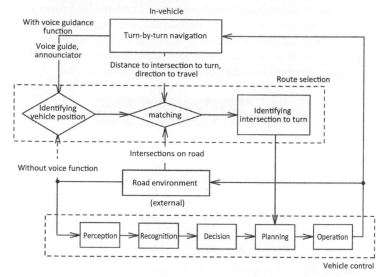

Fig. 6.37: Information-processing model for drivers identifying the next intersection at which to turn when using turn-by-turn navigation system.

way to the intersection. Finally, by comparing the in-vehicle and external information, they determine when to make a turn at the upcoming intersection. Thus, when the navigation system is not equipped with audio functions, drivers have to gather and process information every time they approach an intersection. Consequently, drivers tend to look at the in-vehicle device often. In other words, when audio functions are available, drivers do not have to look at the in-vehicle display very often. When intersections exist at short intervals, drivers may find several intersections along the road ahead of them. In such a situation, it is difficult for them to determine at which intersection they should make a turn. Drivers may encounter similar situations when a car navigation system has poor positioning accuracy due to GPS-related problems.

6.3.2.1.3 Information-processing Models for Drivers using a Navigation System Capable of Displaying an Enlarged View of Intersection

The navigation system capable of displaying an enlarged view of intersections is the most common type of car navigation system used in Japan today. This in-vehicle navigation system (Fig. 6.38) displays an enlarged view of the upcoming intersection where the driver should make a turn and landmarks around the intersection (crossing macrograph, see 5.2.2.9). This function enables drivers to easily identify which intersection to make a turn at. Compared to the turn-by-turn navigation system, this system appears to be useful for areas where there are intersections at short intervals and for low accuracy of positioning the current vehicle location. In most cases, the enlarged view of intersections is displayed with a digital map. When the vehicle approaches the intersection to make a turn, an enlarged view of the intersection appears on the display either by itself or side by side with the digital map.

Figure 6.39 shows an information-processing model for drivers using a navigation system capable of displaying an enlarged view of intersections. The system displays the next intersection at which the drivers should make a turn, the landmarks around the intersection, name and appearance of the intersection, etc. Thus, this route guidance system displays more detailed information than the turn-by-turn route guidance system. Since this system is often equipped with audio functions, the model assumes systems with audio functions including voice and annunciator sounds.

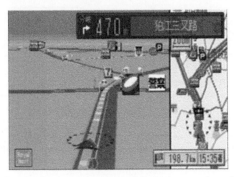

Fig. 6.38: Example of navigation system capable of displaying an enlarged view of intersections.

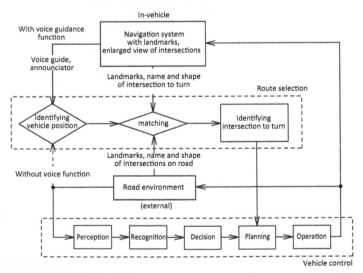

Fig. 6.39: Information-processing model for drivers identifying the next intersection at which to make a turn (navigation system capable of displaying an enlarged view of intersections).

Similar to the turn-by-turn navigation system, when the navigation system capable of displaying an enlarged view of intersections is used, audio information triggers drivers to find the location of the next intersection to make a turn. Then, they visually collect landmark information from the in-vehicle display, which might be helpful in identifying the intersection. If the intersection has a name, drivers recognize it by reading it in the enlarged view of the intersection or by hearing it spoken. And then, using the obtained landmark information, drivers search corresponding structures around upcoming intersections in their front view, and identify the correct intersection to make a turn. When there is no adequate landmark around the intersection and the intersection has no name, drivers have to rely on different information, e.g., the appearance of and the distance to the intersection, shown in the enlarged view of the intersection, as clues. These types of information are similar to the types of information provided by the turn-by-turn navigation system.

When audio functions are unavailable, there are similarities between drivers using the navigation system capable of displaying an enlarged view of intersections and those using the turn-by-turn navigation system. In both cases, drivers typically begin to compare in-vehicle information with external information and identify which intersection to make a turn at, when noticing an intersection ahead. After these drivers obtain information from the in-vehicle display, they process the information in a similar manner compared to drivers who use the navigation system equipped with audio functions.

6.3.2.2 ACT-R (Adaptive Control of Thought-Rational) Model of Driving Behavior

6.3.2.2.1 Driving Behavior and Integrated Driver Models with an ETA Framework Viewpoint

Salvucci has developed a rigorous computational model of driver behavior from the perspective of a cognitive architecture that is a computer-simulation framework based on basic psychological theories that take account of basic characteristics of human systems (Salvucci, 2006). The model used ACT-R architecture to takes account of control, monitoring, and decision-making processes as related to driving on a multiple-lane freeway. A computational model was a powerful tool for studies of complex driving tasks. There have been other cognitive architecture models called Soar and the queuing network-model human processor (QN-MHP) (Laird et al., 1987; Newell, 1990; Liu, 1996). However, the application of Soar and QN-MHP models is limited respectively to the situation where a vehicle is approaching an intersection and where a vehicle is keeping a lane.

The aim of a computational driver model is to simulate and predict actual driver behavior. The application of an ETA framework to the development of driving/driver models is beneficial in terms of understanding advantages and limitations of the model (Byrne, 2001; Gray and Boehm-Davis, 2000). The assumption of this framework is that humans use 'artifacts' when they perform tasks. In doing so, they perceive, think, and interact with the world via artifacts in the process of 'embodied cognition'.

Driving behavior can be regarded as a set of basic tasks that can be integrated and adaptively replaced depending on driving situations. Michon described the following three components in driving tasks: (1) an operational component, which regulates input, enabling a driver to perform stable driving, (2) a strategic component that enables a driver to safely interact with the environment and other vehicles, and (3) a tactic component that enables a driver to make an inference and a plan at a higher level (Michon, 1985) (see 6.1.2.1.5). These three components together contribute to safe and stable navigation.

'Artifacts' relevant to driving behavior are an automobile and the interface between a driver and an automobile. The interface includes the steering wheel, acceleration/brake pedal, clutch pedal, turn signal, wiper, etc. Also there are knobs and buttons of various kinds. 'Embodied cognition' refers to an integration of perceptual, cognitive, and motor processes involved in task implementation while driving. Cognition definitely occurs when a driver performs high-order decision-making. Cognition is also involved, to a lesser degree, in low-order automobile operation and situation awareness. Embodiment exists between human cognition and an automobile being driven. Embodiment takes the form of a perceptual process involving the senses of vision, audition, vestibular, etc., by which external information is obtained, and a motor process, in which a driver performs external actions, e.g., moving hands and feet, based on the result of information processing. In the system that integrates these processes, more than one process may be performed concurrently. Conversely, the performance of these processes may be compromised due to restrictions in the driver's processing capacity and a processing bottleneck.

In the integrated driver model proposed by Salvucci, the three components mentioned above are modeled while taking into consideration many driving-related tasks, realistic vehicle operation and automobile dynamics, and interacting cognitive process. The cognitive architecture deals with two types of knowledge: 'declarative knowledge', which is made up of 'chunks', and 'procedural knowledge', which is made up of production rules. Each rule is a set of conditions and actions, and condition-action rules manipulate declarative knowledge. The architecture integrates human abilities and limits into one system. The abilities include unlimited memory capacity and retrieval of memory in a flexible manner. Human limits include forgetfulness, the inability to acquire detailed visual information from a visual field other than the central vision.

6.3.2.2.2 Integrated Driver Model using the ACT-R Cognitive Architecture

The ACT-R driver model includes three components: control, monitoring, and decision-making. 'Control' refers to mapping of all aspects and characteristics of perceptual variables obtained from the perception of the surrounding environment to operate a vehicle (steering and stepping on the acceleration/brake pedal). 'Monitoring' denotes the continuous recognition of the current situation by a driver through periodical perception of the surrounding environment and the encoding of perceived information. 'Decision-making' means a strategic decision to take a steering action, e.g., changing lanes. These three components are implemented in the ACT-R cognitive architecture, making it possible to develop psychologically plausible driver behavior models.

(1) *Control*: The control component of a driver model conducts perception of low-order visual clues and lateral (steering) and longitudinal (acceleration/deceleration) control of a vehicle. Lateral control of a vehicle is consistent with Salvucci and Gray's steering model (Salvucci and Gray, 2004).

This model employs a two-level control method based on the perception of two visual reference points (Donges, 1978; Land and Horwood, 1995). The first is a near point, which represents the center of the lane the driver is currently traveling in, to estimate how close the vehicle is to the center of the lane. Drivers use the center of the lane 10 m ahead of the vehicle that is perceivable as the closest reference point. The second is a far point, which drivers use as a reference point to estimate the curvature of the upcoming road trajectory, and to decide their next action to keep the vehicle within the current lane (see 6.3.1.2.3).

A far point can be one of the following three targets: (a) the vanishing point of a straight road ahead, which is no farther than the point that can be passed through by the vehicle in 2 seconds, (b) the tangent point of a curve in the road ahead (Land and Horwood, 1995), and (c) the lead vehicle (Fig. 6.40). Information from both the near and far points complementarily support the driver's steering control, with the former guiding to keep the center of lane and the latter to facilitate stability.

According to the model, drivers need to obtain information from these points in order to achieve lateral control of the vehicle. Drivers pay visual attention to a near point first, and then to a far point. They subsequently store the information on visual angles for the near and far points, θ_n and θ_f, respectively, in their memory buffer. The model adjusts the steering angle using these quantities. The following equation expresses the rules for steering angle control:

$$\Delta\Phi = k_f\Delta\theta_f + k_n\Delta\theta_n + k_I\theta_n\Delta t \tag{6.1}$$

Similar rules apply to longitudinal control (speed control). The model encodes the position of the lead vehicle and calculates the time headway of vehicles, thw_v. Then, the model calculates differences from the previous cycle of the control loop in terms of Δthw_v and Δt, and then, using these values, it calculates the acceleration-related value ψ.

$$\Delta\psi = k_v\Delta thw_v + k_{follow}(thw_v - thw_{follow})\Delta t \tag{6.2}$$

This acceleration equation specifies the following two constraints: the time headway of vehicles is stable ($\Delta thw_v = 0$) and equals the time headway required for a vehicle to follow another vehicle ahead of it ($thw_v = thw_{follow}$). The acceleration value ψ represents two kinds of control: its positive and negative values are interpreted as a driver stepping on the acceleration and brake pedals, respectively. Its absolute values range between 0 and 1.

There are certain constraints that apply to the execution of lateral/longitudinal control. Because cognitive processors operate sequentially, they are incapable of continuous processing and their input to control loops is updated discretely. In addition, cognitive mechanisms trigger the firing of production rules every 50 ms. Thus, it takes 150 ms to execute the loop in the production system as it involves the firing of three production rules: encoding near points, encoding far points, and issuing motor commands.

On a highway, most low-order vehicle control occurs while traveling within the current lane and while driving on curves. However, a lane change also takes place frequently when there are multiple lanes. Equation 6.1 can easily be extended to a lane change. If you want to take into account a lane change in the steering control equation, you simply substitute near and far points in the current lane with those in the

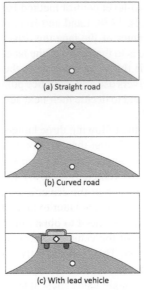

(a) Straight road

(b) Curved road

(c) With lead vehicle

Fig. 6.40: Near and far points.

lane to which the driver intends to go (Salvucci and Liu, 2002). Lateral control rules are expressed by the following equation:

$$\Delta\Phi = k_f\Delta\theta_f + k_n\Delta\theta_n + k_{1min}(\theta_n, \theta_{nmax})\Delta t \qquad (6.3)$$

where parameter θ_{nmax} restricts the effect of the changed steering angle for θ_n.

(2) *Monitoring*: In the driver model, the monitoring component continuously checks the current situation. When driving on a freeway, the most important aspect of monitoring is to find the positions of surrounding vehicles. Accordingly, a random sampling model is used to monitor every direction of the surrounding area—left lane, right lane, front, and back (using side and rear-view mirrors)—to detect other vehicles at the probability of $p_{monitor}$. If a vehicle is nearby, the model stores the information on the lane in which the vehicle is traveling, its direction, and how far the vehicle is in its ACT-R declarative memory.

By encoding the surrounding situations using ACT-R declarative knowledge, the model can quickly predict the potential driver errors. The cognitive process embodies a memory decay process, by which encoded 'chunks' of vehicle positions and distances quickly attenuate. Consequently, memory might be lost unless it is frequently updated. Therefore, if the model employs a strategy that relies on memory, rather than visual check before changing lanes, then the model might 'forget' the fact that it detected a vehicle in a blind spot. Thus, in this way, models can be used to study the difference between accessing 'knowledge in the world' and accessing 'knowledge in the head' (Gray and Fu, 2004).

(3) *Decision-making*: The decision-making component determines whether to formulate a new strategy using the information collected by the control and monitoring components. On highways, a decision is made as to when the vehicle should change

lanes. A lane-changing decision might vary, depending on which lane (driving or passing lane) the vehicle is traveling in. When the subject vehicle is traveling in the driving lane, the model measures the time headway to the preceding vehicle if there is one. If the measured time headway is less than the required time headway thw_{pass}, the model decides that the subject vehicle changes lanes. On the other hand, when the subject vehicle is traveling in the passing lane, the model checks if there is another vehicle ahead of it. When there is no vehicle ahead, the model decides that the subject vehicle shifts to the driving lane.

Once the model decides that the subject vehicle changes lanes, it also decides when to make the change. First, the model attempts to retrieve information from the declarative knowledge on a vehicle closest to the subject vehicle, traveling in the lane to which the subject vehicle intends to shift. If the model detects such information, and if the distance between the nearby vehicle and the subject vehicle is shorter than the safe distance d_{safe}, it decides that the subject vehicle aborts the plan to change lanes. However, as mentioned earlier, in the case where a nearby vehicle travels in a blind spot for a long time, the model might 'forget' the existence of the vehicle due to memory decay. The model also updates its mental model of the environment. After the model confirms that no vehicle is present within safe distance d_{safe} through monitoring, it performs a lane change.

(4) *Component integration and multitasking*: The model includes three components—control, monitoring, and decision-making—and each component executes its own tasks. To run the model, these components need to be integrated through a multitasking method. Because the model is implemented in the ACT-R cognitive architecture, which utilizes sequentially-operating cognitive processers, these tasks are executed sequentially and alternately. The model has a single, tight, main loop and performs tasks step-by-step. The control component implements either lateral or longitudinal control rules; the monitoring component checks the lane selected and traveling direction of the vehicle and the decision-making component decides when to begin or end a lanc change. Drivers deviate from a control task to a monitoring or decision-making task only when they have made sure that the current situation is stable and safe.

Implementation of these three subtasks in the model is important from the perspective of realistically predicting the driver's performance. Although the main loop is tightly linked, it still takes some time to complete the cycle. As a result, even when a driver model operates a vehicle on a straight road without performing any secondary task (or non-driving related activity), it never demonstrates perfect performance, like human drivers never drive perfectly. Moreover, when drivers perform secondary tasks in addition to primary tasks (driving), we expect their performance to worsen. This expectation is the basis for the prediction of driver's distraction.

(5) *Parameter values*: Because driving tasks are complex, it is reasonable to include many domain-specific parameters in driver behavior models. There are two types of parameters. The first is domain specific and related to driving behavior. Among them, there are parameters that were set to be compatible with empirically established 'informal values' that can be obtained from literatures and those to be set by reasonable 'estimated values' to operate the model realistically (Table 6.5).

Table 6.5: Parameters, Values, and Methods Used to Determine these Values for ACT-R Driver Models.

parameter	value	method
k_f	16.0	estimated
k_n	4.0	estimated
k_l	3.0	estimated
θ_{nmax}	0.07 rad	estimated
k_v	3.0	informal
k_{follow}	1.0	informal
thw_{follow}	1.0 s	informal
thw_{pass}	2.0 s	informal
$P_{monitor}$	.20	informal
d_{safe}	40 m	informal
θ_{stable}	0.07 rad ($\approx$ 1/4 lane)	informal
θ_{stable}	0.035 rad/s ($\approx$ 1/8 lane/s)	informal

The second type of parameters is unrelated to the driving domain. Based on numerous modeling studies, the ACT-R community has determined default values for many of the parameters. However, since the model assumes an experienced driver, some parameter values and declarative knowledge chunks change with those representing trained drivers.

6.3.2.2.4 Validation and Application ACT-R Model of Driving Behavior

The major purpose of the ACT-R driver model is to perform rigorous evaluations and prove its validity by making it drive a vehicle in the environment identical to a human driving environment. The driving environment refers to multiple-lane highways generated by a fixed-base driving simulator. When participants drive in the driving simulator, the system produces driving protocols that include vehicle operation data (steering position, accelerator pedal position, etc.), drivers' eye movement data, and environmental information (positions of surrounding vehicles). The ACT-R driver model is fully capable of simulating human driving behavior using a simulator. Quantitative characteristics of the model can be studied, and direct comparisons of driving behavior can be made between the model and humans.

The model can be applied to evaluate in-vehicle devices. Accordingly, the ACT-R driver model can be used as a tool to predict the effect of driver distraction. Because cognitive architecture applies restrictions on a driver, the execution of secondary tasks (use of in-vehicle devices) affects the execution of primary tasks (driving), and vice versa. So far, studies using prototype models have successfully explained driver distraction (Salvucci, 2001).

6.3.3 Statistical Behavior Models

6.3.3.1 Structural Equation Models for Driving Behavior

Drivers perform proper driving tasks—following a lead vehicle, changing lanes, making a turn at an intersection, etc.—in response to continuously changing driving situations. Modeling is a useful tool for understanding driving behavior. Modeling can be viewed as a process of removing unessential components. However, determinations as to which components are essential differ, depending on the person's viewpoint on a given event. In this section, we introduce structural equation models where setting up of a viewpoint plays a major role in modeling.

6.3.3.1.1 Structural Equation Models (SEM)

Structural equation modeling (SEM) is a type of multivariate analysis, and is also called a covariance structure analysis. It is a combination of regression analysis, which is used to estimate a relationship between two directly observed variables (dependent and independent variables), and a factor analysis, which is used to search for common factors hidden between directly observed variables.

Active utilization of SEM began in the 1990s, when software (AMOS and EQS) capable of taking advantage of following characteristics of SEM was developed.

- SEM graphically illustrates estimated relationships between variables [users can readily comprehend results or quantitative relationships between variables once they learn how to interpret the graphical presentations];
- Models describing relationships between variables have large degrees of freedom [users can analyze relationships between different datasets in various ways]; and
- SEM numerically indicates the fit of the model to the data (fit index) [users can quantitatively comprehend the reliability of the models they have built].

Graphic representations that indicate relationships between variables are called path diagrams. Figure 6.41 is an example of a path diagram. Directly measured variables (e.g., vehicle speed, headway distance, pedal stepping force, and numerical responses on questionnaires) are called 'observed' variables, while immeasurable variables that can only be estimated using observed variables are called 'latent' variables. As shown in the figure, a path diagram is a graphic representation of the equations. In this method, users prepare a variance-covariance matrix of observed variables expressed as the function of the parameters indicated in the diagram (i.e., $a_{1A} - a_{6B}, f_{BA}$, independent variables [latent variable A and B], and error variables $[e_1 - e_6$ and $d]$) and a variance-covariance matrix using actual measured data for observed variables. Then, they estimate and adjust parameters, allowing the two sets of variance-covariance to become as close as possible numerically. The method of maximum likelihood is a general parameter estimation method. There are many other estimation methods available in SEM software, including those that do not assume the multivariate normal distribution.

The variance-covariance matrix of observed variables that are expressed as the function of parameters are called 'covariance structure'. For this reason, SEM is also known as 'covariance structure analysis'. To estimate parameters, restrictions need to be applied to the variances of latent variables, and path coefficients that derive from these variances.

One characteristic of SEM is that it has many indices (fit indices) available and they can be used to quantitatively assess how good the estimated parameters are in explaining the data. In other words, these indices are useful in determining the fit of the structural equation model constructed as per the data. This approach allows users to quantitatively compare several models for their suitability, and to quantitatively evaluate how much improvement was made to the model by adding a path (i.e.,

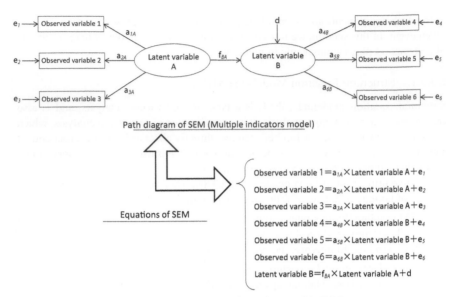

Fig. 6.41: Path diagram and equations used in SEM.

how appropriate the model is in explaining the data). Commonly used fit indices are presented below.

(i) *Chi-square goodness-of-fit test*: The test evaluates discrepancies between the model and data in terms of probability. (Note that when this test is applied to SEM, it works differently from its general application. When the test is performed for SEM, a null hypothesis is 'the model fits the data'. As such, smaller, non-significant chi-square values indicate a better fit of the model to the data.)

(ii) *Root mean square error of approximation (RMSEA)*: The distance between estimated parameters and data points is adjusted by the degree of freedom. (The index value of 0.05 or less indicates good fit, while that of 0.1 or more indicates poor fit.)

(iii) *Goodness-of-fit index (GFI) and adjusted goodness-of-fit index (AGFI)*: The variability of explanatory variables is the measure of their influence on the variability of the dependent variable. GFI and AGFI respectively serve roles equivalent to a multiple correlation coefficient and a multiple correlation coefficient adjusted for the degrees of freedom in a multiple regression analysis. (GFI and AGFI values of at least 0.9 or 0.95 indicate good fit.)

(iv) *Comparative fit index (CFI)*: The index indicates the relative position of an estimated model that falls between the saturated model (the maximum model in which paths are drawn between all variables) and the independence model (the minimum model in which no path is drawn between variables). The best model should have a good fit and the least possible number of paths. (CFI values of at least 0.9 or 0.95 indicate good fit.)

SEM users must keep in mind that a constructed model is not always the optimum one to explain the behavior of interest because there is a great amount of freedom in model construction. In other words, users select the best-fit model out of several different models they have examined. However, a more appropriate model might exist in model structures that they did not consider. It must be pointed out as a research limitation that there may be unnoticed variables that could explain the subject of interest.

6.3.3.1.2 Structural Equation Model of Driving Behavior for Making a Turn

A structural equation model has been constructed using driving behavior data collected in studies conducted on actual roads. The modeling was conducted to study the effect of vehicle speed and position, relative to other vehicles ahead and behind, on a driver preparing to make a side turn (Sato and Akamatsu, 2008). In the study, as a driver approaches an intersection and prepares to make a driver's side turn, measurements were taken on the timing of drivers moving their feet from the accelerator pedal to cover the brake pedal (their feet was above the brake pedal) and the timing of onset of turn signaling. They found that these timings varied depending on measuring dates, even though the same intersection was used throughout the study. Consequently, they investigated the factors contributing to the variation.

Analysis of the collected behavioral data showed linear relationships between the timing of subject drivers covering the brake and vehicle speed. Linear relationships were also observed between the timing of covering the brake and the headway distance to the lead vehicle, which traveled straight through the intersection. Moreover, additional linear relationships were found between the timing of covering the brake and the distance from the subject vehicle to the vehicle behind, before the subject vehicle made the turn. These results indicated that both slower vehicle speed and shorter headway distance to the lead vehicle cause subject drivers to delay covering the brake pedal. It was also found that the three factors measured did not have a major impact on the timing of the onset of turn signaling (Sato and Akamatsu, 2007).

They constructed a structural equation model with five observed variables: the position at which subject drivers started covering the brake, the position at which subject drivers switched on the turn signals, vehicle speed, headway distance to the lead vehicle, and distance to the vehicle behind. The model initially had poor fit when using only five observed variables. However, fitting of the model was improved by adding latent variables to the model. Specifically, 'level of vehicle congestion' was included as a latent variable of the model. It was a latent variable under two observed variables—distance to the vehicle ahead of the subject vehicle and distance to the vehicle behind. Also, 'position at which subject drivers started taking preparatory actions for driver's turn' was another latent variable, which was under two observed variables: position at which subject drivers started covering the brake and position at which they switched on the turn signal. As a result, the model had a good fit to the data collected on actual roads. It was found that headway distance to the lead vehicle and the distance to the vehicle behind interactively, not independently, affect subject drivers' preparatory actions for the driver's side turn. Even if the headway distance is the same, the driver's behavior for turning at an intersection is not the same. The results suggest that when traveling in the group, a driver needs to break away from the group in order to turn at an intersection. Compared to traveling alone, the timing of breaking away from the group gets delayed. As illustrated by this example, drivers' internal aspects (in this example, drivers' perception of traveling with a group of vehicles, which is influenced by their positions relative to surrounding vehicles), which cannot be measured directly, may be expressed in terms of latent variables.

6.3.3.1.3 Application of Structural Equation Model to Theory of Planed Behavior

Another example of SEM includes modeling of relationships among variables obtained from questionnaire surveys. Many studies were conducted in the framework based on the theory of planned behavior, to identify the types of driving attitudes that influence risk-taking behavior, such as speeding and failure to wear the seat belt (Şimşekoğlu and Lajunen, 2008; Warner and Aberg, 2006; Ulleberg and Rundmo, 2003). The theory states that an individual's behavior is influenced by behavioral intentions, which are determined by attitude toward behavior, subjective norms, and perceived behavioral control (Ajzen, 1985). Because factors that influence behavior are conceptual and cannot be observed directly, researchers collected this type of data through questionnaire surveys and estimated relationships between different factors by performing SEM.

6.3.3.2 Bayesian Network Models for Driving Behavior

6.3.3.2.1 Bayesian Network Model

Driving behavior is regarded as a result of information processing (such as perception, decision, and action) and involves many factors, including those that are difficult to observe (see 6.1.2.2). Bayesian network (BN) is a probabilistic model applicable to the modeling of driving behavior when a large amount of behavioral data is available (Daly, 2001; Murphy, 2002; Pew and Mavor, 1998).

BN is a probabilistic graphical model with a directed acyclic graph, and consists of nodes, which represent random variables and links, which indicate conditional dependence relationships between random variables. A simple example is shown in Fig. 6.42, where a joint distribution is set to be $P(A, B, C) = P(A)P(B|A)P(C|B, A)$. The joint distribution of a BN generally takes the following form:

$$P(X_1, \cdots, X_n) = \prod_{i=1}^{n} P(X_i|Pa(X_i)) \tag{6.4}$$

where $Pa(X_i)$ is a set of random variables on which the random variable X_i is dependent. Learning about BNs refers to a selection of BN structure (i.e., $Pa(X_i)$) and determination of its conditional probability distribution $P(X_i|Pa(X_i))$. These are statistical learning techniques based on data (Daly, 2001). A BN structure is also predetermined based on users' knowledge of the subject of interest.

Once the structure of the BN is determined and some of the observed variables or probability distributions are obtained, the probability distributions of other variables can be inferred (Murphy, 2002). To calculate exact solutions, the junction tree algorithm is used to convert a graph into a tree structure, and the belief propagation algorithm is used to obtain marginal distribution. It is important that X_i depends only on $Pa(X_i)$ and does not depend on any of the variables. This fact makes it possible to reduce the amount of calculation required for probabilistic inference and to study the conditional independence of and dependency relationships between variables.

To apply a BN to the modeling of driving behavior, modelers need to (i) select relevant factors and items to be inferred (selection of variables), (ii) learn about the BN from data (selection of structure and parameter learning), and (iii) infer unobserved items using observed items. An appropriate selection of variables and a BN structure are critical for effective modeling and inference.

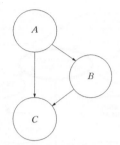

Fig. 6.42: Bayesian network.

BNs can be applied to driving behavior modeling in the following ways.

(i) *Behavior prediction*: Inferring the probability distribution of driving actions based on known environmental factors (Akamatsu et al., 2003; Amata, 2009).

(ii) *Behavior detection*: Detecting certain maneuvering, such as switching lanes and collision avoidance behavior, and detecting behavioral intentions (McCall and Trivedi, 2006; Kasper et al., 2012).

(iii) *State estimation*: Estimating the state of drivers, such as feeling fatigued or irritated (Al-Sultan et al., 2013).

Figure 6.43 shows an example of driving behavior modeling, focusing on the behavior of slowing down to stop at an intersection with stop signs (Akamatsu, 2004). The BN includes variables regarding driving actions (e.g., speed when accelerator is off and TTC of onset of braking), those regarding road structures (e.g., viewing angle at an intersection) and those regarding traffic situation (e.g., a crossing vehicle appears while approaching a subject vehicle). The BN structure and variables were obtained from the data on 2,500 run by 12 driver's using instrumented vehicles (see 2.3.2.1). The relationship between a viewing angle at an intersection and the timing of the release of the acceleration pedal, and the relationships among the viewing angle at an intersection, presence/absence of vehicles traveling through the intersection, the position at which the subject vehicle came to a stop, vehicle speed, etc., can be seen. Also, using the BN, probability distributions of driving actions to be taken in response to environmental factors can be inferred and formulated as an algorithm, enabling them to detect deviation from normal driving behavior.

6.3.3.2.2 Dynamic Bayesian Network Model

A BN whose random variables have dependency between time slices is called a dynamic Bayesian network (DBN) (Murphy, 2002). A DBN can describe the dynamics of the BN and is a probability model capable of describing context-dependent time series. Non-dynamic BNs can be called static Bayesian networks (SBNs).

The structure of a DBN can be selected according to the objectives of the modeling. Appropriate selection of DBN structure and variables is vital for effective modeling and inference.

Like hidden Markov models (HMMs), switching models, including auto-regressive hidden Markov models (AR-HMMs) (Fig. 6.44), a basic type of DBNs, have discrete variables representing unobserved states. An HMM, which activates

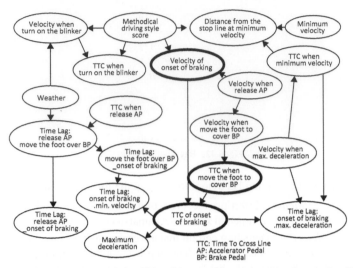

Fig. 6.43: Bayesian network model describing drivers making a stop at an intersection (Akamatsu, 2004).

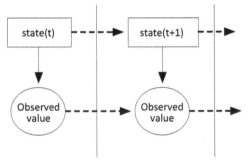

Fig. 6.44: Dynamic Bayesian network (AR-HMM).

different submodels when the state changes, can be considered a DBN with a simple structure. Switching models are used for modeling such maneuvering, as switching lanes, making a right/left turn, and making a stop. In a switching model, individual submodels represent the dynamics of various driving actions, such as stopping and accelerating, while indicating the transition probability between driving actions (Fig.6.45) (Kumagai and Akamatsu, 2006).

A DBN enables users to infer unobserved random variables based on some observed variables (e.g., state of the switching model). It also enables users to infer past and future models through time-series analysis. DBNs have been applied to modeling of driving behavior in the following ways:

(i) *Prediction*: Using observations to a certain point in time, time series of future driving behavior is predicted (Kumagai and Akamatsu, 2006; Kishimoto and Oguri, 2008). Also, assuming a future state, time series are predicted up to that point in time (Kumagai and Akamatsu, 2006).

(ii) *Estimation*: The states of driving behavior are estimated through the inference of unobserved random variables. For example, driving modes can be estimated by inferring the state of a switching model (Hayashi et al., 2006; Hamada et al., 2013). In other studies, fatigue and frustration of drivers were estimated (Malta et al., 2011; Yang et al., 2010). Behavioral data obtained by noisy measurement can be filtered (Gindele et al., 2010).

(iii) *Recognition*: By incorporating observed time series data into a DBN and evaluating likelihood values, researchers can detect drivers' behavior and behavioral intensions (Pentland and Liu, 1999; Sekizawa et al., 2007).

Figure 6.46 shows AR-HMM modeling of vehicle speed when the vehicle is turning to the driver's side at an intersection with the stop sign and the prediction of vehicle speed when the vehicle is about to enter the intersection (time = 0s in the figure). The probability distribution of predicted vehicle speed peaked at around 10 km/h and 0 km/h when the vehicle was traveling in the intersection (time = 4s in the figure). These peaks respectively represent two scenarios: (i) the vehicle yields when entering the intersection, and (ii) the vehicle makes a stop in the middle of the intersection, allowing a vehicle coming from the opposite direction to cross the intersection.

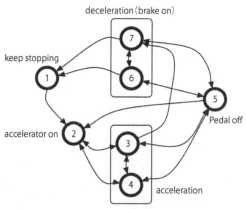

Fig. 6.45: State transition diagram produced by an AR-HMM, which was trained with data on driving behavior during the driver's side turns obtained from 33 reacted trips (Kumagai and Akamatsu, 2006).

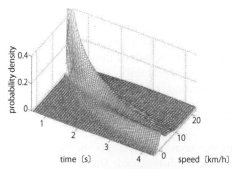

Fig. 6.46: Prediction of vehicle speed using AR-HMM when the vehicle is making a driver's side turn (Kumagai and Akamatsu, 2005).

6.3.3.3 *Modeling Driving Behavior Using Hidden Markov Models*

Hidden Markov models (HMMs) are statistical models commonly used in the field of speech recognition. In recent years, the models have also been used to estimate driver behavior and characteristics of individuals. HMMs are suitable for modeling of drivers' ambiguous behavior, which cannot be adequately reproduced by physical models, complex movement of vehicles, and environmental factors. The models' accuracy can be improved by collecting large amounts of labeled data and training the models with them. On the other hand, HMMs are inadequate for explaining phenomena in a theoretical manner. Accordingly, the models should be used carefully as per the study targets to which models are applied. In this section, we present some examples of modeling drivers' characteristics and driving behavior using HMMs and observing driving signals shown in Fig. 6.47.

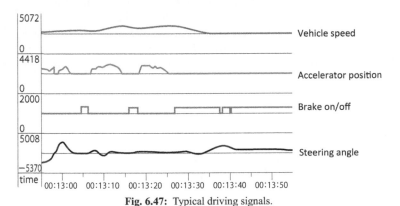

Fig. 6.47: Typical driving signals.

6.3.3.3.1 Theoretical Background of Modeling Driving Behavior Using HMM

Driving behavior can be estimated using label column B (e.g., right/left turn, making a stop, and starting a vehicle), which matches the input pattern vector X, obtained from incoming driving signals under any given situation. Assuming that the time series vector $X = x_1, x_2, ..., x_T$, where T refers to the Tth frame, and the driving behavior label $B = b_1, b_2, ..., b_I$, where I refers to the Ith label, this estimation can be achieved statistically by finding an appropriate label B, which makes the probability, $P(B|X)$ or $P(b_1, b_2, ..., b_I|x_1, x_2, ..., x_T)$, maximum for a given X. It is not practical to solve this problem using a direct calculation approach since there are thousands of different input signal values, and there are over a 100 million combinations of these values to be calculated. This problem can be solved through indirect calculations using prior probability $P(B)$ and conditional probability $P(X|B)$. In this approach, first, $P(B|X)$ is defined in the following equation according to Bayes' theorem:

$$P(B|X) = \frac{P(X|B)P(B)}{P(X|B)} \tag{6.5}$$

Since $P(X)$ in the denominator represents the probability of occurrence and is independent of $P(B)$, it can be disregarded. Thus, the maximum $P(X|B) \, P(B)$ can be calculated by the equation, $B = argmax_B P(X|B) P(B)$. In this equation, $P(B)$ and $P(X|B)$ can respectively be generalized as follows:

$$P(B) = P(b_1, b_2, ..., b_l)$$
$$= P(b_1)P(b_2|b_1)P(b_3|b_1, b_2)$$
$$\cdots P(b_l|b_1, b_2, ...b_{i-1})$$
$$= P(b_1) \prod_{i=2}^{l} P(b_i|b_1, ...b_{i-1}) \tag{6.6}$$

$$P(X|B) = P(x_1, x_2, ..., x_{T1}|b_1, b_2, ..., b_l)$$
$$= \Sigma_{T1<...<TN-1} \, P(x_1, x_2, ..., x_{T1}|b_1)$$
$$P(x_{T1+1}, x_{T1+2}, ..., x_{T2}|b_2) \tag{6.7}$$

Equations 6.6 and 6.7 describe the successive occurrence of driving behavior labels, which are the components of B. Each term in Equation 6.7 represents a conditional probability of observed driving signals when a label is presumably entered. For example, single driving behavior can be detected by calculating $P(x_{Ti+1}, x_{Ti+2}, ..., x_{Tn}|b_{Tn})$. Modeling driving behavior, $P(X|B)$, expressed in terms of HMM, can be called a driving behavior model.

6.3.3.3.2 Example of Constructing a Driving Behavior Model Using Discrete HMM

An example of how an HMM is used to make estimations is shown in this section. For simplicity, we use a left-to-right HMM to estimate a driver's intention to start or stop a vehicle using two types of time series signals: acceleration signal (A) and deceleration signal (D) (Fig. 6.48). Our objective is to automatically identify the intention of the driver—either starting or stopping the vehicle, assuming that the observation signal vector X is '*ADA*' during any given observation time. To make the HMM properly predict the probabilities of occurrence and transition for both the 'start' and 'stop' states (q_{1e} and q_{2e} are the final states) shown in Fig. 6.48, the model is trained with time series signals previously measured. For example, when entering observation signal X, which is equal to '*ADA*', into the model, the 'start' model describes a series of potential transition states q_{ij} in the form of $q_{11} \rightarrow q_{11} \rightarrow q_{12} \rightarrow q_{1e}$ and $q_{11} \rightarrow q_{12} \rightarrow q_{12} \rightarrow q_{1e}$, and the 'stop' model describes a series of potential transition states in the form of $q_{21} \rightarrow q_{21} \rightarrow q_{22} \rightarrow q_{2e}$ and $q_{21} \rightarrow q_{22} \rightarrow q_{22} \rightarrow q_{2e}$. Assuming that probabilities that the driver intends to start (P (start)) and stop (P (stop)) a vehicle are both 0.5, posterior probability when '*ADA*' is observed can be calculated using the following procedure:

$$P(start|ADA) = P(ADA|start) \, P(start)$$
$$= (b_{11\rightarrow A} a_{1,1\rightarrow 1} b_{11\rightarrow D} a_{1,1\rightarrow 2} b_{12\rightarrow A} a_{1,2\rightarrow e}$$
$$+ b_{11\rightarrow A} a_{1,1\rightarrow 2} b_{12\rightarrow D} a_{1,2\rightarrow 2} b_{12\rightarrow A} a_{1,2\rightarrow e})$$
$$P(start)$$
$$= (0.7 \times 0.4 \times 0.3 \times 0.6 \times 0.8 \times 0.7$$
$$+ 0.7 \times 0.6 \times 0.2 \times 0.3 \times 0.8 \times 0.7)$$
$$\times 0.5$$
$$= 0.021168 \tag{6.8}$$

$P(\text{stop}|ADA) = P\,(ADA|\text{stop})\,P(\text{stop})$

$$= (b_{21\to A}\,a_{2,1\to1}\,b_{21\to D}\,a_{2,1\to2}\,b_{22\to A}\,a_{2,2\to e}$$
$$+\,b_{21\to A}\,a_{2,1\to2}\,b_{22\to D}\,a_{2,2\to2}\,b_{22\to A}\,a_{2,2\to e})$$
$$P(\text{stop})$$
$$= (0.4 \times 0.3 \times 0.6 \times 0.7 \times 0.1 \times 0.5$$
$$+\,0.4 \times 0.7 \times 0.9 \times 0.5 \times 0.1 \times 0.5)$$
$$\times 0.5$$
$$= 0.00441 \tag{6.9}$$

Since $P(\text{start}|ADA)$ is greater than $P\,(\text{stop}|ADA)$, the observation signal X, or 'ADA', presumably stands for the driver's intention to start a vehicle, assuming that higher posterior probability of the two is more likely to represent the driver's intention.

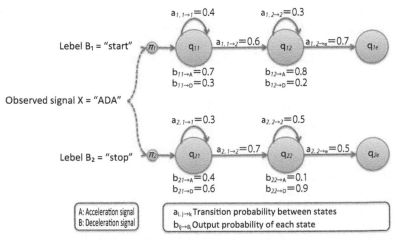

Fig. 6.48: Example of estimating driver's intention to start and stop a vehicle using HMM.

6.3.3.3.3 Estimation of Road Shape and Driving Behavior Using Continuous HMM

Continuous signals observed in a real-world situation can be described with continuous HMM, which estimates output probability of each state q_{ij} of Fig. 6.48 by performing a clustering method using a Gaussian mixture model (GMM). We present an example of constructing a GMM-HMM, capable of estimating driving behavior performed on a specific road and based solely on continuous driving signals entered into the GMM.

(1) *Collection of driving signals and creation of corpus*: To produce appropriate labels and a corpus including relevant driving behavior signals using a GMM-HMM, the model needs to be trained with driving behavior labels and a corpus including relevant driving behavior signals. Kaminuma et al. collected driving signal data (vehicle speed V, yaw rate Y, accelerator position A, brake B, steering angle S, and first/second derivatives of these variables) from 18 people, who drove through a predetermined route on a public road (Fig. 6.49a). The data were collected at 19 locations along the route and were labeled (Kaminuma and Nankaku, 2013).

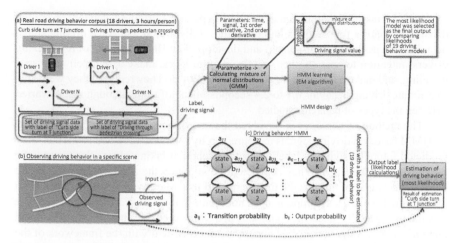

Fig. 6.49: Construction of HMM for estimating driving behavior and steps involved in the estimation.

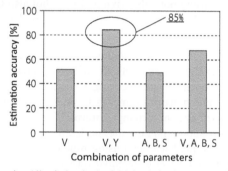

Fig. 6.50: Estimation (discrimination) of driving behavior measured at 19 locations.

(2) *Estimation of driving behavior in relation to specific road shape*: We trained the GMM-HMM using the EM algorithm obtained from the abovementioned corpus (Fig. 6.49a). Computation was performed using 2 to 16 GMM mixtures and 5 and 10 HMM states (Fig. 6.49c). In addition, we differentiated driving behavior by 19 measurement locations by entering non-learning data into the Hidden Markov Model Toolkit (HTK) (Fig. 6.49b) (http://htk.eng.cam.ac.uk). In this experiment, we achieved 85 per cent estimation accuracy using vehicle speed and yaw rate (Fig. 6.50). It is feasible to improve discrimination performance by selecting signals that optimally represent the driving behavior of interest. The performance can also be improved by assessing appropriate numbers of GMM mixtures and HMM states in relation to change in input signals and observed sections, and to avoid excessive quantization of data and over-training of the model.

6.3.3.3.4 Estimating Driving Behavior Using HMM and other Applications

(1) *Prediction of driving behavior*: It is vital for the driver-assistance technology to smoothly facilitate the driver's vehicle operation. Some studies have been tackling

this issue by taking the approach of predicting people's driving behavior. To predict driving behavior a few seconds into the future, Oliver et al. collected driving signal data from 70 people (1 hour/person) and developed an HMM capable of recognizing drivers' intention to pass another vehicle, change lanes, make a right/left turn, start traveling, and make a stop (Oliver and Pentland, 2000). They reported they were able to estimate driving behavior 0.1 to 2.4 seconds ahead of time although results varied between the types of behavior studied.

The method of Oliver et al. involved the use of posterior probability computed by HMM based on driving signals measured up to a specific time of a day. Then, they predicted future behavior by labeling driving behavior taking place after the specific time and using the HMM. They conducted the study on the assumption that a future event occurring after the specific time is related to a Markov chain taking back from the specific time. This assumption does not mean that the past chain of events to the specific time actually meets the conditions of a Markov chain. Thus, it should be noted that Oliver et al.'s theoretical explanation is not consistent with the framework of the Bayes' theorem.

Straub et al. proposed a method to estimate destinations of and typical routes taken by subject drivers using not only driving signals but also GPS signals and the orientation of subject vehicles (Straub et al., 2014). They used a hierarchical HMM to optimize model parameters.

(2) *Estimating characteristics of individuals*: By estimating individuals' characteristics, it is feasible to comprehend drivers' preferences and habits, and provide them with personalized interfaces and services. Meng et al. proposed a method to differentiate individuals as a means to prevent vehicle theft (Meng et al., 2006). With the proposed system, driving patterns of individuals can be estimated by modeling each driver and optimizing the model, using an HMM. To differentiate seven drivers, Meng et al. modeled them using nine-dimensional data space and six HMM states, and conducted driver differentiation with the constructed system. As a result, they achieved differentiation with 80 per cent accuracy on average (ranging from 40 to 100 per cent). Kaminuma et al. trained an HMM capable of discriminating between empathic and non-empathic behavior using a system similar to that shown in Fig. 6.48, and conducted estimation experiments to automatically optimize the response guidance of an in-vehicle voice interface. The study demonstrated discrimination accuracy of up to about 70 per cent (Kaminuma and Nankaku, 2014).

(3) *Future direction and issues*: In the field of voice recognition, a Gaussian mixture model in combination with HMM (GMM-HMM) had been commonly used. At the end of the 2000s, the use of deep neural networks (DNNs) in combination with HMM (DNN-HMM) was proposed to calculate the output probability and transition probability in voice recognition research (Hinton et al., 2012). Compared to GMM-HMM, DNN-HMM had enhanced voice recognition performance by at least 10 points even when the corpora of the same condition were used with these models. DNN-HMM is now widely used for public services, including Google Voice. DNN-HMM used for voice recognition is trained with large amounts of data, about 10 times greater than the conventional amount of data, collected before and after the arbitrarily determined time. This method is compatible with driving behavior estimation without producing noise signals, and its application to driving behavior should be considered.

The accuracy of both GMM and DNN can be enhanced by training these models with large amounts of corpora. From this viewpoint, it is important to effectively collect corpora at low cost. The function to upload vehicle signals to internet storage, used for various maintenance purposes and services (so-called connected vehicles), may be a promising source of corpora. However, from the viewpoint of personal information protection, it may be difficult to obtain driving signal information of the general public.

References

Ajzen, I. (1985). From intentions to actions: A theory of planned behavior. pp. 11–39. *In*: J. Kuhl and J. Beckmann (eds.). Action Control: From Cognition to Behavior, Springer, Heidelberg, Germany.

Akamatsu, M., M. Yoshioa, N. Imacho, T. Daimon and H. Kawashima. (1997). Analysis of driving a car with a navigation system in an urban area. pp. 85–95. *In*: Y.I. Noy (ed.). Ergonomics and Safety of Intelligent Driver Interfaces, Lawrence Erlbaum Associates, New Jersey, USA.

Akamatsu, M., Y. Sakaguchi and M. Okuwa. (2003). Modeling of driving behavior when approaching an intersection based on measured behavioral dada on an actual road. Proceedings of Human Factors and Ergonomics Society 47th Annual Meeting, 1895–1899.

Akamatsu, M. (2004). Evaluating driver while driving real road environment using equipped vehicles. Journal of Society of Automotive Engineers of Japan, 58(12): 53–59 (in Japanese with English abstract).

Akamatsu, M., K. Hayama, A. Iwasaki, J. Takahashi and H. Daigo. (2006). Cognitive and physical factors in changes to the automobile driving ability of elderly people and their mobility life. Questionnaire survey in various regions of Japan, IATSS Research, 30(1): 38–51.

Akamatsu, M. (2008). Japanese approaches to principles, codes, guidelines, and checklists for in-vehicle HMI. pp. 420–438. *In*: M.A. Regan, J.D. Lee and K.L. Young (eds.). Driver Distraction, Theory, Effects, and Mitigation, CRC Press, London, UK.

Al-Sultan, S., A.H. Al-Bayatti and H. Zedan. (2013). Context-aware driver behavior detection system in intelligent transportation systems. IEEE Transactions on Vehicular Technology, 62(9): 4264–4275.

Amata, H., C. Miyajima, T. Nishino, N. Kitaoka and K. Takeda. (2009). Prediction model of driving behavior based on traffic conditions and driver types. Proceedings of the 12th International IEEE Conference on Intelligent Transportation System, 747–752.

Anstey, K.J., J. Wood, S. Lord and J.G. Walker. (2005). Cognitive, sensory and physical factors enabling driving safety in older adults. Clinical Psychology Review, 25(1): 45–65.

Anstey, K.J. and J. Wood. (2011). Chronological age and age-related cognitive deficits are associated with an increase in multiple types of driving errors in late life. Neuropsychology, 25(5): 613–621.

Anstey, K.J., M.S. Horswill, J.M. Wood and C. Hatherly. (2012). The role of cognitive and visual abilities as predictors. The multi-factorial model of driving safety, Accident Analysis and Prevention, 45: 766–774.

Asoh, T., T. Muraki and T. Ito. (2000). Comparative study of effectiveness of maps and turn-by-turn navigation systems. JARI Research Journal, 22(29): 82–85 (in Japanese with English abstract).

Ball, K. and C. Owsley. (1993). The useful field of view test: A new technique for evaluating age-related declines in visual function. Journal of the American Optometric Association, 64(1): 71–79.

Ball, K.K., B.L. Beard, D.L. Roenker, R.L. Miller and D.S. Griggs. (1988). Age and visual search: Expanding the useful field of view. Journal of the Optical Society of America, Optics and Image Science, 5(12): 2210–2219.

Barber, P. and N. Clarke. (1998). Advanced collision warning systems. IEEE Colloquim on Industrial Automation and Control: Application in the Automotive Ind., 234: 2/1–2/9.

Boer, E.R. (1996). Tangent point-oriented curve negotiation. Proceedings of the 1996 IEEE Intelligent Vehicles Symposium, 7–12.

Boer, E.R., M.E. Rakauskas, N.J. Ward and M.A. Goodrich. (2005). Steering entropy revisited. Proceedings of the 3rd International Driving Symposium on Human Factors in Driver Assessment, Training and Vehicle Design, 25–32.

Byrne, M.D. (2001). ACT-R/PM and menu selection: Applying a cognitive architecture to HCI. International Journal of Human-Computer Studies, 55(1): 41–84.

Cheng, B., M. Hashimoto and T. Suetomi. (2002). Analysis of driver response to collision warning during car following. JSAE Review, 23(2): 231–237.

Christ, S., P. Delhomme, A. Kaba, T. Mäkinen, F. Sagberg, H. Schulze and S. Siegrist. (2000). Investigations on influences upon driver behavior—Safety approaches in comparison and combination. GADGET Project: Final Report, 2000.

Crowell, J.A. and M.S. Banks. (1996). Ideal observer for heading judgments. Vision Research, 36(3): 471–490.

Daimon, T. and H. Kawashima. (1996). A study of analysis and evaluation of in-vehicle information systems. Transaction of Automotive Engineers of Japan, 27(39): 113–118 (in Japanese with English abstract).

Daly, R. (2001). Learning Bayesian networks: Approaches and issues. The Knowledge Engineering Review, 26(2): 99–157.

Devos, H., A. Nieuwboer, W. Vandenberghe, M. Tant, W.D. Weerdt and E. Uc. (2013). Validation of driving simulation to assess on-road performance in huntington disease. Proceedings of the Seventh International Driving Symposium on Human Factors in Driver Assessment, 241–247.

Donders, F.C. (1868). Over de smelheid van psychische processen. Onderzoekingen gedaan in het Physiologisch Laboratorium der Utrechtsche Hoogeschool, 1858–1869, Tweeds Reeks, 11: 92–120 (translated by Koch, On the speed of mental processes. Acta Psychologica, 1969, 30: 412–431).

Donges, E. (1978). A two-level model of driver steering behavior. Human Factors, 20(6): 691–707.

Donorfio, L.K.M., L.A. D'Ambrosio, J.F. Coughlin and M. Mohyde. (2008). Health, safety, self-regulation and the older driver: It's not just a matter of age. Journal of Safety Research, 39(6): 555–561.

Endsley, M.R. (1995). Toward a theory of situation awareness in dynamic systems. Human Factors, 37: 32–64.

Endsley, M.R. (2000). Direct Measurement of situation awareness: Validity and use of SAGAT. pp. 147–173. In: M.R. Endsley and D.J. Garland (eds.). Situation Awareness Analysis and Measurement, Lawrence Erlbaum, Mahwah, NJ, USA.

Engström, J. and E. Hollnagel. (2007). A general conceptual framework for modeling behavioral effects of driver support functions. pp. 61–84. In: P. Carlo Caccaiabue (ed.). Modelling Driver Behavior in Automotive Environments, Springer, London, UK.

Ericsson, K.A. and H.A. Simon. (1993). Protocol Analysis: Verbal Reports as Data (revised edition), MIT Press.

Franz, B., J. Haccius, D. Stelzig-Krombholz and B. Abendroth. (2015). Evaluation of the SAGAT method for highly automated driving. Proceedings of 19th Triennial Congress of the International Ergonomics Association.

Fujioka, T. (2007). Theoretical research on the stability of the closed system composed of a look-ahead driver and a planer vehicle. JSAE Annual Congress, 11-7: 29–34 (in Japanese).

Fuller, R. and J.A. Santos. (2002). Psychology and the highway engineer. pp. 1–10. In: R. Fuller and J.A. Santos (eds.). .Human Factors for Highway Engineers, Elsevier, Oxford, UK.

Fuller, R. (2005). Towards a general theory of driver behavior. Accident Analysis and Prevention, 37(3): 461–472.

Fuller, R. (2011). Driver control theory. pp. 13–26. In: B.E. Porter (ed.). Handbook of Traffic Psychology, Elsevier, Oxford, UK.

Gawron, V.J. (2008). Human Performance, Workload, and Situational Awareness Measures Handbook, CRC Press, 68–75.

Gibson, J.J. (1950). The Perception of the Visual World, Houghton Mifflin.

Gibson, J.J. (1966). The Senses Considered as Perceptual Systems, Houghton Mifflin.

Gindele, T., S. Brechtel and R. Dillmann. (2010). A probabilistic model for estimating driver behaviors and vehicle trajectories in traffic environments. Annual Conference on Intelligent Transportation Systems, 1625–1631.

Gray, W.D. and D.A. Boehm-Davis. (2000). Milliseconds matter: An introduction to micro strategies and to their use in describing and predicting interactive behavior. Journal of Experiment Psychology: Applied, 6(4): 322–335.

Gray, W.D. and W.T. Fu. (2004). Soft constraints in interactive behavior: The case of ignoring perfect knowledge in-the-world for imperfect knowledge in-the-head. Cognitive Science, 28(3): 359–382.

Green, P., B. Cullinane, B. Zylstra and D. Smith. (2003). Typical values for driving performance with emphasis on the standard deviation of lane position: A summary of literature. Technical Report UMTRI-2003-42.

Green, P. (2013). Standard definitions for driving measures and statistics: Overview and status of Recommended Practice. J2944, Proceedings of the 5th International Conference on Automotive User Interfaces and Interactive Vehicular Applications, 184–191.

Hamada, R., T. Kubo, K. Ikeda, Z. Zhang, T. Bando and M. Egawa. (2013). A comparative study of time series modeling for driving behavior towards prediction. Asia-Pacific Signal and Information Processing Association Annual Summit and Conference (Apsipa).

Hayashi, K., Y. Kojima, K. Abe and K. Oguri. (2006). Prediction of stopping maneuver considering driver's state. pp. 1191–1196. *In*: Intelligent Transportation Systems Conference, 2006. ITSC'06. IEEE.

Hayhoe, M. and D. Ballard. (2005). Eye movements in natural behavior. Trends in Cognitive Sciences, 9(4): 188–194.

Hildreth, E. and C.S. Royden. (1998). Computing observer motion from optical flow. pp. 269–293. *In*: T. Watanabe (ed.). High-level Motion Processing, Bradford Book, Cambridge, MA, USA.

Hildreth, E.C., J.M. Beusmans, E.R. Boer and C.S. Royden. (2000). From vision to action: Experiments and models of steering control during driving. Journal of Experimental Psychology, Human Perception and Performance, 26(3): 1106–1132.

Hillyard, S.A., E.-K. Vogel and S.J. Luck. (1998). Sensory gain control (amplification) as a mechanism of selective attention: Electro-physiological and neuro-imaging evidence. Philosophical Transactions of the Royal Society B, 353(1373): 1257–1270.

Hinton, G., L. Deng, D. Yu, G.E. Dahl, A.R. Mohamed, N. Jaitly, A. Senior, V. Vanhoucke, P. Nguyen, T. Sainath and B. Kingsbury. (2012). Deep neural networks for acoustic modeling in speech recognition. IEEE Signal Processing Magazine, 29(6): 82–97.

Hollnagel, E. and D.D. Woods. (2005). Joint cognitive systems. Foundations of Cognitive Systems Engineering, Taylor & Francis, Boca Raton, FL, USA.

Horswill, M.S., K.J. Anstey, C. Hatherly, J.M. Wood and N.A. Pachana. (2011). Older driver's insight into their hazard perception ability. Accident Analysis and Prevention, 43(6): 2121–2127.

http://htk.eng.cam.ac.uk.

http://www.oecd.org, OECD Guidelines on the Protection of Privacy and Transborder Flows of Personal Data, http://www.jipdec.or.jp/publications/oecd/2013/index.html.

Innes, C., R. Jones, T. Anderson, J. Dalrymple-Alford, S. Hayes, S. Hollobon, J. Severinsen, G. Smith and A. Nicholls. (2007). Prediction of driving ability in persons with brain disorders using sensory-motor and cognitive tests. Journal of the Neurological Sciences, 260: 188–198.

Itti, L. and C. Koch. (2001). Computational modeling of visual attention. Nature Reviews Neuroscience, 2(3): 194–203.

Kahneman, D. (1973). Attention and Effort, Prentice Hall, New York, USA.

Kaminuma, A. and Y. Nankaku. (2013). Automatic character and action recognition of the drivers' using driving signal. Proceeding of JSAE Fall Meeting, 20135857. 128-13: 1–7 (in Japanese with English abstract).

Kaminuma, A. and Y. Nankaku. (2014). Automatic Driver Characteristics Estimation Using Driving Signals, FISITA2014, F2014-AHF-023.

Kasper, D., G. Weidl, T. Dang, G. Breuel, A. Tamke, A. Wedel and W. Rosenstiel. (2012). Object-oriented Bayesian networks for detection of lane change maneuvers. Intelligent Transportation Systems Magazine, 4(3): 19–31.

Kishida, E., N. Matsuzaki, K. Uenuma, H. Shigematsu, M. Kitazaki and K. Iwao. (2008). Study on device controlling visual information to improve driver's performance. Transaction of Society of Japan Mechanical Engineering (C), 74(745): 2254–2263 (in Japanese).

Kishimoto, Y. and K. Oguri. (2008). A modeling method for predicting driving behavior concerning with driver's past movements. *In*: Vehicular Electronics and Safety, 2008, ICVES 2008, IEEE International Conference on Vehicle Electronics and Safety, 132–136.

Kitamura, A., H. Naito, T. Kimura, K. Shinohara, T. Sasaki and H. Okumura. (2014). Distribution of attention in augmented reality: Comparison between binocular and monocular presentation. IEICE Transactions on Electronics, E97-C(11): 1081–1088.

Koch, C. and S. Ullman. (1985). Shifts in selective visual attention: Towards the underlying neural circuitry. Human Neurobiology, 4: 219–227.

Kondo, M. (1958). On fundamental relations existing between steering and vehicle motion. Transaction of Society of Automobile Engineers of Japan, 5: 40–43 (in Japanese).

Kondoh, T., T. Yamamura, S. Kitazaki, N. Kuge and E.R. Boer. (2008). Identification of visual cues and quantification of drivers' perception of proximity risk to the lead vehicle in car-following situations. Journal of Mechanical Systems for Transportation and Logistics, 1(2): 170–180.

Kondoh, T., T. Yamamura, N. Kuge, P. Miguel and T. Sunda. (2015). Development of a real-time steering entropy method for quantifying driver's workload. Transactions of Society of Automotive Engineers of Japan, 46(1): 167–172 (in Japanese, English abstract is available).

Konz, S. and D. McDougal. (1968). The effect of background music on the control activity of an automobile driver. Human Factors, 10(3): 233–244.

Kramer, A.F., L.J. Trejo and D. Humphrey. (1995). Assessment of mental workload with task-irrelevant auditory probes. Biological Psychology, 40(1-2): 83–100.

Kumagai, T. and M. Akamatsu. (2005). Human driving behavior prediction using dynamic Bayesian networks. Proceedings of JSAE Annual Congress, Paper #20055492, 64-05: 13–18.

Kumagai, T. and M. Akamatsu. (2006). Prediction of human driving behavior using dynamic Bayesian networks. IEICE Transactions on Information and Systems, 89D(2): 857–860.

Laird, J.E., A. Newell and P.S. Rosenbloom. (1987). Soar: An architecture for general intelligence. Artificial Intelligence, 33(1): 1–64.

Land, M. and J. Horwood. (1995). Which parts of the road guide steering? Nature, 377(6547): 339–340.

Land, M.F. and D.N. Lee. (1994). Where we look when we steer. Nature, 369(6483): 742–744.

Lee, D.N. (1976). A theory of visual control of braking based on information about time to collision. Perception, 5(4): 437–459.

Liu, Y. (1996). Queuing network modeling of elementary mental processes. Psychological Review, 103(1): 116–136.

Loren, S., H.L. Kathy, W.G. Kenneth and E.D. Lawrence. (2003a). Model driver screening and evaluation program final technical report. Vol. 1, Project Summary and Model Program Recommendations, HS-809 582. NHTSA.

Loren, S., K.H. Lococo, K.W. Gish and L.E. Decina. (2003b). Model driver screening and evaluation program final technical report. Vol. II, Maryland Pilot Older Study, HS-809 583, NHTSA.

MacAdam, C.C. (1981). Application of an optimal preview control for simulation of closed-loop automobile driving. IEEE Transactions on Systems, Man, and Cybernetics, SMC-11(6): 393–399.

Malta, L., C. Miyajima, N. Kitaoka and K. Takeda. (2011). Analysis of real world driver's frustration. IEEE Transactions on Intelligent Transportation Systems, 12(1): 109–118.

Marat, S., T.H. Phuoc, L. Granjon, N. Guyader, D. Pellerin and A. Guérin-Dugué. (2009). Modelling spatio-temporal saliency to predict gaze direction for short videos. International Journal of Computer Vision, 82(3): 231–243.

Mars, F. (2008). Driving around bends with manipulated eye-steering coordination. Journal of Vision, 8(11). Article 10: 1–11.

Marshall, S.C., F. Molnar, M. Man-Son-Hing, R. Blair, L. Brosseau, H.M. Finestone, C. Lamothe, N. Korner-Bitensky and K.G. Wilson. (2007). Predictors of driving ability following stroke: A systematic review. Topics in Stroke Rehabilitation, 14(1): 98–114.

McCall, J.C. and M.M. Trivedi. (2006). Human behavior based predictive brake assistance. Intelligent Vehicles Symposium 2006: 8–12.

McGehee, D.V., J.D. Lee, M. Rizzo, J. Dawson and K. Bateman. (2004). Quantitative analysis of steering adaptation on a high performance fixed-base driving simulator. Transportation Research Part F, 7(3): 181–196.

McKnight, A.J. and A.S. McKnight. (1999). Multivariate analysis of age-related driver ability and performance deficits. Accident Analysis & Prevention, 31(5): 445–454.

McRuer, D.T., R.W. Allen, D.H. Weir and R.H. Klein. (1977). New results in driver steering control models. Human Factors, 19(4): 381–397.

Meng, X., K.K. Lee and Y. Xu. (2006). Human driving behavior recognition based on hidden markov models. Proceedings of IEEE International Conference on Robotics and Biomimetics, 274–279.

Michon, J.A. (1985). A critical view of driver behavior models. What do we know, what should we do? pp. 516–520. In: L. Evans and R. Schwing (eds.). Human Behavior and Traffic Safety, Plenum Press, New York, USA.

Miura, T. and K. Shinohara. (1998). Characteristics of visual attention related to ITS: Focusing on the temporal characteristics in observing car navigation display. Journal of the Traffic Science Society of Osaka, 28(1/2): 53–59 (in Japanese with English abstract).

Morita, K., M. Sekine and T. Okada. (2006, October). Factors with the greatest influence on drivers' judgment of when to apply brakes. pp. 5044–5049. In: SICE-ICASE, 2006, International Joint Conference, IEEE.

Muira, T. (ed.) (2012). Visual Attention and Behavior—Bridging the Gap between Basic and Practical Research. Kazama Shobo, Tokyo, JAPAN.

Mullen, N.W., M. Bedard, J.A. Riendeau and T.J. Rosenthal. (2010). Simulated lane departure warning system reduces the width of lane that drivers use. Advances in Transportation Studies, 33–44.

Murphy, K.P. (2002). Dynamic Bayesian Networks Representation, Inference and Learning, Ph.D. thesis, University of California, Berkeley.

Nakayama, O., T. Futami, T. Nakamura and E.R. Boer. (1999). Development of a steering entropy method for evaluating driver workload. SAE Technical Paper, 1999-01-0892.

National Police Agency (2009). Guidelines for testing equipment for senior driver training. Police Agency, Traffic License Division, Notice No. 52 (in Japanese).

Navalpakkam, V. and L. Itti. (2006). An integrated model of top-down and bottom-up attention for optimizing detection speed. Computer Vision and Pattern Recognition, IEEE Computer Society Conference on Computer Vision and Pattern Recognition (CVPR '06), Vol. 2: 2049–2056.

Newell, A. (1990). Unified Theories of Cognition, The William James Lectures, 1987, Harvard University Press, Cambridge, MA.

Nishizaki, Y., M. Nagai, J.I. Kawahara, T. Sato and H. Nemoto. (2013). Individual differences in drivers' cognitive functions in backward manoever and merging behaviour. Transactions of Society of Automotive Engineers of Japan, 44(4): 1059–1065 (in Japanese with English abstract).

Norman, D.A. (1986). Cognitive Engineering. pp. 31–61. In: D.A. Norman and S.W. Draper (eds.). User Centered System Design, Lawrence, Erlbaum, NJ, USA.

Oliva, A. and A. Torralba. (2006). Building the gist of a scene: The role of global image features in recognition. Progress in Brain Research, 155: 23–36.

Oliver, N. and A. Pentland. (2000). Graphical models for driver behavior recognition in a smart car. Proceedings of the IEEE Intelligent Vehicles Symposium, 7–12.

Ostlund, J. et al. (2004). Driving performance assessment—methods and metrics. Available at http://www.aide-eu.org/pdf/sp2_deliv_new/aide_d2_2_5.pdf.

Ostlund, J., L. Nilsson, J. Tornros and A. Forsman. (2006). Effects of cognitive and visual load in real and simulated driving. VTI Report 533A.

Owsley, C. and K. Ball. (1993). Assessing visual function in the older driver. Clinics in Geriatric Medicine, 9(2): 389–401.

Pentland, A. and A. Liu. (1999). Modeling and prediction of human behavior. Neural Computation, 11(1): 229–242.

Pew, R.W. and A.S. Mavor (eds.) (1998). Modeling Human and Organizational Behavior, Academic Press, Washington, USA.

Plöchl, M. and J. Edelmann. (2007). Driver models in automobile dynamics application. Vehicle System Dynamics, 45(Nos. 7-8): 699–741.

Posner, M.I. (1980). Orienting of attention. Quarterly Journal of Experimental Psychology, 32(1): 3–25.

Ranney, T.A. (1994). Models of driving behavior: A review of their evolution. Accident Analysis and Prevention, 26(6): 733–750.

Rapoport, M.J., G. Naglie, K. Weegar, A. Myers, D. Cameron, A. Crizzle, N. Korner-Bitensky, H. Tuokko, B. Vrkljan, M. Bédard, M.M. Porter, B. Mazer, I. Gélinas, M. Man-Son-Hing and S. Marshall. (2013). Relationship between cognitive performance, perceptions of driving comfort and abilities, and self-reported driving restrictions among healthy older drivers. Accident Analysis & Prevention, 61: 288–295.

Rasmussen, J. (1990). The role of error in organizing behavior. Ergonomics, 33: 1185–1200.

Readinger, W.O., A. Chatziastros, D.W. Cunningham, H.H. Bülthoff and J.E. Cutting. (2002). Gaze-eccentricity effects on road position and steering. Journal of Experimental Psychology: Applied, 8(4): 247–258.

Reger, M.A., R.K. Welsh, G. Watson, B. Cholerton, L.D. Baker and S. Craft. (2004). The relationship between neuropsychological functioning and driving ability in dementia: A meta-analysis. Neuropsychology, 18(1): 85–93.

Reitan, R.M. (1986). Trail Making Test: Manual for Administration and Scoring, Tucson, AZ, Reitan Neuropsychology Laboratory.

Rensink, R.A. (2000). Seeing, sensing, and scrutinizing. Vision Research, 40(10-12): 1469–1487.

Rey, A. (1941). L'examen psychologique dans les cas d'encephalopathie tramatique, Les problems. Archives de Psychologie, 28: 286–340.

Rizzolatti, G., L. Riggio, I. Dascola and C. Umiltá. (1987). Reorienting attention across the horizontal and vertical meridians: Evidence in favor of a pre-motor theory of attention. Neuropsychologia, 25(1): 31–40.

Robertshaw, K.D. and R.M. Wilkie. (2008). Does gaze influence steering around a bend? Journal of Vision, 8(4), Article 18: 1–13.

Rosenholtz, R., A. Dorai and R. Freeman. (2011). Do predictions of visual perception aid design? ACM Transactions on Applied Perception, 8(2): 12:1–12:20.

Royden, C.S., M.S. Banks and J.A. Crowell. (1992). The perception of heading during eye movements. Nature, 360(6404): 583–585.

Royden, C.S. and E.C. Hildreth. (1996). Human heading judgments in the presence of moving objects. Perception & Psychophysics, 58(6): 836–856.

SAE J2944 (2015). Surface Vehicle Recommended Practice. Operational Definitions of Driving Performance Measures and Statistics, 2015.

Salvucci, D.D. (2001). Predicting the effects of in-car interface use on driver performance: An integrated model approach. Journal of Human-Computer Studies, 55(1): 85–107.

Salvucci, D.D. and A. Liu. (2002). The time course of a lane change: Driver control and eye-movement behavior. Transportation Research Part F, 5(2): 123–132.

Salvucci, D.D. and R. Gray. (2004). A two-point visual control model of steering. Perception, 33(10): 1233–1248.

Salvucci, D.D. (2006). Modeling driver behavior in a cognitive architecture. Human Factors, 48(2): 362–380.

Sato, T. and M. Akamatsu. (2007). Influence of traffic conditions on driver behavior before making a right turn at an intersection: Analysis of driver behavior based on measured data on an actual road. Transportation Research Part F, 10(5): 397–413.

Sato, T., M. Akamatsu, A. Iwasaki, H. Imaizumi and H. Daigo. (2007). Change in driving behavior by elderly (second report)—Analysis of driving behavior to adapt to age-related changes of cognitive and physical functions. Transaction of Automotive Engineers of Japan, 38(4): 209–214 (in Japanese with English abstract).

Sato, T. and M. Akamatsu. (2008). Modeling and prediction of driver preparations for making a right turn based on vehicle velocity and traffic conditions while approaching an intersection. Transportation Research Part F, 11(4): 242–258.

Sato, T., M. Akamatsu, Y. Miyazaki, N. Nakamura, K. Hiramatsu and S. Fukuzumi. (2010). Controlling task demands while approaching intersections. Proceedings of European Conference on Human Centred Design for Intelligent Transport Systems, 603–614.

Sato, T., M. Akamatsu, A. Tanaka, J. Hatada, Y. Denda and T. Ishii. (2013). Estimation method of task demands in road traffic environments. Journal of Mobile Interactions, 3(1): 13–19 (in Japanese with English abstract).

Sato, T., Y. Takeda, S. Iwaki and M. Akamatsu. (2016). Relationship between allowable task demand while driving and allocation of attentional resources. Transaction of Automotive Engineers of Japan, 47(2): 503–507 (in Japanese with English abstract).

Sekizawa, S., S. Inagaki, T. Suzuki, S. Hayakawa, N. Tsuchida, T. Tsuda and H. Fujinami. (2007). Modeling and recognition of driving behavior based on stochastic swithed ARX model. IEEE Transactions on Intelligent Transportation Systems, 8(4): 593–606.

Sharp, R.S., Daniele Casanova and P. Symonds. (2000). A mathematical model for driver steering control, with design, tuning and performance result. Vehicle System Dynamics, 33(5): 289–326.

Shepard, R.N. and J. Metzler. (1971). Mental rotation of three-dimensional objects. Science, 171(3972): 701–703.

Şimşekoğlu, Ö. and T. Lajunen. (2008). Social psychology of seat belt use: A comparison of theory of planned behavior and health belief model. Transportation Research Part F, 11(3): 181–191.

Smiley, A. (2004). Adaptive strategies of older drivers. *In*: Transportation in an Aging Society: A Decade of Experience, Transportation Research Board Conference Proceedings 27, Washington, D.C., Transportation Research Board, 36–43.

Sommer, S.M., O. Arno, M. Strypsten, G. Eeckhout and S. Rothermel. (2003). D4.3: On-road assessment methodology and reference road test. AGILE Project Deliverable 4.3, QLRT-2001-00118.

Staplin, L., K.H. Lococo, K.W. Gish and L.E. Decina. (2003). Model driver screening and evaluation program. Final Technical Report, Vol. I: Project Summary and Model Program Recommendations, DOT HS 809 582.

Sternberg, S. (1969). The discovery of processing stages: Extensions of Donders' method. Act Psychologica, 30: 276–315.

Straub, J., S. Zheng and J.W. Fisher. (2014). Bayesian nonparametric modeling of driver behavior. IEEE Intelligent Vehicles Symposium (IV): 932–938.

Strayer, D.L., F.A. Drews and W.A. Johnston. (2003). Cell phone-induced failures of visual attention during simulated driving. Journal of Experimental Psychology, 9(1): 23–32.

Sugimoto, F. and J.I. Katayama. (2013). Somatosensory P2 reflects resource allocation in a game task: Assessment with an irrelevant probe technique using electrical probe stimuli to shoulders. International Journal of Psychophysiology, 87(2): 200–204.

Summala, H. (1988). Risk control is not adjustment: The zero-risk theory of driver behavior and its implications. Ergonomics, 31(4): 491–506.

Takeda, Y., M. Sugai and A. Yagi. (2001). Eye fixation related potentials in a proof reading task. International Journal of Psychophysiology, 40(3): 181–186.

Takeda, Y. and N. Yoshitsugu, K. Itoh and N. Kanamori. (2012). Assessment of attentional workload while driving by eye-fixation-related potentials. Kansei Engineering International Journal, 11(3): 121–126.

Taylor, R.M. and S.S. Selcon. (1991). Subjective measurement of situational awareness. pp. 789–791. *In*: Y. Queinnec and F. Daniellou (eds.). Designing for Everyone, Proceedings of the 11th Congress of the International Ergonomics Association, Taylor & Francis, London, UK.

Theeuwes, J. (2001). The effects of road design on driving. pp. 241–263. *In*: Pierre-Emmanuel Barjonet (ed.). Traffic Psychology Today, Kluwer, Boston, MA, USA.

Treisman, A.M. and G. Gelade. (1980). A feature-integration theory of attention. Cognitive Psychology, 12(1): 97–136.

Ulleberg, P. and T. Rundmo. (2003). Personality, attitudes and risk perception as predictors of risky driving behavior among young drivers. Safety Science, 41(5): 427–443.

Vogel, K. (2003). A comparison of headway and time to collision as safety indicators. Accident Analysis and Prevention, 35(3): 427–433.

Wada, T., S.I. Doi, K. Imai, N. Tsuru, K. Isaji and H. Kaneko. (2007). Analysis of drivers' behaviors in car following based on performance index for approach and alienation (No. 2007-01-0440). SAE Technical Paper.

Warner, H.W. and L. Aberg. (2006). Drivers' decision to speed: A study inspired by the theory of planned behavior. Transportation Research Part F, 9(6): 427–433.

Warren, W.H., M.W. Morris and M. Kalish. (1988). Perception of translational heading from optical flow. Journal of Experimental Psychology: Human Perception and Performance, 14(4): 646–660.

Wickens, C.D. (1992). Introduction to engineering psychology and human performance. pp. 1–17. *In*: C.D. Wickens and J.G. Hollands (eds.). Engineering Psychology and Human Performance (3rd ed.), Prentice Hall, New Jersey, USA.

Wickens, C.D., J. Goh, J. Helleberg, W.J. Horrey and D.A. Talleur (2003). Attentional models of multitask pilot performance using advanced display technology. Human Factors, 45(3): 360–380.

Wilde, G.J.S. (1982). The theory of risk homeostasis: Implications for safety and health. Risk Analysis, 2(4): 209–225.

Wilkie, R.M. and J.P. Wann. (2002). Driving as night falls: The contribution of retinal flow and visual direction to the control of steering. Current Biology, 12(23): 2014–2017.

Winsum, W.V., K.A. Brookhuis and D.D. Waard. (2000). A comparison of different ways to approximate time-to-line crossing (TLC) during car driving. Accident Analysis and Prevention, 32(1): 47–56.

Wood, J.M., M.S. Horswill, P.R. Lacherez and K.J. Anstey. (2013). Evaluation of screening tests for predicting older driver performance and safety assessed by an on-road test. Accident Analysis and Prevention, 50: 1161–1168.

Wortelen, B., M. Baumann and A. Lüdtke. (2013). Dynamic simulation and prediction of drivers' attention distribution. Transportation Research Part F, 21: 278–294.

Yamanaka, K., K. Morishima and H. Daimoto. (2014). A study of measuring and evaluating useful field of view while driving. Journal of Society of Automotive Engineers of Japan, 68(3): 92–96 (in Japanese with English abstract).

Yang, G., Y. Lin and P. Bhattacharya. (2010). A driver fatigue recognition model based on information fusion and dynamic Bayesian network. Information Science, 180(10): 1942–1954.

Index

A

acceleration 2, 14, 45, 46, 48–50, 54, 56, 62–64, 66, 68–70, 72, 73, 75, 77, 111, 190, 204, 217, 232, 234, 249, 250, 276, 278, 283, 294, 308, 309, 317, 319, 321, 322
ACC 13, 15, 16, 27, 54, 226, 238, 244–246, 248, 284
ACT-R 307, 308, 310–312
ADAS 15, 33, 44, 190, 216, 221, 239
air quality 83
arousal enhancing 114, 125
arousal level 9, 11, 17, 87, 104, 107, 112, 115–119, 121–123, 126–130, 133, 134, 137, 143–145, 147, 148, 150, 177, 195–197, 227, 248, 262, 265
attention 1, 5, 8, 11, 12, 14, 17, 49, 59, 87, 104, 107, 109, 116, 121, 133, 134, 136, 137, 139–141, 143, 144, 146, 149, 154, 165–167, 169, 170, 181, 195–197, 199, 209, 210, 214, 217–219, 222, 234, 241, 245, 248, 257–262, 265, 267–270, 281, 284, 286–290, 292, 293, 298, 309
automated driving system 15–17, 237, 242, 246, 248–250, 284

B

Bayesian network model 277, 316–318
biochemical reaction 152
biosignal measurement 130–133
blood pressure 87, 145, 146, 153
body movement 4, 66, 68, 121
body size 3, 4, 34, 69

C

cabin comfort 9, 10, 83
car navigation system 6, 10–12, 164, 171, 178, 183–188, 190, 192, 194, 268, 274, 282, 301, 302, 305
cellular phone 169, 175, 178, 181, 213, 214, 284
CFF 133, 139
checklists 35
compatibility 7, 21, 200, 244, 290

compliance 215, 216, 223, 224
Cooper-Harper scale 163
criticality 217, 219–221, 226–228, 240, 241

D

DALI 164
design stage 34, 36, 37, 45
development process 32, 33, 35, 45, 165, 180
distraction 11–13, 15, 46, 47, 49, 140, 169, 193–210, 212, 213, 230, 281, 282,
    293, 311, 312
door closing 79–81, 83
driver distraction 11, 15, 140, 194, 198, 213, 230, 312
driver model 13, 16, 293, 296, 297, 307, 308, 310–312
driver state 102, 130, 132, 139, 148, 152, 247, 276
driver steering control model 293–295
driving behavior 13–20, 44–46, 49–54, 56, 59, 121, 154, 169, 172, 201, 242–246,
    272, 274–276, 285–289, 291–294, 299, 307, 308, 311, 312, 315–324
driving enjoyment 18–20
driving fatigue 8, 87, 102
driving performance 7, 8, 19, 70, 103, 104, 106, 107, 121, 124, 127, 170–173,
    178, 197, 201–205, 209, 245, 275–277, 278, 280–282, 287, 299
driving simulator 14, 34, 35, 44–49, 73, 87, 104, 123, 139, 140, 154, 171, 175,
    201, 202, 208, 213, 214, 221, 222, 259, 262, 272, 277, 279, 280, 282, 285,
    299–301, 312
drowsiness 87, 116, 117, 124, 125, 247, 262, 265
DRT 172, 202, 208, 209

E

EEG 9, 87, 103, 116, 118, 120, 121, 126, 133–138, 210, 214, 262–264
elderly driver 4, 17–19, 261, 275, 276, 287–289, 291–293
electrodermal activity 146
engine sound 45, 47, 75
enjoyment 18–20, 109–111, 113, 114
ethnography 38, 39
exhaust sound 75, 77
external communication 178
eye movement 122, 123, 128, 133, 136, 140–142, 201, 257, 263, 264, 298, 301,
    312

F

facial expression 9, 121, 125, 148–151, 152
fatigue 7–9, 87, 90, 102–104, 106, 107, 109, 116, 120, 121, 128, 130, 133, 134,
    139, 143, 145, 148, 151–154, 165, 197, 262, 263, 265, 288, 317, 319

feedback 6, 8, 13, 14, 19, 20, 47, 110, 118, 123, 136, 151, 177, 212, 267, 269, 270, 273, 292, 293, 295
field operational test (FOT) 53
flow theory 19, 109, 110–113
fragrance 86–88

H

heart rate 7, 9, 103, 104, 126, 144, 145, 147, 173, 202, 210, 214
hidden Markov model 317, 320, 323
hierarchical model 272–275
history 9, 39, 217
HMI 6, 10, 16, 27, 33, 37, 44, 46, 47, 58, 145, 178, 213, 239–242, 247–250, 266
HMM 317–324
HRV 104, 145, 146, 210

I

information-processing 12, 138, 167, 197, 267–270, 272, 299–304
instrumented vehicle 50–53, 279, 280, 282, 317
interior lighting 88, 89, 91, 92
interior material 94, 96, 97
in-vehicle system 7, 10, 18, 130, 162, 210, 225, 227, 230, 233, 304

L

LCT 202, 206–209, 212, 213
level of automation 235, 236, 246

M

map display 181, 188
mental map 181, 182
mental resource 11, 12, 104, 107, 141, 162, 229, 277
mental workload 12, 17, 33, 102, 104, 105, 107, 140, 142, 144, 145, 148, 162–165, 168, 169, 171, 172, 210, 214, 239
menu 6, 179, 180, 191–194
meters 4, 5, 26, 31, 40, 85, 147, 178–180, 219, 229

N

NASA-TLX 104, 105, 162–164, 172, 213
naturalistic driving study (NDS) 53
N-back 209–211
near and far 299, 309, 310

O

occlusion method 11, 12, 30, 202, 205, 206, 277
odor 83–86

P

parking maneuver 283, 284
pedals 2, 51, 54, 69, 70, 94, 167, 204, 232, 246, 249, 273, 294, 309
PERCLOS 9, 117, 123–126
priority 169, 178, 183, 198, 207, 215, 223–227

Q

questionnaire 19, 22, 34, 36, 38, 41–43, 48, 87, 162, 172, 202, 222, 233, 245

R

respiration 9, 145–147, 153
riding comfort 7, 8, 27, 62–64, 66, 110
RSME 167, 168, 172

S

SAGAT 272, 284–286
saliency 170, 266, 267, 268
SART 284–286
SDLP 280, 282
seat 2, 3, 10, 14, 25, 37, 48, 50, 62–71, 89, 95–98, 129, 223–225, 228, 316
service 21, 22, 31, 36, 38, 39–41, 43, 44, 51, 154, 190, 324, 325
situation awareness 5, 230, 237, 248, 271, 272, 275, 284–286, 308
sleepiness 9, 104, 115–130, 134, 140, 143, 145, 147, 148, 152
smells 9, 83–88, 215, 216
SRK model 273, 274
statistical model 320
steering entropy 173, 175, 204, 280, 281
steering reversal 173, 280, 281
steering wheel 2–5, 19, 47, 69–71, 89, 94, 95, 178, 201, 204, 273, 283, 288, 294, 301, 308
stress 4, 20, 68, 87, 88, 102–109, 121, 126, 130, 148, 151–154, 164–166, 195, 271
structural equation model 312, 313, 315, 316
subsidiary task method 13, 30, 168, 169, 171–173, 261
SuRT 212, 213
SWAT 164–166, 168, 213
symbols 5, 6, 184, 186, 191, 193, 218, 222, 224, 290

T

tangent point 298–301, 309, 310
target user 33, 37
task analysis 40
task demand 19, 20, 104, 107, 141, 162, 178, 197, 207, 209, 210, 229–231, 261,
    268, 275–277
task-capability interface model 19, 20, 275, 276
texture 94–96, 98, 184
TLC 204, 280, 282, 283

U

understanding 3, 16, 17, 21, 32, 37–40, 42, 44, 60, 130, 135, 195, 210, 217, 218,
    222, 231, 232, 236, 238–242, 247, 248, 250, 265, 268, 271, 275, 293, 307,
    312
urgency 57, 216, 217, 219–221, 226, 227, 241
useful field of view 142, 143, 257, 259–262, 289, 291, 292
user requirements 33, 34, 36–38, 40, 42
user's manual 36

V

vanity lamp 89–91
vibration 7–9, 45, 47, 62–71, 74, 76, 78, 82, 126, 129, 154, 215, 281
vigilance 17, 103, 106, 116–121, 123–126, 133, 215, 219, 248

W

warning signal 14, 44, 215–219, 221–223, 228, 243, 279, 284
workload 12, 13, 17, 33, 43, 44, 102, 104–107, 130–132, 140, 142–145, 148 ,
    162–169, 171–176, 178, 196–198, 200, 210, 211, 213, 214, 229–233, 239,
    245, 271, 282, 293, 300, 301
workload manager 229
WP 166, 167

Printed and bound by CPI Group (UK) Ltd, Croydon, CR0 4YY

24/10/2024

01778307-0012